GEERDT MAGIELS

FROM SUNLIGHT TO INSIGHT

Jan IngenHousz,
the discovery of photosynthesis
& science in the light of ecology

Editing: Carla Aerts
Cover picture: Koen de Waal
Cover design: Frisco, Oostende
Book Design: Geerdt Magiels
Print: Silhouet, Maldegem
© 2010 Geerdt Magiels

© 2010 Uitgeverij VUBPRESS Brussels University Press
VUBPRESS is an imprint of ASP nv
(Academic and Scientific Publishers nv)
Ravensteingalerij 28
B-1000 Brussels
Tel. ++32 (0)2 289 26 50
Fax ++32 (0)2 289 26 59
E-mail: info@vubpress.be
www.vubpress.be

ISBN 978 90 5487 645 8
NUR 738 / 681 / 922
Legal Deposit D/2009/11.161/103

"A true lover of natural knowledge never looks with indifference on any phenomenon which seems somewhat extraordinary though he does not understand at first sight neither the cause by which it is produced, nor the utility which might be derived from it. He delights first of all to trace the cause without any regard to the benefit, which he may derive from it either for himself or for mankind. ...

There are indeed very few new experiments or discoveries, which are capable of being turned to any immediate advantage except perhaps that of surprise or admiration, and in the discoverer a kind of delight, unknown to any but true rational minds, mixed with a kind satisfaction bordering on a degree of irresistible pride inseparable from the consciousness of having enlarged the boundaries of human knowledge."

Jan IngenHousz

A first speck of light

The humble beginnings of this book was a very simple question, asked more than ten years ago by a good friend who knew that I was a biologist and had done my graduate studies in botany and plant ecology. "Do you know who discovered photosynthesis?" I had to admit that I did not know. Since then I asked every biologist I met the same question. Nobody was able to give me the answer.

This is an astonishing fact. Photosynthesis is the single most important biochemical process on this planet, making animal life and human culture, including you reading this text, possible. Yet we don't know who discovered this natural phenomenon. And it is not just the lay person who does not know, even plant physiologists, biologists and ecologists don't know. One would think that the person who first described photosynthesis would be on a par with the discoverers of the place of the earth in the solar system, of gravity, of evolution, relativity or DNA. But, no.

Jan IngenHousz. That was the name my friend mentioned. And he happened to know the name by coincidence because one of his grandfathers had worked at the estate of the family IngenHousz, an old and respectable Dutch lineage, just like his grandfather from Brabant in the South of the country.

When I first looked up IngenHousz, the internet was in its infancy and it was hard to find much information on this illustrious unknown man. But what began with a trickle turned into a flood. Unexpected and unstudied archives turned up and a wonderful story unfolded.

This story will intertwine two story lines. The first is the reconstruction of Jan IngenHousz' life and work in the second half of the eighteenth century. This was a tumultuous period, both politically and intellectually, and that will resonate as the story of IngenHousz unfolds, a man who can be justifiably considered a prototype of a man of the Enlightenment. The story will reconstruct how he contributed to the discovery of photosynthesis, within the wider context of medicine, chemistry and physics, at a moment in time when the foundations of modern European society and culture were being laid.

The second is a reflection on the human endeavour of acquiring trustworthy knowledge about the world. That is to say: the mundane and pragmatic but also exciting and captivating business of performing science. At the time Jan IngenHousz tried to understand what plants did with sunlight, his activities would not be called 'science', but were known as 'natural philosophy'. Still, from the way IngenHousz gathered data, ran experiments and tested hypotheses, we would now definitely recognise him as a contemporary scientist. His case is a largely unknown and definitely under-researched example of scientific enquiry and as such it will be very useful to throw some fresh light on the phenomenon 'science'. It will be an exercise in the philosophy of science, albeit philosophy of science with a slant. Being trained in ecology in the first place and some postgraduate philosophy on top of that, I will develop an unusual ecological perspective on science. Trying to answer the question why IngenHousz is not a household name leads into a complex reality for which the existing approaches in the philosophy of science can offer no real satisfactory clarification. An ecological perspective might be able to do just that. And at the same time it may offer a novel approach to some dead-end avenues in the philosophy of science of the last couple of decades.

In a way this story will mingle two different strands of thinking. On the one hand it will fill in an unwritten chapter in the history of science. On the other, it will shed some light on this story with the tools of philosophy of science. These two disciplines have not been the best of friends, as the former tries to reconstruct the unique particular characteristics of once-only events, while the latter tries to understand all things under the sun (and the sun and the understanding itself) in one grand theoretical framework.

Here is another point in which the proposed ecological perspective provides a useful and elucidating approach. Ecology is about complex systems, evolving in dynamic equilibrium over time. History is therefore a fundamental ingredient of ecology. Looking at science in an ecological way might offer inroads to reunite the history and the philosophy of science in the most natural of ways. How Jan IngenHousz worked may help to understand how science works in general.

History and philosophy meet at the point where they illustrate and illuminate the way people throughout the ages have been trying to collect trustworthy knowledge about the world in which they live. Jan IngenHousz was one of them. He is the protagonist of the following story.

The first decades of curiosity
1730-1779

Jan IngenHousz was born on 8 December 1730 in Breda, at 17 in the Eindstraat.[1] At that time, in the first decennia of the eighteenth century, Breda was an industrious provincial town, an important hub in the trade between England and the Southern Netherlands. It was a fortified town, with a garrison and an illustrious past. It was here that the peace treaty of Breda had been signed on 31 July 1667, between England, the Republic of the United Provinces, France and Denmark, which called an end to the Second Anglo-Dutch War. As a result, New Amsterdam was handed to the English, while the Republic could keep Surinam, a handful of tropical islands in the West Indies and some slave trade forts on the West African coast. Breda was not unknown to the English. Charles II lived for some time in Breda after fleeing England for Oliver Cromwell, staying at the castle of Mary Henrietta Stewart, the widow of Willem II of Orange.

The English involvement in the region did not end there. In 1743, when young Jan was 13, the English fleet sailed again to the Low Lands to help defend southern Holland against the French. They were to join the 'pragmatic army' coalition of English, Hanoverian and Austrian troops who were to stop the French army's expansionist march to the north. Louis XV's French troops had been predating on Flanders, which was part of the Austrian Empire. The defeat of the French in the battle of Dettingen reinforced the status of Maria-Theresia as the Archduchess of Austria, which had been in jeopardy after the death of her father Charles VI. The military victory also confirmed her as the undisputed ruler of the empire, including Brabant, the region now known as a part of Belgium. After the battle, the English troops retreated strategically to Flanders and many English regiments were camped at Terheijden, just north of Breda.[2]

The chief medical officer of the English army was John Pringle. He had studied medicine in Leyden, under the world-famous Boerhaave. He was fluent in Dutch and most probably sought contact with the medical community in

---

1   The richest single source of biographical knowledge about IngenHousz is Wiesner 1905
2   Browning 1993

Breda. One of its members was Arnold IngenHousz[3], merchant in leather and skins and one of the pharmacists in town. They became close friends and Pringle got to know the two teenage sons of the household, Ludovic born in 1729 and Jan born in 1730. Their mother however, he would not get to know, as she died just before Jan's first birthday.

A doctor in the making

Jan was a clever boy with a talent for languages and a natural curiosity. He was exceptionally good at Latin and Greek, two languages he spoke and wrote as fluently as his mother tongue.[4] At sixteen he went to the university of Louvain, stimulated by his teacher Hoogeveen.[5] It should not come as a surprise that Jan went south to study medicine. Leyden, with its world wide reputation in medicine was no option for a catholic boy from a mainly catholic region. North Brabant lay as a buffer zone between the catholic south and the protestant north. Louvain was (and still is) a catholic university and was the natural destination for someone like Jan IngenHousz. He graduated on 24 July 1753. The poetic declaration still exists[6] and begins with this opening:

A MONSIEUR

MONSIEUR

JEAN INGENHOUSZ

DE BREDA

AU JOUR QU'IL PREND LES DÉGRÉS DE LICENCES

EN MEDECINE

EN LA TRES CELEBRE UNIVERSITE DE LOUVAIN

LE XXIV. DE JUILLET M.D.C.C.LIII.

---

3   The family name IngenHousz is the Brabant version of the Limburg family name Ingenhuys. 'Ingen' stands for ' in de' (in the) which has the meaning of 'from the', while 'huys' stands for 'house'. It might be referring to a stone house in a town which was in the fifteenth century mostly build from wood. The Ingenhuys lineage stretches back to Venlo and Limburg in the fifteenth century. Descendants have come to Brabant. Louis IngenHousz was born in Zaltbommel in 1661. His son Arnoldus was born there in 1693, married Maria Beckers and moved to Breda where he became pharmacist.

The name IngenHousz can be found in many spelling forms such as Ingen-Housz, Ingenhus, Ingenousz, Ingenhousz and others. Jan IngenHousz signed his letters with his name spelled in the form that has been chosen here.

4   Breda Gemeentearchief A5

5   These and many of the details of IngenHousz' biography are known thanks to the short biography written by Dr MJ Godefroi, first published 75 years after his death. This story was based on the testimonial and documentation provided by Jonkheer HF de Grez, son to Theresia IngenHousz, the sixth child of Louis, Jan's brother. This biography has been updated, supplemented and annotated in Ingen Housz J 2005. About the young age at which Jan IngenHousz went to Louvain there is no definite consensus.

6   Breda Gemeentearchief A14

After that he travelled Europe for another four years and attended courses at the universities of Paris, Leyden and Edinburgh, not only in medicine but also in physics and chemistry. He spent some two years at the university of Leyden where he had tutorials with the renowned experimental physicist Van Musschenbroeck and the chemist Gaubius. In Edinburgh he specialised in gynaecology. Edinburgh was at that time one of the centres of the Enlightenment. The Scottish Enlightenment was a remarkable intellectual flourish that lasted for much of the eighteenth century and an underappreciated event in the history of western culture.[7] Edinburgh was a meltingpot of intellect where scientific, economic and philosophical advances were made, which had an immediate impact in Europe and beyond. IngenHousz must have felt at home as it was not only a place where his curiosity could be quenched, but also where his social qualities could flourish. The Scottish Enlightenment was a highly social affair in which writers and thinkers were kept together by bonds of friendship and where discussions were facilitated in numerous clubs and societies. Science and medicine were at the core of the movement, inspired by men such as William Cullen, James Watt and Joseph Black who were driven by the distinct need to better the material conditions of the country, which at the beginning ot the eighteenth century, were in a desperate need for improvement. Scientific method in all its aspects permeated the whole range of intellectual disciplines. The interdisciplinary and holistic approach to intellectual problems was totally in line with IngenHousz' approach. His multicultural background must have given him a distinct sense of entering a very inspiring arena of knowledge. The Edinburgh medical faculty was probably one of the best places to be at that time, because it injected the art of medicine with expertise from other scientific disciplines, not in the least chemistry.[8] Botany too was a natural part of the curriculum, as the the vegetable kingdom was practically the sole source of medicines.

Joseph Black was one of the professors. Black was famous for being the first to descibe fixed air, a kind of air that could be expelled from solid rock by heating limestone (and now known as carbon dioxide). He was involved in the research into the true nature of acidic and alkaline liquids in relationship to fixed air, with the underlying idea that this could result in a more effective internal solvent for dissolving urinary stones. This was fundamental research in the sense of solving real world problems, an approach that fitted nicely in IngenHousz' line. Black's description of fixed air also initiated a new departure in the study of airs as it illustrated the complexity of the atmosphere. Black's teachings and writings were guided by an underlying conception of general laws which link heat, attraction and change of state. Black was one of the most famous pupils of William Cullen, one of the other key figures at the Edinburgh

---

7   Broadie 2003
8   Donovan 1983

medical school. Cullen counted Hume and Smith as colleagues and friends. He not only promulgated clinical expertise, but was also deeply concerned with research in the function of the body and with physiology, or, as it was then known, "animal oeconomy".[9] At the same time he was also giving lectures on agriculture, focusing on its chemical aspects.[10] Both agriculture and medicine were considered a natural knowledge that could be seen as for the public good and conducted within the realm of explicit economical relationships. They were both disciplines that acquired 'applied' knowledge. IngenHousz absorbed these new practical and theoretical approaches as a dry sponge. All his interests, in medicine, chemistry and electricity, were coming together in one conceptually interrelated field of research. He was to take all this knowledge and inspiration back to his hometown.

Back in Breda

In 1757 he settled in Breda as a medical doctor, only a couple of houses down the road from where he was born. Being a wellknown and respected man, his medical practice turned into a flourishing business. It left him little time for his experimental work in physics and chemistry. Electricity was one of his favourite study objects and after the patients had been seen and treated, he disappeared in his make-shift laboratory to perform experiments with 'elektriseermachines' of his own design, an electrical machine using disks instead of cylinders. This was an innovation which would later be claimed by Ramsden.[11] This work did not result in publications but traces of his work can be found in his lively and intense correspondence with his former teachers in the Low Countries and in England and with friends, such as Pringle who had always encouraged the young and promising IngenHousz to follow the paths of knowledge.[12]

An academic career was no option as all Dutch (protestant) universities were out of bound for the catholic intellectual from the south. The same was true for any official government position. This situation was partly the reason for a brain drain of catholic intellectuals to catholic countries and universities abroad. The famous physician Gerard van Swieten was a wellknown example. He had left Leyden and gone to Vienna. IngenHousz would finally do the same and leave his home country, but only after his father had died.

Jan IngenHousz arrived in London in 1764, by invitation of his patron John Pringle. After his army career he had returned to civilian medical practice and

9   Rocca 2007
10   Fissell 2003
11   McConnell 2007
12   Wiesner 1905 is until now the only comprehensive biographical work on IngenHousz, written at the occasion of the International Botanical Congres, held in Vienna May 1905, when also a statue of IngenHousz was unveiled. This biography refers to these letters which have as yet not been located.

became famous because of his book *Diseases of the Army*. He was knighted, became member of the Royal Society and was appointed royal physician to George III after his accession to the throne. Sir Pringle was a man of means and prestige. He was IngenHousz' perfect introduction to the world of science and culture in London. IngenHousz would quickly get to know people such as the physicians William Hunter, George Armstrong, John Hunter and Alexander Monro I. They would become friends and intellectual companions, introducing him to anatomical collections, surgical techniques, preparation methods and new medical procedures, such as inoculation against smallpox. In 1765 he returned to Edinburgh where he was very much impressed by the teachings of William Cullen. He admired Cullen's almost universal knowledge, combining chemistry, pharmacology and physiology. Cullen was also one of the men who tried to find applications for chemistry in other fields of activity, such as agriculture. IngenHousz writes to his friend Dr Deckers in Holland: "I am every day at the hospital where I admire the care and attention of the physicians, all doing their best both for the patients and for the instruction of the students. Day after day one is describing here all symptoms of diseases and the activities of drugs in all details."[13]

The speckled monster

Next, back in London, he would turn his medical attention to the battle against smallpox. Smallpox was considered the most terrible of all the scourges of death. Because it spreads through infected saliva droplets, smallpox is easily transmitted by everyday human contact.

After exposure, the Variola virus incubates in the body for about twelve days. Then fever, fatigue, head- and backache set in, followed within two to three days by a prominent rash on the face, arms and legs. This will form lesions that fill with pus and then begin to crust developing scabs which fall off after a few weeks. One third of all victims die. Many are left blind. Most of those who recover are scarred, hence its nickname 'speckled monster'.

Smallpox is one of humankind's oldest and most dreadful diseases. It was present throughout the Old World, and particularly dangerous in crowded cities, recurring regularly in deathly epidemics. The history of human struggle with the virus goes back a long time. Already around 1000 AD, stories are told about the practice of 'inoculation' in China and India. The first written accounts date from the sixteenth century in Chinese books on medicine. Powder from dried crusts of smallpox pustules was stuffed up the nose with impregnated plugs of cotton or blown into the nostrils or applied to a scratch on the skin. Around the fifteenth century, a practice of applying powdered smallpox crusts,

---

13   quoted by Godefroi 1875, the correspondence with Deckers has not been recovered

pushed with a pin or poking device into the skin, became quite common in the Middle East. This technique was known as 'variolation'. The primary intent of variolation was that of preserving the beauty of wives and daughters without being too preoccupied with saving lives.

But saving lives it did, although it is not exactly clear how. Maybe the virus was not as virulent because it did not enter the body in the usual way. The variolation powder was also dried and kept for months at body temperature, so most of the viruses may have been dead by the time they reached a new breeding ground, with their infectious potency reduced. Reports of these practices in India and the Far East found their way to the learned societies in the West.[14] The Royal Society in London received a letter from the Italian physician Timoni, reporting on inoculation in Constantinople. In Boston the reverend Cotton Mather read Timoni's paper and he must immediately have realised that this confirmed the story told to him by his slave Oneseimus. Having recently arrived from Tripoli, this man had an inoculation scar. In his homeland a little pus from an infected person was applied to a cut made in the skin of a healthy person. The healthy person usually got a mild case of the disease and soon recovered; exposure to smallpox later in life had no ill effect. Mather had seen three of his children nearly die during earlier smallpox epidemics and became very interested in his slave's testimony.

Mather sent his account to the Royal Society in London in an essay 'Curiosities of Smallpox' in 1716 in which he asks "How does it come to pass that no more is done to bring this operation into experiment and into Fashion? ... For my own part if I should live to see the smallpox again enter into our city, I would immediately procure a consult of our physicians, to introduce a practise which may be of so very happy a tendency."[15]

So, inoculation was not something unheard of when Lady Mary Mortley Montagu, the wife of the British Ambassador to Turkey, saw the ancient inoculation customs in Turkey and wrote about it in letters to friends back home. When she was back in England she persuaded the physician to the Turkish embassy, Charles Maitland, to inoculate her three-year old daughter in 1718. She must have been the first patient to experience the procedure in England, performed by a professional physician. Lady Montagu made great efforts to promote inoculation in England. She did score some success, but had to fight fierce opposition.

At the time smallpox invaded London, the same bitter controversy raged on the other side of the ocean. There had been no smallpox in the Americas until Europeans brought the virus to the New World. Native people had hardly a chance to build immunity against the new disease and the hypothesis goes that the indigenous people of the Americans succumbed to smallpox

14 Glynn 2004 p 60
15 Silverman 1984 p338-9

in great numbers; providing the conquistadors an easy advance. The English colonists tried to keep smallpox out of their part of the world. Boston inspected all incoming ships and if one arrived with smallpox on board, it was forced to fly a quarantine flag and remain in isolation until the disease had passed. People afflicted by smallpox were sent to stay and often die on Spectacle Island in Boston Harbour, tucked away from the city's population. Still, smallpox outbrakes occured at regular intervals, killing numerous people while others survived and became immune against the disease. In the spring of 1721 the virus struck once more. A ship sailed past the quarantine barrier and a number of smallpox infected sailors landed ashore in Boston. As soon as the first cases appeared, the town took dramatic measures to isolate the infected men. But it was too late: by May the city was in the grip of a virulent epidemic. Mather tried to tell doctors what he knew about inoculation; yet all but one ignored him. The surgeon Zabdiel Boylston found Mather's arguments persuasive and inoculated his own son and two of his slaves. Controversy errupted the minute word got out. Although the Boston clergy backed up Mather and Boylston, the town's physicians and the common people considered the practice barbaric. They called his technique "poisoning" and urged the authorities to charge the doctor with murder if the experiment failed.

The New England Courant published by James Franklin, Benjamin's older brother, offered a polemical platform to writers who opposed inoculation. Those in favour found themselves accused of trying to subvert the will of God. Smallpox, many people claimed, was a judgment of God on the sins of the people. Averting this and other diseases was deemed provoking Him. Actions were undertaken to reinforce the opinions of those who tried to stop the inoculators. Cotton Mather's house was set to fire. Zabdiel Boylston had to stay put in his house to avoid being assaulted.

When news spread that Boylston's son and slaves had fully recovered, people began to defy the ban, visiting the doctor to get inoculated. The clinical results that were gradually assembled refuted the critics. But nevertheless, scripture resonated more convincingly among many believers. The discussion about the good and evil of variolation raged in all countries where the new procedure was introduced. And many of the opponents' arguments sounded pretty similar to the ones that are still being expressed against vaccination today.

But after overcoming considerable difficulty and achieving notable success, Boylston travelled to London in 1724, where he published his results and was elected to the Royal Society in 1726. The results of the first clinical study in the history of medicine were clear: 6 of the 247 (2.4%) inoculated inhabitants of Boston died compared with 844 of 5980 people (14%) who died after contracting the disease naturally and only accepting the usual medical treatment.[16]

---

16   Boylston 1726

A meeting in London

IngenHousz arrived in London fourty years after Boylston. He was warmly welcomed by Pringle and this friendship must have been the perfect platform on which to build a wide network of knowledge. He kept an address book which reads like a 'who's who' in London of the world of science, technology and politics. Only a handful of loose pages still exist, fallen apart, filled with *Directions in London nobility & Learned*[17] in alphabetical order. One can find over 450 names and addresses beginning with

> *Ash Dr N 88 New Bond Street*
> *[...]*
> *Cavallo Mr 51 Wells Street Oxford Road*
> *Cavendish Mr Bedford Square corner of gower Street*
> *Combe Dr*
> *[...]*
> *Dimsdale Baron 50 Cornhill & N 6 Saville Row*
> *[...]*
> *Earl of Warwick*
> *Earl of leicester*
> *Mr Bauer*
> *Mr De Le Blancherie*
> *[...]*
> *Franklin governor 87 Northon Str Portland Place*
> *[...]*
> *Kirwan N 4 Cecil Street Strand*
> *[...]*
> *Thomas Mr (who was in China)*

and under Marchands & artificers, he lists among others:

> *Arnold watchmaker*
> *Bickley stove grate maker*
> *Bins makes portable water closet*
> *Bird locksmith*
> *Stancliff eminent quadrant maker*
> *Blanchard steel pen mmaker*
> *Blunt instrument maker to the king*

---

17   Breda Collectie Ingen Housz A16

Brookman pensilmaker[18]
Cullingworth sells the very finest flanel
Cuthbertson john mathem. instrum maker ~~Holland Street~~ 56 Shoe Lane Holborne
Cuthbertson mathem instrum maker Poland street
Dr Simmons author of the London Medical Journal
Haley watch maker
Jackson&Moser, have their emyreal stove warehouse in Soho
james Woodmason stationer
Kelly shoemaker
Kimbel opticien
le chevalier Mollinedo secrd'embassadeur d'Hispagne
Long operator in medical electricity
Peacock, architect makes philosophical percolators
Pearson painter on glas
Pepys cutler, poultry counter near the mansion
Pether, makes telescopes
Richard makes artificial mineral waters, like schweppe
Richardson, ivory turner
Rundell&Bridge, jewellers
Schweppe makes mineral water King Street Holbin 75 Margaret Str Cov g
Smith botanist
Smith button maker
Stainford wine merchant
Turner eminent practical chymist

Amongst all these ambassadors, earls and doctors, clockmakers, chymists and other craftsmen, one name definitely sticks out: Benjamin Franklin. Franklin was a bit of a polymath: printer and publisher, scientist and philosopher, revolutionary, statesman, and diplomat, a function he took up in London in 1757. Having lost a son to smallpox in 1736, he published in 1759 a pamphlet on smallpox inoculation as it was practised in the American colony.[19] He wrote the preface and co-authored the booklet with the physician William Heberden. Thinking beyond orthodoxy he defended the practice on the basis of the hard facts that mortality was much lower than in smallpox through naturally infection. Around 1500 copies were printed for free distribution overseas. Although Franklin had no formal medical training, medicine was one of his great interests. He did not set out to behave as a physician, but was continually consulted about remedies, especially as he was continually asked for advice.

---

18   IngenHousz uses a peculiar idiosyncratic spelling, and especially so in his handwritten notes. I have tried to be careful to reproduce all texts as he wrote them. I have chosen for not indicating all spelling or other errors or deviations with [sic] as that would clutter the text and render it less readable.
19   Franklin 1759

His interest in medicine stemmed from his insatiable curiosity about many aspects of life and how to find solutions for so many ailments and disorders.[20]

Pringle was the obvious link between IngenHousz and Franklin, who were likeminded spirits in many respects. Both shared so many interests that the mutual understanding must have been natural and instantaneous. Both had been travelling extensively and were not in their mother country, still feeling perfectly at home in England, at that time the centre of the world, economically and culturally. Electricity and chemistry, health and disease, physics and geology, were the wide-ranging topics they showed more than a superficial interest in. The diverse interests and activities of both were not unusual for the men of their time. Many educated people were active in and interested in politics as that was a means for civil and social improvement, and scientific exploration was just an efficient tool to better understand the world. IngenHousz' and Franklin's engagements - such as their advocacy for smallpox variolation - arose from the social consciousness typical for researchers in the era of the Enlightenment. It was a time when reason and observation were highly appreciated as a means to knowledge and to improve the organisation of society. They shared the high regard for rationality, (self) discipline, dissemination of knowledge and social consciousness.

Electricity must have made sparks fly between the two men. IngenHousz was interested in the phenomenon since he heard the lectures of Petrus Van Musschenbroeck, back in Leyden in 1754. Franklin's scientific reputation was made after his 1752 experiments with the kite demonstrating the electrical nature of lightning. With his pragmatic mind Franklin had been speculating about the medical use of electricity, as can be seen in his 1757 letter to John Pringle:

> I never knew any advantage from Electricity in Palsies that was permanent. And how far the apparent temporary Advantage might arise from the Exercise in the Patients Journey and coming daily to my house, or from the Spirits given by the Hope of Success, enabling them to exert more Strength in moving their Limbs, I will not pretend to say.[21]

Medicine and electricity made IngenHousz' heart beat a little faster and he translated Franklin's work on electricity in Latin. Long lost, but referred to by Joseph Priestley, IngenHousz must have offered to translate Franklin's letters on electricity at one of their first meetings, so Van der Pas reconstructed the history of this paper and the first encounters between these men.[22] Priestley and IngenHousz probably met on 28 January 1768 when Priestley was in London

20   Gensel 2005
21   Franklin 1757
22   Van der Pas 1978

on one of the very few occasions he visited the capital. But that day he had to be there to sign the admission bylaws of the Royal Society. Priestley mentions IngenHousz in his later *History of electricity*:

> *The ingenious Dr. Ingenhaus [sic] has constructed a machine in which friction is not given to any kind of hollow glass vessel whatever; but to a circular plate of glass, generally about nine inches in diameter... In the first edition of this work, I mentioned of this work one Mr. Ramsden as the inventor of this construction; a mistake, which he himself led me into.*[23]

According to Van der Pas Priestley must have obtained this information from IngenHousz himself, for at that time, IngenHousz had published nothing on the subject yet. And also the translated letters by Franklin would never see print. IngenHousz was too busy travelling and inoculating. A handwritten fragment however was identified by Van der Pas in the National Library of Austria. Priestley after all was right when he wrote about Franklin's letters in his *History and present state of electricity*:

> *Nothing was written upon the subject of electricity which was more generally read and admired in all parts of Europe than these letters. There is hardly any European language into which they have not been translated; and if this were not sufficient to make them properly known, a translation of them has lately been made into Latin.*[24]

But unknown to Priestley and Franklin, it became a "bibliographical rarissimum", as Cohen described it succinctly.[25] But it illustrates nicely the place IngenHousz was taking in a network of knowledge that spread all over Europe, extended into the New World and had one of its hubs in London.

The inoculators

By the time IngenHousz and Franklin meet in London with their common interest in the inoculation against smallpox, the practice was wide spread. It had been given official approval in 1755 by the College of Physicians: "That in their Opinion the Objections made at first to it have been refuted by experience, and that it is at present more generally esteemed and Practised in England than ever, and that they Judge it to be a Practice of the utmost benefit to Mankind."

---

23   Priestley 1769
24   Priestley 1767
25   Cohen 1941

This did not prevent that in England and on the continent, many were not yet convinced, as can be seen in Leopold Mozart's letter to Lorenz Hagenauer:

> *Do you know what people are always wanting? They are trying to persuade me to let my boy be inoculated with smallpox. .. for my part, I leave the matter to the grace of God. It depends on His grace whether He wishes to keep this prodigy of nature in the world in which He has placed it, or to take it to Himself.*[26]

Whatever its success and the fact that many doctors approved, the method was far from a standardised procedure. Some said the inoculated patient had to rest and stay inside, others added complementary remedies such as quina bark or antimony. It was not common practice to isolate the patient for a few days, which meant that infected patients could easily infect others. Jan IngenHousz studied the inoculation method made famous by Sutton and Dimsdale, who made the technique more acceptable and less dangerous by using only superficial scarification with only small amounts of lymph. IngenHousz learned the application of inoculation from Dr William Watson, one of the best exponents of the technique, who further simplified the method by dispensing with the need for residential treatment.[27] Watson was also an expert in electricity, having received the prestigious Copley Medal of the Royal Society in 1745, which must have given these two doctors something to talk about apart from discussing the best possible inoculation method.

When IngenHousz was appointed as doctor to the London Foundling Hospital in 1766 to assist Watson, he was given the task of inoculating all children that were admitted. He writes enthusiastically about the new method to his friend Deckers describing the dry but convincing facts: at the hospital (an institution to receive unwanted and often illegitimate newborn children of the poor) every day four to five children used to die from the small pox, but after inoculation was introduced, scarcely one in ever three hundred died recently.[28] Sutton inoculated 40,000 people without losing a single life, Dimsdale inoculated more than 1,500 without losing a soul to the illness.[29]

Together with Dimsdale he inoculated all the inhabitants of two Hertfordshire parishes, Little Berkhampstead and Bayford, to prevent a recurrence of the small pox epidemic, which had devastated these communities three years earlier resulting in a 20% mortality. IngenHousz describes his experiences in a letter to the pastor of the Wallonian community in The Hague, which was published in 1768.[30] It was just one of the many instances of intense correspondence with colleagues and friends in Holland to convince them of the

---

26 Anderson 1985
27 Miller 1957
28 Godefroi 1875
29 Jenkins 1999
30 IngenHousz 1768

importance of inoculation for public health. He wrote to Deckers: "I hope the eyes of my compatriots will be opened at last on this subject and that they will accept a practice that is so helpful for the human race."

There was, after all, a lot of opposition against the practice, all over Europe. But the mighty and wealthy were often the first to let themselves and their family be inoculated. Not only did they have access to the information about the new medical approach, but they also had the money to pay for it. Smallpox hit Vienna hard, like many other cities in Europe. And it affected all classes of society indiscriminately, from low to upper. In 1761 Maria Theresia lost her sixteen year old son Karl Joseph to the pox, followed a year later by her twelve year old daughter Johanna Gabriele. In 1767, the fifty year old Empress herself almost died from the pox virus, but survived with scarred skin. Her daughter Maria Josepha died from smallpox later the same year, just before she was due to get married to Ferdinand VI of Naples. Smallpox was an omnipresent threat. Leopold Mozart was in town with his family and changed his mind. He writes once more :

> In the whole of Vienna, nothing was spoken of except smallpox. If 10 children were on the death register, 9 of them had died from smallpox. You can easily imagine how I felt; whole nights went by without sleep, and I had no peace by day.[31]

The Empress herself became disillusioned with Gerard van Swieten, her Dutch doctor, who advised the 'cold treatment': bleeding, slightly acidic drinks and lots of fresh cool air. Van Swieten had followed the same academic path as IngenHousz. He first studied medicine in Louvain, before continuing in Leyden with Boerhaave. Being Roman Catholic, a professorship at Leyden was out of bounds to him and in 1745 he went to Vienna, invited by the Empress to become her physician and to reform the medical faculty at the Viennese university.[32] The combination of being a student of the world famous Boerhaave and being a Catholic were the essential characteristics that were high in demand at the Austrian court. Van Swieten would not only attend to all matters medical but also manage the imperial library (and therefore take care of what was considered necessary censorship) and contribute to the establishment of the botanical garden, where another Dutchman Nicolaus Jacquin, botanist and chemist, was to take up the post of head.

On the medical front, he had stood powerless at many death beds. Maria Theresia's husband Joseph I died unexpectedly in 1765. His son Joseph II lost his first wife Isabella from Parma in 1763 and his second spouse Jozefa

31   ibid, letter of 10 November 1767
32   Van der Korst 2003

from Bayern only four years later. Both women succumbed to the speckled death. Both daughters from his first marriage died in Van Swieten's hands, the youngest from smallpox.[33]

But even many years later, the new revolutionary practice of inoculation was still not his favourite technique. He was not so much a theoretical sceptic, he was just very cautious, preferring to err on the side of safety. He was afraid of the possible side effects of the treatment, as some patients developed full-blown smallpox after inoculation and died. Maria Theresia however was converted to the idea of inoculation. She wanted all her younger children who had not been affected by the terrible disease to be inoculated, fully supported in this strategy by her sons Joseph, the new Emperor, and Leopold, Archduke of Tuscany. Some trials were set up to run clinical tests on Austrian soil, as it was not a common treatment. This was in sharp contrast to other countries such as the United Provinces of the Netherlands, Geneva or England. She invited the Genevan specialist Tissot to variolate her children, but he declined the invitation to come to Vienna. Then she decided to ask for advice in England, inspired by Catherine of Russia who had been inoculated by Thomas Dimsdale. The Empress' question was delivered via Count Seilern, the Austrian ambassador in London, to land on the desk of three Royal Physicians: Pringle, Middleton and Duncan. When Middleton himself declined, Pringle thought about Jan IngenHousz, who had just returned from a successful inoculation campaign in Hertfordshire with Dimsdale. His friend was only 37 and had successfully inoculated tree- to four hundred patients. And he was a Catholic, not an unfavourable argument when sending someone to the Viennese court. Count Seilern invited IngenHousz for dinner on 15 March 1768 and requested kindly that he go to Vienna to inoculate the Imperial family. There was little time left to ponder the proposal. He left for Vienna on 1 April 1768.

On the road

IngenHousz took the boat from Harwich to The Hague. There he is welcomed by the Austrian envoy Baron Reischach who hands him his passport. He is invited by the Prince of Orange and visits his old teacher Hoogeveen in Delft. After a short stop in Breda to be with his brother's family, he travels to Brussels, where he is welcomed by Prince Karl Von Lotharingen, the brother-in-law of Maria Theresia, and by Count Ludwig Cobenzl, Minister for the Austrian Netherlands. He is treated as a high status traveller, as his luggage is exempt from customs, on command of his highly esteemed hosts.

---

33 Arneth A Ritter von 1876

In a notebook[34] that doubles as a financial logbook and travel diary, he wrote, in his typical mix of languages and eclectic spelling:

> *je partois le 22 d'avril 1768 de Bruxels pour Vienne dans une Bironge [?] de voyage que j'avois acheté à Bruxelles pour ducats. j'arrivois a Ratisbone le 7 de May au matin ayant passé 4 nuites entieres dans le voyage sans coucher dans une maison. je partois de ratisbone en batau Dimanche le 8 a midi; le batailer m'avoit promis de me livrer a Vienne en 4 ou 5 jours. nous n'y arrivames que samedis le 14 à midi. Sa Majesté Imperiale et Royale ayant donné ordres que quelqu'un seroit à la porte pour me recevoir et me conduit dans des appartements meublés pour moi et tout pret.*
>
> *le baron Van Swieten m'acceuillis trèes poliment, ainsi que le Prince Kaunitz, Prince Strarrenberg, Baron Roschach chambelland, den grave Ferdinand van Harrach en den grave Ernest van Harrach, den grave van Degensfeld Hollandsche envoye &c.*
>
> *Den 22 May op Pinxterendag had ik de eer van een particuliere audientie te hebben met hare Majestyd. Hare hand kussende met een gebogen knie, sprak zy my aanstonds aan op de allervriendelykste wyze, betuygende verblyd te syn van my te zien. Zy toonde my aanstonds haare wezen en armen die door de kinderziekte geschonden waren, zeggende dat deze ziekte zo vreesselyk gewoed had in haare familie, en dat zy hoopte dat ik haare 2 aertshertogen door de inenting behouden zou, dog dat dit afhing van het goed dunken van den Baron van Swieten (die by ons was) zy vroeg my hoe hoe sig den Prince van Walles had en de rest der koninklyke familie, of zy alle nog sterk waren, of den jonge prince van Bruynswyk geinoculeerd is, hoe den prince van orange met zyne princes het maakt, of zyn wezen na dat van zyn vader gelykte &c. of ik van de preparatie van de inenting kon, of ik de patienten in de open air laat komen. Daarna zeide zy nog <u>voulez vous voir ma famille</u>, waarop zy my in het naaste vertrek geleyde dara ik vier aertshertoginnen en de 2 aertshertogen vond, aan wien De Kyserin my selfs presenteerde, en van wie ik alle de eer had regter hand te kussen met een kniebuyging.*

He departed from Brussels on 22 April 1768 in a carriage that he had bought in Brussels. He arrived in Regensburg on 7 May, after having spent four nights on the road without sleeping under a roof. From Regensburg his travels continued by boat over the Donau to Vienna, leaving on Sunday 8th at noon. The boatman promised him that it would take 4 or 5 days to Vienna, but they arrived only on Saturday 14th. Her Majesty had given orders for somebody to attend him on his arrival and take him to his apartment.

---

34   Breda, Collectie Ingen Housz, A7

He was warmly received by Van Swieten, by Prince Kaunitz, Prince Starrenberg, Baron Roschach, Duke Ferdinand, Duke van Harrach, Duke van Degensfeld, the Dutch Envoy and many others. On 22 May, Pentecost, he had the honour to be received by the Empress in audience. He kissed her hand while genuflecting and she spoke to him instantaneously in a most affectionate manner, stressing how glad she was to see him. She showed him her face and arms mutilated by the illness, and told him how badly her family had been hit by this disease. She expressed her dearest hope that he would be able to save her two archdukes through inoculation, although that would depend on the opinion of Van Swieten (who was with them). She asked him if the Prince of Wales and his family were well, if they were all good health and whether the young Prince of Brunswick had already been inoculated. And how was the Prince of Orange with his princess, and did he look like his father? Etc. And she asked about the preparation of the inoculation and if the patients could take the open air. Then she said would you like to see my family (in French) and she opened the door to the next room where four archduchesses and two archdukes were waiting, of which he was honoured to kiss their right hands with a genuflection. (One of the archduchesses was the young Marie Antoinette.)

IngenHousz was first to demonstrate his mastery of the method. He inoculated sixteen children in the presence of Maria Theresia and Van Swieten. Later he inoculated another thirteen children while Emperor Joseph looked on. One of the children died, but not as a consequence of the treatment. All went convincingly well and IngenHousz was able to proceed, as described by Arneth:

> Es ist wohl nicht zu bezweifeln, dass es gleichfalls auf van Swietens betreiben geschah, wenn Maria Theresia sich im September 1768 entschloss, die Inokulation bei ihren beiden jüngeren Söhnen vornehmen zu lassen. Und Jozef befahl, dass das gleiche Verfahren auch bei seinem eigenen Töchterchen angewendet werde.
>
> Der berühmten Artz und Chemiker Johann Ingen-Housz, welcher zu diesem Ende eigens nach Wien berufen wurde, nahm am 10. september die Inokulation an den Erzherzogen Ferdinand und Maximilian und an der Erzhogin Therese vor. Mit höchster Spannung verfolgten nicht nur Maria Theresia und Jozef, sondern alle, die um den Vorgang wussten, den process; Fünf Tage nach der Einimpfung machte sich etwas fieber bemerkbar und am 18. zeigten sich die erste Blattern. Bis zum 24. hielt die Eiterung an und am 29. begann das Verdorren der Blattern. Mit grosser Verwunderung sah man, wie die Geimpften sich nicht zu Bette legen brauchten, sondern entweder zu Wagen oder zu Fuss sich häufig im Freien bewegten. Und so ungemeinen Wert legte man auf die Sache, dass nach ihrer glücklichen beendigung in der Schlosskapelle den Te Deum gesungen wurden.[35]

---

35   Arneth A Ritter von 1876

Maria Theresia decided to proceed, supported by Van Swieten. On 10 September 1768 the two young archdukes Ferdinand and Maximillian, fourteen and twelve years old, were variolated by IngenHousz. At the same time the oldest daughter of Joseph II was treated. That made headline news. Especially when everything went according to plan. After five days the fever started, followed three days later by the pustules, which started to dry on day nineteen. To the amazement of many, the inoculated children did not have to take bed rest, but could walk or drive around. To celebrate the happy endings, a Te Deum was sung in the palace chapel. On 29 September a big garden party was thrown in the Schlosspark of Schönbrunn to celebrate the successful treatment. Entrance was free to all. The Viennese newspapers prominently featured the news.[36] It was the best propaganda for the new preventive therapy one could hope for. Variolation was the talk of the day.

The Empress was extremely thankful towards the Dutch physician. A medal was minted, with on the front the image of Maria Theresia and Joseph II and on the flip side the text:

> Ferdinandus, Maximilanus eorumque neptis
> Theresia, archiduces Austriae, de insertis
> variolis restituti, 29. Sept. 1768.

IngenHousz received two golden and thirty silver reproductions of the medal, with the message to please his friends with these memorabilia. And he got more than just memory medallions. As compensation for his travel expenses he received 200 Ducats and was given 1200 Gilders for his stay in Vienna. He was asked to remain in Vienna as physician to the royal household, with the official title of Hofrat and was awarded an annual pension for life of 5000 Gilders, about 40,000 euro at present values. He was a well known man at the time; his name and his method were the talk of the day. Leopold Mozart writes to Lorenz Hagenauer in Salzburg on 6 August 1768:

> Do you know that inoculation for small pox is proceeding steadily here? The Empress has engaged the English inoculator to perform in a house near the Schöbrunn palace. Already 40 poor children have been inoculated satisfactorily. The Emperor and the Empress come to the centre almost daily and are completely won over by it.[37]

IngenHousz hit the headlines. The Gazette des Pays-Bas, a newspaper published 'avec privilege de sa majesté impériale apostolique', publishes an item on its front page of Thursday 19 May 1768 on the introduction of the new medical inoculation method in Brussels. Mr. Ingenhousse [sic] is described as one of

---

36   Wienerisches Diarium 1. Weinmonat (Oktober) 1768, Nr 79
37   cited by Jenkins 1999

the "Inoculateurs des plus éclairés & des plus habiles du continent ... qui vient d'être appellé à Vienne pour y établir ce salutaire préservatif contre le danger & la malignité de la petite vérole naturelle."[38]

On the road again

And it did not end there. The Empress' twenty-two year old son Leopold II, Archduke of Tuscany, had as yet been unexposed to smallpox. Maria Theresia asked her new Royal Physician to go and inoculate him. The entries in his diary give an inkling of the life he was leading:

> *Louis Mathi in dienst als knecht op 10 mei 1768 voor 12 fl / maandag Francois Ernest Manchet de 31 mei 1768 voor zelfde bedrag*

Two servants entering service, he notes every two months the wages he pays them, while in between he notes the expenses for the reparations of old shirts or the making of new ones: fine shirts with cuffs.

> *23 december heeft de Kyzerin my vereert met 30 silver en twe goude medailles wegens de inenting der 3 vorstelyke kinderen, verder rysgeld 200 ducaten voor de postpaarden en 1200 guldens voor 4 maanden rysvertering 't zamen 2100 florynen, alle in kremnitsche duc.*

The Emperess gives him money for the inoculation of the three emperial children, a travel stipend and enough money to cover his expenses during the voyage to come.

> *le domestique Antoin Skitard entre en service le premier de 1769*
> *op 30 december heeft hy 1500 florynen gegeven aan de Heer Stametz om het op de bank te plaatzen, moetende ik by myne terugkomst de obligatie tegen zyne quitantie te verwisselen.*

Another servant takes up service on the first day of 1769.

> *1769*
> *te Venetien*
> *kleed en lorgnet a ressort voor 3 sequinen alsook een silveren lepel vork en mes kostende 91 ivres of 4 sequinen en 3 livres, den koker kost 4 livres*
> *Napels*
> *snuyfdoosje van wit schilpazt met gout ingelegt, en witte en zwarte gebreyde zyde koussens*

---

38   Hasquin 1987

geeft op 3de paasdag een pakje mee van zyde koussen naar Breda met schipper van galy dat naar Amsterdam vaart voor zyn nigtje en zyn soeur.

In 1769 he arrives in Venice where he buys a lorgnet, as well as a silver spoon. Later in Naples he buys a sniff box, made of white turtle shell, with golden inlays. he also buys white and black silk stockings. On the third day after Easter he hands a parcel with stockings for his niece and her sister in Breda to a captain set to sail with his gallion for Amsterdam.

eygenhandige brief van Kyserin met een bankbrief van 8250 florijnen No32787 lopende den intrest van den 20 may 1769
te Luca gekogt de encyclopedie, de 12 tomen van stoffen en de 4 tomen van platen. de tome des matières kosten 2 gilliati per stuk, een toom met platen 3 gilliati., betaald totaal 47 seguinen ook voor inbinden en verpakken

In Luca he buys an encyclopedia (12 volumes and 4 volumes of plates) and pays for the binding and packaging.

10 juillet 1769 Florence koopt by Vincenzo Landi libraire voor de rekening van Sign Vincenzo Giatini in Luca
j'ai acheté de l'Abbé Fontana touts les tomes de l'ouvrage in folio de mr Manetti des oiseaux qui font à present 5 tomes, que j'ai payé contant a mr l'abbe.

On July 10th 1769 he buys in Firenze all volumes of the book on birds by Manetti for Abbé Fontana

4 nov. ben ik te wenen terug gekoemen. ik ontfing aanstonds een brief van Hoogeveen waar in een assignatie op den boekverkoper Greefer van 16 gl 16 st voor rekening van Hoogeveen
den Heer Stametz heeft my myne obligatie voor het een hem gelatene geld by myn vertrek overgegeven namentlyk 1500 Gl de nummer van de obligatie is 29645, den datum den 3 jan. 1769
11angenomen voor keukemeyd anna marie ornoerin
14 nov strosak, matras en bed voor de meyd betaalt 28 gl 10 kr

In a letter to his friend Deckers, back in Holland, IngenHousz expressed his thoughts and doubts about what he is doing:[39]

Die Inokulation des Grossherzogs von Toskana war mein grösster ärtzlicher Erfolg; aber er ruft mir auch die aussprechliche Unruhe ins Gedächtnis, in welche ich gerate war und mich lebhaft fühlen liess, dass meine Gesundheit

---

39   Letter to Deckers 10 May 1769, transcribed from French in German by Wiesner

angegriffen ist. Obschon die der Impfung gefolgte Erkrankung des Grossherzogs nur eine leichte war uns Seine königliche Hoheit nicht einen tag im Bette zuzubringen genötigt war, so musste ich doch stets mit der Möglichkeit rechnen, dass irgendein Zufall störend einwirken könnte. So musste ein Mann zittern, welcher berufen war, eine Sache von solcher Wichtigkeit auf sich zu nehmen. Ich bin der einzige Mensch auf der Welt, welcher vier der illustersten und teuersten Personen von Europa das leben gerettet hat. Ich habe nun nicht mehr die Ambition, noch andere Prinzen zu Impfen and hege nur den Wunsch, meine Tage in Ruhe zu beenden...

He starts to realise what he has achieved. He treated some of the most important men and women of Europe, and luckily all went well. But the responsability is devastating. He does not want to inoculate any princes anymore and wants to spend his days in peace and quiet.

Not only the inoculation went well, the stay in Tuscany was inspiring. He enjoyed a stay at the Archduke palace Palazzo Pitti in town and in the countryside at the Villa Poggio Imperiale. He met with several likeminded people, such as Abbé Felice Fontana. Fontana was only a couple of months younger than IngenHousz and they shared the same interests. Fontana was a physicist who taught at the University of Pisa and later became the director of the Museum of Natural History in Florence.

When autumn arrived, he was back in Vienna. There he would continue his lifelong advocacy of inoculation, inoculating - on his many travels - children and adults, often the poor of the countryside and battling the prejudices that he encountered almost everywhere. Instigated by his enthusiasm, an inoculation centre for the children of the Austrian aristocracy was established at Schloss Sankt Veit, the summer residence of the Archbishop. For the rest, he enjoyed freedom and independence through the massive stipend he received from the court. He had the material means and the intellectual freedom to pursue his curiosity about all things natural.

And in the mean time, he tried to handle his quite affluent income and build it into a little financial capital, buying bonds and shares, which were handled through his banker Stametz in Vienna. He travelled widely, visiting other 'savants' and 'natural philosophers', exchanging ideas, experimental techniques and theoretical notions. And nowhere did he forget to go shopping. He seemed to enjoy the luxury of buying quality materials and interesting gadgets, as he wrote in his diary:

1770
29 may te Praag
5 july koopt een slot met een gehym om den sleutel terug te trekken wanneer hy een vierde part van een cirkel heeft gedaan voor 35 grossen en een kleyn met

een selfde gehym voor 2 guld.
21 aout koopt aandelen in goud en silver myn in Hongaryen by de grens met
Transsylvanien

While in Prague, he buys a lock with a secret mechanism, and a smaller lock of the same type. The big one costs 35 grossen, the small one 2 gilders.

vertrekt op reis 21 september, laat kostbaarheden en geld en sleutel van
appartement aan jean henry Stametz
krijgt koets en reisgeld van den graave van Metternich,
arriveert 23 october te strasburg, daar laten maken een rondingot met kamisool

He goes on a trip to Strasbourg on 21 September, leaves his valuables, money and the key to his appartment with Stametz. He gets a carriage and travel money from Duke Metternich. When he arrives on the 23rd he gets the tailor make him a rondingot with a camisole.

13 nov in Baal in Zwitserland fyn lynwaad gekogt
17 dec ontfangt van mr De Saussure professeur en phil a geneve six ducats
kremnitz pour les deux premiers tomes du Hortus botan; vindobom. pour le
comte de mr Jaquin
1771
25 jan gekogt by Ferdinand Berthoud geleerd en befaamd horlogie maker te
Parys een repetitie horloge met een eenvoudge kas voor 850 livres. schenkt ook
horloge aan syn knegt voor de ongemakken van de rys.
febr schenkt duyzend guldens aan zyn broer in breda

On 13 November he arrives in Basle where he buys fine linen. In Geneva he hands over a botanical book from Jacquin to De Saussure, receiving six ducats for it. By January 1771 he is in Paris where he buys a watch from the famous watchmaker Berthoud. He also gives a watch as a present to his servant, as compensation for the discomforts of the voyage. He sends a thousand gilders to his brother in Breda.

There might have been another, secret, reason why IngenHousz travelled to Paris. New evidence[40] reveals that he had been deputised by Maria Theresia to investigate why her daughter, Marie Antoinette, who married the French dauphin on 16 May 1770, had not yet conceived a child. IngenHousz was unable to get near the young lady, despite the explicit mandate from her mother. Etiquette at the French court must have prevented him, his rank being too inferior. Neither his diary, nor any letter mentions this. The diary continues on arrival in London.

---

40   Arneth 1874, quoted by IngenHousz 2005

maart 13 te Londen gearriveerd en by mr Pitters gaan woonen in
Northhumberland court charing cross voor 20 sch 2 weeks
stof voor rok broek en camisool, ook voor syn knegt, en voor bath coat

On 13 March he arrives in London and stays with Mr Pitters in Northhumberland Court at Charing Cross, for 20 shillings for two weeks. He buys more cloth for clothing for himself and his servant.

Curiously enough he does not mention in his diary that on 21 March 1771 he is admitted as a new fellow to the Royal Society. He had been elected on 25th May 1769. The certificate at the Royal Society dated 16th Nov 1769 reads:

John Ingen-housz Doctor of Physic, now residing at Vienna, & Physician to
her Imperial majesty, A gentleman of great merit, & well versed in natural
knowledge, being desirous of being elected a fellow of the Royal Society upon the
inland list, is recom[m]ended by us on our personal acquaintance with him, as
highly deserving that honour

The promotors were W. Watson, Rd. Huck, Jno Blair, G. Baker, John Pringle, W Watson Junr, W. Heberden, B. Franklin, Gowin Knight, James Parsons and M. Maty. This was quite an honour for a foreigner who were not easily admitted to the Royal Society. It was an extra reason for joyful reunions and toasting with old friends, who congratulated him for his succesful activities in Vienna and his promotion to imperial physician. In his diary one finds only very prozaic notes on everyday business.

2 may koop pensylvania fire place met appolo's head on front voor Prince Carel
van Lotteringen
8 june op zondag aan den heer Franklin betaalt voor plated silver werk van cheffield
20 april briefwisseling van Stametz dat hij interest heeft ontvangen voor zijn
kapitaal op de bank
in july ontfangen de eerste zes maanden lyfrent en 1770 action des fermes

On 2 May he buys a fire place for Prince Van Lotheringen. On 8 June he pays Benjamin Franklin for plated silver work. On the 20th he receives a message from Stametz from Vienna about interests on his capital in the bank. And in July he receives his first six months of annuity.

In May he travelled through the towns of Northern England and visited several industries, together with Benjamin Franklin, accompanied by two businessmen from the American colonies John Canton and Jonathan Williams. They travelled via Matlock and Bakewell, where they visited a marble mill, examining the water-driven saws and polishers. In Castleton they explored the Peak Cavern. They visited Manchester, saw the Bridgewater Canal and mines before they doubled back across the Pennines towards Leeds. There they met Joseph Priestley who showed them some experiments. They took in silver-plating in Sheffield and iron-smelting at Rotherham. From there, they headed for Burton-on-Trent, "remarkable for good Ale" and journeyed on to Erasmus Darwin at Lichfield to end their packed itinerary via Birmingham, Boulton and Soho.[41] The industrial revolution was building up steam and for both IngenHousz and Franklin everything must have been extremely fascinating. They fostered a curious mind and anything must have been of interest to these men. Everything was relevant, the division between pure and applied science, between research and development, had yet to be made. The whole world was their laboratory.

Back in Vienna his diary picks up the pace of life.

> 1772
> 14 Mars aangekomen a Vienne
> rekent af met knegt, speciaal ook voor de rys, en betaalt 14 april tegen quitantie voor een maand koetzhuur 6 fl en aan den koetzier 3 fl 20 kr t zamen 5 souverainen

On 14 March 1772 he is back in Vienna, pays his servant, also extra for the long travelling, and pays in exchange for an invoice for the carriage and the driver.

> juliet
> maria catharina geboren 20 maart 1765 N; ik ben peter van arnoldus en Theresia
> Petrus josephus guilielmus 4 may 1766
> Henricus Fernandinus 22 april 1767
> Joannes 13 july 1768
> Ludovicus 23 aug 1770 gestorven den 2 nov 1774
> Theresia 2 dec 1771
> Gasper 10 jan 1773 gestorven den 16 sept 1783
> Elisabetha 4 febr 1774
> Joanna Maria 17 maart 1775 gestorven den 3 nov 1775
> josina agatha 21 octobre 1776
> Ludovicus 28 novem 1778

In July he lists the names and ages of his cousins in Breda. He is godfather to Arnoldus and Theresia. They were born in the natural quick succesion of the big families of the time, three out of eleven expired. IngenHousz must have added some of these dates retrospectively after 1772.

---

41   Uglow 2002

july krijgt voorschot voor onder andere de onkosten voor de postpaarden van Wenen naar Florence en wederom terug, zullende ik daarna rekening aan de heer van Meyer doen

In July, he receives an advance on his expenses for another trip to Italy. There was more inoculating to be done. Maria Theresia asked him to take care of the inoculation of some more nephews and nieces in Florence.

aout 30 arrivé a Florence
den negende september heb ik op het palys van Poggio Imperiale by Florence de kinderpokjes ingeent aan hare Koninklyke Hoogheden den Aertshertog Franciscus erfprins van Toscanen en syne suster de Aertshertogin Maria Anna. In de selvde maand heb ik ook de volgende personen ingeent Josepha Deblands 18 jaren out, geboren in Holmutz, Elisabth Francois oud negentien jaren, kamerist van de Aertshertogin Maria Anna. Joannes Hopf out 10 jaren, Joseph Hopf out 8 jaren, Maria Francisca Rosa out 4 jaren, Maria Clara Politi out elf jaren, maria anna politi out 7 jaren, joseph politi out 9 jaren, Franciscus Politi out 5 jaren, Ludovicus Politi out 2 jaren, Leopoldus Politi out elf maanden. Allen hebben de ziekte bekomen behalve drie, namelijk Josepha Deblands, Elisabeth Francois en maria anna politi. Deze twee hadden de kinderziekte jong zynde gehad, maar waren bedugt dat dezelve terug konde komen, omdat zy slegs wynig pokjes gehad hadden.

On 30th he arrives in Florence. On 9 September he grafts the smallpox on Archduke Fransiscus and his sister Archduchess Maria Anna. Later in the same month he inoculates eleven more children of the household, the youngest only eleven months old. All became ill as expected, two of the three others had had the illness when they were young, but were afraid of recurrence as they only had developed a mild form.

Electrifying experiments

On 1 January 1773 on his way North, he arrived in Livorno, where he performed an experiment on an electric ray or 'torpedo'.[42] This ray is a fish known since the Antiquities for its electrical capacities. The Roman physician Scribonius Largus used this swimming battery in the first century to treat a variety of diseases. He placed weakened fish over the brows of people suffering severe headaches. Livorno was one of the hotspots for torpedo research in the seventeenth century, together with Pisa and Firenze. After the middle of the eighteenth century evidence was accumulating to suggest the possible electric nature of the shock

42   IngenHousz 1775

produced by some species of fish. John Walsh, who frequented the circle of the Royal Society related philosophers in London interested in electric phenomena such as Benjamin Franklin and Joseph Priestley, had been doing research on these fish in the summer of 1772.[43] The results were only published in the Transactions in 1773, after IngenHousz set up his experiments in the Adriatic. But IngenHousz, with his 'electric' connections, probably knew about them. His fellows at the Royal Society such as John Hunter and John Walsh were among those fascinated by electricity and its possible application in health care. When passing through Livorno, he could not let go of this occasion to experiment on the famous fish himself.

IngenHousz hired a fishing boat and a crew of eighteen and sailed out to catch five rays. He compared the electric shock one gets from a charged leyden jar, which he took with him, to the discharge from the fishes. He compared the shocks he got of the fish in or out of the water, compared between big and small fish, varied the ways he held them to see if that changes the charges felt and puts a brass chain to the back of the fish to find out if that would give a difference. Suspending it by a dry silk ribbon, he examined if the fish attracted light bodies. He tried to charge a bottle by contact with the fish. In the dark he observed no sparks, and heard no crackling noises. He put some men holding hands in a line to see how the charge carried on through these conducting bodies. After these in vivo experiments, he dissected the fishes to find the nervous tissue that appears to be the reservoirs of the electric power. He concluded:

> I waited for a better opportunity to make a good many more experiments, and repeat, with more care, those already made. But though I have not been fortunate enough to find any other than what is above mentioned, I thought it my duty to inform you what I have attempted, howsoever incomplete my researches have been. And if ever any further opportunity should offer, and in a better season, I shall not fail to make all the experiments I can think of, for illustrating so curious and interesting a subject. I am, &c.

He wrote this report in Salzburg on 27 March and sent it to the Royal Society in London, where it is read 10 November 1774. IngenHousz was on his way back home in Vienna where to take up his work, experiments in various fields of investigation, combined with his advocacy work for smallpox inoculation. And he was in close contact with the imperial family, discussing topics of scientific and governemental importance. As is testified by Benjamin Franklin in his 1775 letter to Philip Mazzei, an Italian physician and promotor of liberty:

> The Congress have not yet extended their views much in relation to foreign owers. They are nevertheless obliged by your kind offers of your service, which

---

perhaps in a year or two more may become very useful to them. I am myself much pleased, that you have sent a Translation of our Declaration to the Grand Duke; because having a high esteem of the Character of that Prince, and of the whole Imperial Family, from the accounts given me of them by my friend Dr. Ingenhousz and yourself.[44]

On 3 November 1775 he wrote a letter to John Pringle, which is read to the Royal Society on 15th February 1776: "Early methods of measuring the diminution of Bulk, taking place upon the Mixture of common Air and nitrous Air; together with experiments on Platina."[45] It opens with:

Some time ago I amused myself with some experiments relating to nitrous air. Having received from the learned Abbé Fontana a copy of a pamphlet, which he published this year under the title, Decrizione e usi di alcuni stromenti per misurare la salubrita del aria, di felice Fontana. In Firenze, l'anno MDCCLXXV.

As IngenHousz wrote, research was fun for him. He enjoyed the freedom to explore the wonders of the world, be it electrical, chemical or medical. With these first recorded experiments on gasses he opened up a field of investigation that was new to him. He must have learned about Fontana's experiments on various kinds of air when the two met on IngenHousz' last trip to Italy. Fontana's experiments built onto the findings of Priestley who found that mixing nitrous air (what we now would call nitric oxide) and common air leads to a diminishing of the total quantity of air. The amount of this diminishing is according to Priestley a criterion for the degree of salubrity of the common air tested. The salubrity or goodness of the common air was something of particular interest to IngenHousz, the doctor. Priestley and Fontana were the forerunners of the 'pneumatic physicians'. These were the men who had realised that there were different kinds of 'air', an invisible fluid that filled the atmosphere and which could have profound influence on the well-being of human and animal life.

In search of good air

Priestley stood at the origin of this field of investigation.[46] In 1767 he had started to study fixed air (carbon dioxide). He showed that this air was soluble in water and that its solubility increased with pressure. This led to the publication in the summer of 1772 of a pamphlet[47] on the making of seltzer water, previously

---

44   Franklin 1775
45   IngenHousz 1776
46   Schofield 1997
47   Priestley 1772b

only available from some natural mineral springs. This new water was supposed to be able to deliver medical benefits (possibly preventing scurvy) and drew wide attention, also overseas. While manipulating all these fluids and gases, Priestley mastered the use of the 'pneumatic through'. This device was in essence not much more than an inverted glass vessel, in which you could mix various gases over a tub filled with water or mercury, sealing its contents from the atmosphere. In the course of 1772 Priestley started presenting the results of his pneumatic experiments to the Royal Society, resulting later that year in the article "Observations on different kinds of air"[48] which he sent to Sweden and which was translated in French by mid-1773.

This experimental work and especially his artificial sparkling water made quite an impression and was a direct cause for being awarded the prestigious Copley Medal from the Royal Society. In his speech at the award ceremony in November 1773, John Pringle painted a heroic future for pneumatic medicine. Pringle being a medical doctor himself was in an ideal position to paint the scope and importance of this new discipline. He pointed out that Priestley's test for the salubrity of the air had environmental and therapeutic implications. As he was speaking, Pringles words were coming true. Priestley's apparatus and manipulative techniques brought chemical, physiological and medical questions within the realm of empirical verification. His publications were inspiring and influenced many investigators all over Europe. Fontana and IngenHousz were just two of them.

IngenHousz' 1775 paper is his first written account of pneumatic experiments. No matter the fun he may have had, he found it difficult to perform the experiment properly. One has to mix two samples of air and "I found it a difficult matter to force always the same quantity of nitrous air into the vessel… If this quantity is not always the same, some variety must happen in every experiment; and thus an exact valuation of the quantity absorbed cannot well be made." Therefore he describes how he contrived an instrument by which application he hopes to improve the experimental setup.

He was not the only one struggling with this problem. Priestley's paper had been translated into Italian, and Fontana in Tuscany and Landriani in Milan were at work improving his methods. In 1775, Landriani would coin the word 'eudiometer' for the instrument he had developed to measure the "good quantity of the weather" or the goodness of common air. This tool would allow a standardised administration of Priestley's nitrous air test. Fontana developed his version of the instrument, while IngenHousz was contributing his suggestions for improvement. As indicated by Schaffer, the discussions among the pneumatic physicians dramatized the problems of replication of the test

---

48   Priestley 1772. As Schofield has indicated, this is a brilliant paper, but confusing in form and content. Though read officially to the Royal Society in November 1772, it is endorsed at its beginning with 'Read March 5, 12, 19, 26, 1772', while it clearly contains editing and additions added througout the summer of 1772

and the manner in which the transmission of standard instrumentation also involved the extension and transmission of Priestley's research programme.[49]

IngenHousz also had a go at perfecting the way of measuring the goodness of the air. In his article he makes it explicit that it is very important to make sure that the amount of air used is always the same. He has constructed a device with separate chambers, tubes and vessels, that are separated by airtight cocks, to be opened and shut at will to make fluids or gases enter the desired recipient. He even designs and tests two different installations. The glass tube in which he collects the mix of common air and nitrous air is divided into a number of equal parts, as to be able to measure its contents. Still, he concludes: "It requires some practice to perform this experiment with dexterity." He and his colleagues were developing an experimental methodology and they knew their procedures were far from perfect. Communicating about them, on something like the platform of the *Philosophical Transactions* of the Royal Society, was enhancing the process of dissemination of knowledge.

By the time this report is read in London, IngenHousz takes a big step in life. Life is more than work, even for a man as tirelessly curious as IngenHousz. There is evidence from letters written soon after IngenHousz first arrived in Vienna that, as a court official, he would be expected to marry.[50] But he had grown up in a man's world, brought up by his widowed father, with just a brother and never having known his sister who died very young. He went to a single-sex school and university was an exclusive male environment too. But at the age of 44, on 25 November 1775, he writes two letters. One to his brother in Breda, one to his friend Pringle in London, two letters that are summarized in his letter-logbook[51] as follows:

> Broer dat ik heden getrowd ben met juffrouw Agatha Maria Jacquin Van Luyden, vyf of zes jaren jonger dan ik
> Pringle that I am married today with Agatha maria Jacquin
> send him some pieces of opal from Hungary

The day before on the 24th (so says the official document in the Vienna archives[52]) he married Agatha Maria Jacquin, 40 years old, the sister of another Dutch expat in Vienna, Nicolaas Jacquin. He and Sebastian Jacquet were the witnesses at the wedding. Jacquin was professor in botany at the Vienna University and, like Van Swieten and IngenHousz, had left Leyden behind as a Catholic without a future in that inhospitable protestant world.

---

49 Schaffer 1984
50 Beale 1999
51 Breda Collectie Ingen Housz, A6
52 Wiesner 1905 Anhang 1, p 237

A fresh wind

Vienna was going through a rejuvenation phase. With the death of Joseph I
and his son Joseph II taking the throne as co-regent with his widow-mother,
a fresh ideological breeze was blowing through Austria. Inoculation was just
one of these new ideas blowing in from the west. The older generation, such
as that of Van Swieten, was having trouble to accommodate to the signs of
modernisation. But the young Emperor did not hide his keen interest in this
progressive body of thought and openly declared his admiration for French
enlightened thinkers such as Voltaire. Not only in and around the palace were
signs of cultural and social emancipation emerging. Coffee houses started to play
a role of small scale centres where news and ideas could be exchanged. Coffee
offered an alternative to the drinking of alcohol from early morning. It was a
pharmacological stimulant that must have acted as a booster for intellectual
activity of those gathered in the European coffeehouses.[53] Coffeehouses were just
one of the new reading spaces, together with the salon, the private library and
the cabinet were new reading materials with new genres, increased print runs
and smaller and cheaper formats were instigating new reading practices and
spreading knowledge on a wider scale than ever before.[54]

Travelling around and meeting the many intellectuals, politicians and natural
philosophers all over Europe became part of IngenHousz' natural environment.
He did have enough money to do it, got the permission from his bosses and the
introductory letters that opened many doors in remote towns and palaces. His
proficiency in a handful of languages was a bonus for communication. And as
far as one can deduce anything about his character from his writings, he must
have been able to blend in easily with new people in ever fresh circumstances.
In 1778 he traveled back to London. His diary, in his peculiar Frenglish, says:

London 1778
jan. Le 12 arrived in London
le 13 Thursday, I took a lodging at mrs jamison palmal court, palmal N2 for
20 sch a week

at Dollond bought a pair of spectacles with double joints in silver
a case to put them in
a little glass tube with two balls in a wooden case
with one end in amber, the other in ivory for the position and negative electricity

at Parker & croskeys fleet street
three ink glasses plain of the new invention at six pence a piece

53   Standage 2005
54   Johns 2003

> *at the same 12 pieces of yellow glass*
> *at Dollond*
> *a little hand microscope with two plano convex glasses in bras*
> *J Tashe Compton street Soho*
> *24 artificial gems for to make impressions at 10 pence a piece*
> *one in relief in a red stone greater*

In the following pages he noted his purchases of clothes, tickets for a masked ball at the Pantheon (it was not all serious science), glassware for his experiments, a refraction telescope with a 20 inch mirror, among many other household items and apparel.

In the mean time Felice Fontana had arrived in London too. On 11 March 1779 he read his article tot the Royal Society. Entitled "Experiments and observations on the inflammable air breathed by various animals"[55] he contributed to the raging debate about the supposed qualities of various gases. Inflammable air is what now is called hydrogen gas. Contradicting opinions were circulating about this air. Priestley believed it was noxious for animals; Scheele was convinced of the opposite. Fontana began to suspect that maybe both could be right and that "their so contradictory effects might be owing to some circumstance not yet attended to."

He performed not only experiments on birds and quadrupeds that would not pass a twentieth century ethical commission, but tried also some hydrogen gas on himself. Sometimes he observed ill effects, sometimes not. After some experimenting he hypothesized that the final effect of the inhaled hydrogen gas was dependent on the ratio between the inhaled volume gas and the respiratory volume of the organism. Filling the lungs completely and exclusively with hydrogen gas was noxious. He concluded that organisms need common air to support life and that this air, after being inspired, exits from the lungs less suitable to be breathed in a second time.

Central in his argument were the results of the nitrous test, which he used constantly to monitor the quality of the airs involved. 'My method' as he called it, had no name yet and he briefly explained it, promising to come back to it more extensively in a later paper.

But first it was IngenHousz' turn to climb the rostrum. On 25 March he delivered a lecture to the Royal Society in which he contributed to the ongoing debate about the characteristics and the possible advantages of various gases.[56] He opened with what can be seen as almost a declaration of love to this field of investigation:

---

55   Fontana 1779
56   IngenHousz 1779

The important discoveries on different kinds of air have opened a new field for one of the most pleasing and interesting scenes which ever were exposed to the contemplation of philosophers, and therefore could not fail of exciting in almost every lover of natural knowledge a decided curiosity to see the pursuit of such striking novelties, and in many almost irresistable temptation to imitate them, and to pursue farther the road already so far opened by the Rev. Dr. PRIESTLEY, Abbe FONTANA, Mr. LAVOISIER, and many other learned and ingenious men.

He referred to Priestley as the man who first discovered this wonderful pure aerial fluid and called it 'dephlogisticated air', which can be extracted from various substances and materials. He referred to Fontana, whom he assisted in October 1776 in Paris in his extraction experiments with water. And he looked forward to the discovery of new methods to easily produce copious quantities of this beneficial air "as will serve to cure several diseases which resist the power of all remedies, and so prolong, as it were, human life."

As a doctor he had some working hypothesis on how that could work. He suspected that too much phlogiston could accumulate in the blood, from which it might be released and perhaps cause fevers and other symptoms. (The causes of which had not yet been clearly elucidated by the best medical writers.) The benefits we derive from respiration is exactly the ventilation of phlogistic particles from the blood.

intermezzo

Understanding oxygen in the era of phlogiston, 18th century chemistry for the 21th century reader

Until the end of the eighteenth century, the kinds of matter known to man were solids or fluids. It was supposed that air was something material and that there were different kinds of air, some good and some bad, because some supported life, others were deadly or at least noxious. But not much was known at all. Not being able to separate, transfer, mix or weigh them explains why the chemists before 1750 were rather more like alchemists. Everybody could see that iron rusts. But how could that phenomenon be explained, not knowing that rusting is the process of oxygen from the air binding to the metal. Alchemists thought that in becoming rusty, and thus losing their value, metals must lose something. And the thing they lost was baptised "phlogiston".

Rusting, burning and all other processes that we now call oxidations were according to those pionering pneumochemists[57] to be explained as a loss of phlogiston (from the Greek for 'setting on fire'). Burning charcoal was to them a process in which phlogiston escaped and became combined with the air. The fact that combustion sooner or later ceased in an enclosed space was the evidence for the fact that air could only absorb a limited amount of phlogiston. Air saturated with phlogiston could not support life, for respiration removes phlogiston from the body with the help of air.

But when in 1770 Lavoisier started using his very sensitive balances, he found that rusted metal actually weighs more than the metal before the rust. Phlogistonists then proclaimed that phlogiston therefore had to have a negative weight. But at about the same time, in the years between 1770 and 1785, chemists all over Europe started catching, isolating and weighing different gases. Atmospheric air was found to be composed of two kinds of air: oxygen and nitrogen. The name for 'oxygen' was derived from the Greek 'acid-former' in the mistaken belief that oxygen was necessary for the formation of all acids. It is indeed necessary for some, the ones Lavoisier & co knew at that time such as sulphuric acid and nitric acid. But others, such as hydrochloric acid, do not. Oxygen is chemically very reactive and consumed in burning and respiration. The other gas was nitrogen or niter-generating gas, and was then called azote, from the Greek 'azoe', meaning 'not sustaining life'.

Not only gases were dissected in the new chemistry. Water was found to be a combination of oxygen and hydrogen (literally "the water-generating gas"). The unmasking of water as a composite was the last nail in the coffin of the Aristotelian doctrine of the four elements.[58] Another gas, formerly known as 'fixed air' was the result of animal respiration, burning wood or heating chalk turned out to be a combination of carbon and oxygen. It is what we now call carbon dioxide. These discoveries of new and hitherto unseen elements delivered new pieces to put the puzzle of nature together in another way. Many processes in nature turned out to be a reshuffling of the basic building blocks. Another insight was that the weight of all products after a reaction must be equal to the weight of the reactants one had started with. This law of conservation of matter became one of the basic insights of the modern chemistry. Phlogiston was no more the mysterious and powerful principle that flew unhindered from one substance to another. From then on "phlogiston" was simply equal to "minus oxygen". "Dephlogisticated air" was therefore air to which oxygen had been added.

Phlogiston had been an invisible 'fluid'. One could not hold a handful of it, just as one could not hold a handful of magnetism or electricity. One could only see these powers of nature in the effects they produced.[59] Combustion

57 Rabinovitch 1969
58 Strathern 2000
59 Crosland 2000

and respiration were some of the processes in which one could see its effects. Priestley's famous experiments showed that mice live longer in what he called "dephlogisticated air". Depriving such air of more than its normal share of phlogiston would explain why it took longer for the air to be used up or to be "phlogisticated". This phlogisticated air was the gas that was left over after the mouse had exhausted all dephlogisticated air. (We would now say that only nitrogen gas is left after all oxygen has been respired and replaced with carbon dioxide, then called fixed air.) The atmosphere consisted of phlogisticated and dephlogisticated air and as IngenHousz would say that air could depart from its simple nature by the addition or the substraction of something.[60] Priestley would see plants as taking something (phlogiston) away from the air, IngenHousz interpreted the same as plants adding something (dephlogisticated air) to the air.

Another merit of Priestley was his discovery of how the addition of nitrous test leads to the absorbtion or destruction of the tested respirable air, giving a degree of goodness. For the moment IngenHousz himself was not adding to that research but proposed an easy method of producing inflammable air, as well as some considerations on the "wonderful substance of gunpowder", both good for some nice philosophical amusements.

He reported on some curious experiments he attended in November 1777 in Amsterdam where Aeneae and Cuthbertson demonstrated explosive and inflammable airs of different kinds. They had developed a sort of pistol that could fire a bullet into an oak board by exploding a mixture of common and inflammable air detonated by an electrical spark. This was based on an invention by Alessandro Volta after experiments with flammable air (methane) from marshes in 1776. When IngenHousz arrived back in London in January he pursued some ideas on how to produce amounts of explosive gas to fire this "chemical pistol" reliably. The use of "vitriolic aether" turned out to be rather useful, despite its volatility, concluding "the ready production of such inflammable air always ready for use, when an explosion is intended to be produced, may be of some importance to philosophers whose time must be sparingly taken up."

In an appendix he described how he, assisted by Fontana, mixed dephlogisticated air with inflammable air, which produced loud explosions when ignited by a flame. Laconically he noted that if they would not have used a strong glass phial, it would have been shattered to pieces. The two men were playing around like two little boys in a toy shop. They abandoned the glass tube pistol and acquired a brass one, which burst apart at the first trial. Luckily everything happened without serious accidents or bodily harm. These exciting capacities of

---

60   IngenHousz 1787

this salubrious gas made him dream of the wonderful and advantageous effects this gas could have when produced in great quantities in the near future.

Fontana and IngenHousz seemed to be working as a team and revealed their results in tandem. In his lecture[61] on 26th April Fontana gave the audience an account "of some experiments which I made at Paris in the years 1777 and 1778 on the air extracted from various kinds of waters", the experiments IngenHousz alluded to in his previous lecture. Fontana boiled various sorts of water (from wells, from the Seine, &c) and put these airs to the nitrous test. He not only extracted air from water, but also exposed waters deprived of air which consequently imbibe the atmospherical air. Clearly, gases can be taken up and given off by fluids like water. These findings were a suitable occasion to expand again on the nitrous test. He stressed that this method had its flaws and that one should be very careful in performing it. He expressed his doubts about all these tests of the salubrity of common air that had been performed by so many authors in so many places.

> Because the method they used was far from being exact, the elements or ingredients for the experiment were unknown and uncertain, and the results very different from one another. When all the errors are corrected, it will be found, that the difference between the air of one country and that of another, at different times, is much less than what commonly believed, and that the great differences found by various observers are owing to the fallacious effects of uncertain methods. This I advance from experience: for when I was in the same error, I found very great differences between the results of the experiments of this nature which ought to have been similiar; which diversities I attributed to myself rather than to the method I used.

He recounted how he examined the air at different locations in the city and on a hill outside the centre, on places abundant with putrid substances and impure exhalations, and he found no big differences. In London he measured the air at different heights (ground level, first floor, rooftop, at the gallery of St Paul's, in the street) and found scarcely any sensible difference. To be on the safe side, he took Mr Cavallo to assist him, "so that a mistake can hardly be suspected". Still, he believed that

> ... it to be a very useful inquiry for mankind, because we do not yet know how far one kind of air more than another may contribute to a perfect state of health; nor at what time small differences may become very considerable, when one continues to breath the same kind of air for whole years, especially in some kind of diseases. An exact method of examining the goodness of common air may even be useful to posterity, in order to ascertain whether our atmosphere

---

61   Fontana 1779b

degenerates in a length of time. This curious inquiry, together with the method,
&c. are the production of this eighteenth century; and our descendants must
have some gratitude for the philosophers who found out, as well as for those
who improved it. If our ancestors had known and transmitted it to us, we
should, perhaps, at present be able to judge of one of the greatest changes of our
globe, a change which very nearly interests human life.

Fontana and IngenHousz, being close friends, like-minded philosophers
and collaborators, must have discussed the nitrous test extensively, probably
performing the test together to master and refine the technique. To IngenHousz
it was the ideal warm-up for a series of experiments he was thinking about and
probably planning to execute in the near future. And thinking about the future,
he wrote to his wife on 14 May to ask her if she would mind when he stayed on
a little longer. He had things to do.

Still, before starting his own research into the air that plants produce, he
delivered a talk on another of his favoured topics. On 4 June 1778 IngenHousz
presented the Baker Lecture to the Royal Society. The Baker lecture was a
yearly lecture set up by Henry Baker, son-in-law of Daniel Defoe (who wrote
Robinson Crusoe in 1719). These lectures were to be dedicated to the advances in
experimental philosophy and the lecturer had to demonstrate some genuinely
new discovery. To be asked to deliver the Baker Lecture was an honour and
a privilege. IngenHousz' talk concerned his experiments with electricity and
used the occasion to elaborate on the theory of his friend Franklin: "Electrical
experiments, to explain how far the phenomenon of the electrophorus may be
accounted for by Dr. Franklin's theory of positive and negative electricity; being
the annual lecture instituted by the will of Henry Baker".[62]

The electrophorus was an instrument developed by Alessandro Volta.
IngenHousz got one from the archduke Ferdinand. He described the
experiments he and other 'electricians' have performed with it. Some thought
their findings were overturning the theory of Benjamin Franklin. IngenHousz
set out to explain that that is not the case.

His Bakerian lecture was only a little excursion on the side, he was making
plans to retreat to a newly rented country house and on June 1779 he entered
the following purchases in his log book. The first column gives the pounds, the
second the shillings, the third the pennies.

juin			
to some elastic gumbottles	*0*	*10*	*0*
to some glasses for experiments	*0*	*12*	*0*
to the brass pistol of my invention with some			
other things of mr Nairn	*6*	*6*	*0*

---

62   IngenHousz 1778

to three deal tables for the country house	0	12	0
to half a dozin table knifes and forkes	0	7	0
to a cushion for the arm chair	0	2	0
to a mahogany dining table	0	18	0
for a brass [sukot?] of the invention of abbe Fontana for to contain the glass measure of air	0	9	0
for another dito thikker and stronger	0	10	0
for a tub for experiments on air and a stand to put it upon	0	18	0
to different articles of linen for the house keeping at southall green	14	6	8
july			
to some glasses for experiments on air	0	18	6

He moved to a house in the nearby country side in Southall Green, some ten miles from London and made preparations for performing some experiments on plants and airs. He didn't only buy the necessary laboratory equipment, but also the things he needed for moving into a new house. "I disengaged myself from the noise of the metropolis, and retired to a small villa, where I was out of the way of being interrupted by any body in the contemplation of nature."[63] By the end of July he is all set up and starts a series of experiments, of which he keeps a meticulously detailed logbook.[64] The first page reads:

> 1779 july 27
> 1. about an ounce M of infl. air was put in a jar containing about 4 ? with a few plants of lenticula aquatica during 24 hours within the house. the same was done in another similar vessel. the plants were about all under or upon the surface of the water, almost none above it in the air.
> 2. a great quantity of lenticula was put in a similar jar so as to be almost all above the water, or in the air it self;
> 3. a portion of the same inflammable air was put in a bottel with an entire plant of piperitis, so that the root was in the air, and the plant in the water.
> 4. a portion of the same air was put in a seperate vessel for to serve as a standart.
> 5. some pieces of salvia were cut of an immediately put underwater in a cylind. jarr and exported to the open air in the sun shine during 48 hours.

[In the margin he notes]

> these experiments are repeated afterwards better, the tryals according the method of A Fontana

63   as quoted in Beale 1999
64   Breda Collectie IngenHousz A11

as the <u>lenticula</u> has corrected the inflammable air I wanted to try whether it would yield good air or correct also common air, therefore:

6. a good deal of lenticula was put in a bottle full of water ... mooted in water[?]... without any air in it and exposed to the air during 48 hours (the sun made those two days very seldom her appearance)

7. another portion of lenticula was put in an ? bottel and a portion of air was let in it, about 3 ounce m. it stood as the former N.G. [in the margin: 251, 351, it extinguished the candle]

8. a good deal of leaves and buts of <u>lauroceratus</u> were put in a bottel without air, in the full day. it gave 178 161 202 247 294 la flame y brilloit

9. an other portion of the same plants was put as in (8.) with about 3 measures of common air. it gave 188 240 336

10. a new portion of salvia was put in a vase of water without air, and kept 24 hours 183 164 146 227 325

11. an other portion of salvia was put in a bottle with about 3 ounces measures of common air

12. good portion (handful) of gras was put (as the salvia in e. 10) without air

13. an other portion of gras was put as in e.11 with 3 ounces M. of Com. air

14. a good portion of pipritis with roots was put in a bottle without air as in e. 10

15. an other portion of Pipritis with roots was put in with some air

16. a portion of a plant as gras which grows under water in the reat pond upon our green was put without water as in e.10

[in the margin in front of e.10]

air from persilly taken in the middel of the day gave july 29 194 180 176 230
air from lauroceratus standing 24 hours gave 178 161 202 297 394 july 30
air from gras kept 24 hours (july 30) 186 170 154 228 326
air from different plants taken early in the morning about 7 or 8 gave (july 30) 186 166 145 171 269
31 july air which is naturally contained in the vagina or hollow leaves of oignon, gave 193 213 312
31 july air taken in a white jar in 2 hours of the middel of a fine day from the leaves of salvia potatoes currents, maliva gave in two different trials
179 159 166 260 355
186 173 168 262 355
NB the second trial was made with new nitrous air
air from salvia taken the same day in the sunshine between 2 and 5
182 176 157 170 265
air from the leaves of potato taken in the same time as the above jar
182 163 146 238 336

First page of IngenHousz' logbook with notes on the experiments in the summer of 1779, beginning with a few plants of *Lenticula aquatica*. He describes which plants or plant parts he is using, how long he submerges them and for how long they stay in the sunlight before he measures the quality of the air they produce.

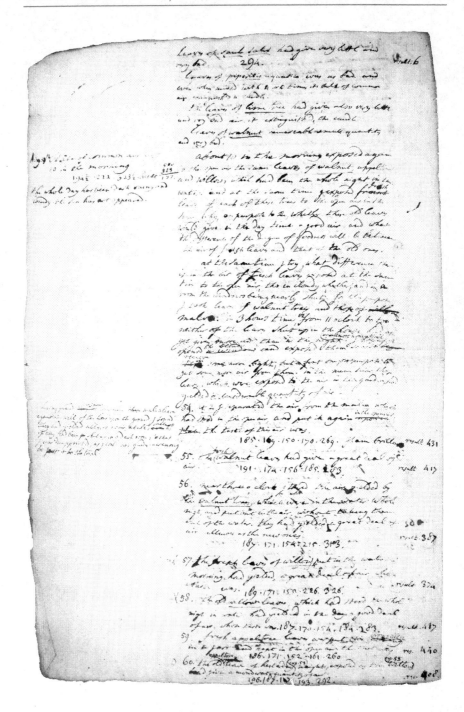

Page 8 with notes on the experiments of 8 August. In the margin he notes: "The whole day has been dark rainy and windy, the sun has not appeared". The jar with malva leaves does yield very little air, " as if they had been shut up in a dark room".

What IngenHousz did was terribly simple. He put some plants in an inverted jar, floating in water. He probably used the plants he had to hand: some duck-weed (*Lenticula aquatica*,[65] as he calls it, since then renamed to *Lemna trisulca*), peppermint (*piperitis*), sage (*salvia*), common laurel (*lauroceratus*), ordinary grass, a portion of a grass-like plant from the reed pond in the garden, parsley and potato.

At first he put in inflammable air with the plant in the vessel. He finds with the nitrous test that the air has improved after 48 hours. That leads to the next step, leaving no air above the plant. So he finds that air starts accumulating above the water. Then he puts some common air with the plants in the vessel. Each time he measures the quality of the air left above the plants, the air has improved.

IngenHousz is not doing anything surprisingly new here. The experiments he performs nor his approach come out of thin air. He steps into a research programme that had been instigated by two of the most influential chemists of that time, Lavoisier and Priestley, who had picked up an old trail that started more than hundred years earlier.

65 Buys 1776

rewind

The first decades of curiosity 1730-1779

Looking back on the first 49 years of IngenHousz' life, in all its details, his many adventures, his intellectual persuits, experiments and publications, with its personal biographical details and psychological intricacies, it is obvious that this man is by no means an ordinary guy.

Two things are showing up in this slowly appearing historical landscape. The development of IngenHousz' scientific thoughts and undertakings is the result of the complex interaction between the idiosyncrasies of his life, personal characteristics, upbringing and the way he picks up the trail of his endeavours. These are a mix of chance events and choices based on his personality. But he is definitely not working on his own. On the contrary. Despite retreating in a country house to perform his experiments, he has been working closely together with Fontana, discussing with people such as Priestley and Franklin, and in correspondence with a wide web of natural philosophers all over Europe. What he knows and does is greatly influenced by the long intellectual history of the western world, which he started moving around in when studying in foreign countries. And just as much through family events. The fact that his brother-in-law was one of the most famous botanists of Europe must have inspired him to start studying plants.

Secondly, it shows how the hypotheses he is developing, the experiments he is setting up, the theories he is slowly building in his mind, fit neatly within the scientific world view which was being developed in these times of Enlightenment, a movement that was gathering full speed at the time of IngenHousz. Therefore he may well be called a prototype of an Enlightenment scientist and intellectual. He was at the front line of a truly modern way of thinking about the world in which evidence from controlled experiments and a critical attitude towards claims of all kinds became paramount in the quest for trustworthy knowledge.

It also shows that science is a human business, driven by individuals with all their personal potentials and limitations, influenced by the group of peers that have taken the same route eagerly working on constructing a better knowledge of the world, embedded in a society that can and will make and break hopes and dreams, using and misusing whatever knowledge the natural philosphers are delivering.

IngenHousz' portrait, on frontispiece of *Experiments upon plants*, 1779

The quest
for the secret of the sunlight
1779-1780

When IngenHousz began his experiments in the summer of 1779, he stepped into a 'research programme' that was already more than a century old. It started in the early seventeenth century when Johannes Baptist van Helmont, one of the last alchemists, who was born and died in Brussels, performed a hands-on experiment on a willow tree that became justifiably famous and would open up a whole new field of research.[66] It might have been preceded by De Cusa's 1450 (thought) experiment, but Van Helmont was the first to try to deduct what plants are built from and how they grow. This is how he describes it himself in the posthumous English translations of his work from 1662:

> But I have learned by this handicraft-operation, that all Vegetables do immediately, and materially proceed out of the Element of water onely. For I took an Earthen Vessel, in which I put 200 pounds of Earth that had been dried in a Furnace, which I moystened with Rain-water, and I implanted therein the Trunk or Stem of a Willow Tree, weighing five pounds; and at length, five years being finished, the Tree sprung from thence, did weigh 169 pounds, and about three ounces: But I moystened the Earthen Vessel with Rain-water or distilled water (alwayes when there was need) and it was large, and implanted into the Earth, and least the dust that flew about should be co-mingled with the Earth, I covered the lip or mouth of the Vessel, with an Iron-Plate covered with Tin, and easily passable with many holes. I computed not the weight of the leaves that fell off in the four Autumnes. At length, I again dried the Earth of the Vessel, and there were found the same 200 pounds, wanting about two ounces. Therefore 164 pounds of Wood, Barks, and Roots, arose out of water onely.[67]

It is would no longer be considered cutting-edge science, but for a natural philosopher it was quite something. The gained mass would come from the water, in a true transmutation of this fundamental element. The real process of

---

66   Ducheyne 2006
67   van Helmont 1662

plant growth and physiology remained obscure, but at least it demonstrated that people started to realise that plants were doing something out there in nature and that one could look for ways of understanding what this could be. Van Helmont also found that there were various kinds of air-like substances, some of them dissolving in water, others smelling repugnant, some inflammable. He concluded, without any means of investigating these damps and fluids better, that these substances without form or colour or sometimes even smell, were in fact a form of proto-matter, the formless substance out of which all matter was assembled. Inspired by Greek mythology in which the cosmos had been created out of chaos, he decided to baptise these vapours "chaos". So he coined the word "gas", based on the Flemish pronounciation with a soft ch-sound of the word "chaos".[68]

By the end of the century John Woodward would disprove Van Helmont's experiment by measuring transpiration from plants, but his work would be largely overlooked, overshadowed by Boyle's musings on plant life. He, like Van Helmont, regarded the substance of plants as little more than transmuted water.

Around that time the microscope entered the field of botany. Nehemiah Grew and Marcello Malpighi described the little mouths or stomata on the leaves' surfaces, the gateway between structures within the plant and the external atmosphere. But they were unable to decide on the fact whether these pores were for excretory or assimilative functions.

The next major step came with *Vegetable Staticks* in which Stephen Hales in 1727 studied the hydrostatics of plants, how their saps, containing air bubbles, flow through their stems and leaves, to be exchanged with the atmosphere. Therefore he suggested that plants breathe air. Even more importantly, he developed an instrument that would turn out to be crucial for the further study of plant-atmosphere interactions: the 'pneumatic trough'. With this device it was possible to collect samples of gases over water or another liquid in an upturned glass vessel. Primitive though they may have been, these experiments supported Hales in his intuitions that plants breathe in and absorb some nourishment from the air through their leaves, and furthermore that light might play a role in that process. Methodically he may have been even more significant as he implemented controlled experiments. And no matter his accomplishments, he made clear beyond any reasonable doubt that there is an important interaction between plants and atmosphere.

This much was clear for Joseph Bonnet who published his *Recherches sur l'usage des feuilles des plantes* in 1754. In a work that is more like poetry than exact science, he described how plants absorb the dew that rises from the soil through the underside of their leaves. It was clear that an interaction between plants and their environment took place, but these pioneering physiologists at that time did not have the right conceptual framework, nor the workable experimental methods to fathom the mystery of plant life.

---

68   Strathern 2000

Something is in the air

By 1770, early in his career, Antoine Lavoisier published a 'mémoire' on the nature of water "Sur la nature de l'eau et sur les expériences par lesquelles on a prétendu prouver la possibilité de son changement en terre",[69] in which he performed painstaking experiments demonstrating that Van Helmont's experiments do not prove the transmutation of water, reemphasising the role of the atmosphere in the economy of plants.

> *Indépendamment de ces différentes substances, qui sont étrangères à l'air, on ne saurait douter que ce fluide lui-même n'entre en proportion très-considérable dans la texture des végétaux, et qu'il ne contribue pour beaucoup à en constituer les parties solides. Il résulte des expériences de M. Hales, et d'un grand nombre d'autres qui ont été faites en ce genre, que l'air existe de deux façons dans la nature : tantôt il se présente sous la figure d'un fluide très-rare, très-dilatable. très-élastique, tel est celui que nous respirons ; tantôt il se fixe dans les corps, il s'y combine intimement ; il perd alors toutes les propriétés qu'il avait auparavant ; l'air dans cet état n'est plus un fluide, il fait l'office d'un solide, et ce n'est que par la destruction même des corps dans la composition desquels il entrait, qu'il revient à son premier état de fluidité. On peut voir, à cet égard, les expériences très-ingénieuses rapportées dans la Statique des Végétaux ; il en résulte que le bois de chêne donne, par l'analyse, environ le tiers de son poids d'air ; que le bois de gaïac en donne encore davantage, et que la quantité de ce fluide contenue dans les différentes espèces de bois est toujours à peu près proportionnelle à leur densité. Ces expériences sont trop constantes pour pouvoir être révoquées en doute, elles ont, d'ailleurs, été répétées un grand nombre de fois, aux yeux de tout le public, dans les leçons de M. Rouelle.*
>
> *Voilà donc deux sources dont les végétaux élevés dans l'eau seule ont pu tirer les principes qu'ils ont donnés par l'analyse : premièrement de l'eau même et de la petite portion de terre étrangère qui se trouvait nécessairement dans toutes celles qui ont été employées, secondement de l'air et des substances de toute nature dont il est chargé.*
>
> *Les expériences faites sur la végétation des plantes par l'eau ne prouvent donc en rien la possibilité du changement d'eau en terre.*

Independently from these substances, which are different from the air, one can not doubt that this fluid itself enters into the texture of the plants in considerable quantities and that it contributes more than a little to the formation of solid parts. It is a consequence of M. Hales' experiments and from those of many others: this air can take two forms in nature. Either it presents itself as a fluid which is very rare, very dilutable, very elastic,

---

69   Lavoisier 1770

which is how we breathe it. Either it takes a fixed form in the bodies with which it combines very intimately. So it looses all its former properties: air in this form is no more fluid but solid and it only regains its former fluid state after the destruction of the body in which it was incorporated. Look at the very ingenious experiments reported in the Vegetable Staticks: oak wood gives after analysis half of its weight in air, while the wood of the guajac gives even more. The quantity of air in various kinds of wood is more or less proportional to their density. These results have been too stable to be doubted as they have been repeated numerous times, even in front of the eyes of the public, such as during the lessons by M. Rouelle. So these are the two sources from which plants growing in water can derive their principles that have been demonstrated in analysis: first of all the water itself and the little portion of strange earth which is hidden in this water, secondly the air and all the substances that the air naturally carries. The experiments on plants growing in water therefore disprove by no means the possibility of a transformation of water in earth.

He continued performing some very precise distillation experiments on different kinds of water, weighing the water before and after the distillation, sometimes repeated several times. He obtained highly purified water, which left behind a small but obvious amount of mineral matter. Then he began to measure the relative densities of the various distillates. By progressively removing the dissoluted salts from the water, the density of the liquid progressively diminished, or so one would expect. It did not happen and Lavoisier continued his experiments in a closed vessel. He finally concluded that the accumulation of earthly matter in the vessels, in the experimentts by himself and by predecessors such as Boyle, was due to the solution of material from the vessel into the water.

In his discussion on how earth (solid matter) could originate from water (fluid matter) Lavoisier readily admitted that he had no problems with the fact as such but with the explanation that had been given to them. He found in this no convincing argument for the transmutation of water into earth. This experimental work by Lavoisier might look of little relevance to IngenHousz' work, but it was important for the natural philosophers of the second half of the eighteenth century because it laid a renewed emphasis on the role of the atmosphere in the economy of plants. He belittled the simple explanation of Van Helmont's experiments in terms of an easy yet mysterious transmutation of water. Still, Lavoisier was totally unaware of the true character of the process at hand.

In search of sweet air

Two years later Joseph Priestley published the results of experiments which were preparing the steps into a new direction. "Observations of different kinds of air"[70] is a long paper, delivered in different chapters and read to the Royal Society on 5, 11, 19, 26 March 1771. It was to be one of the first publications on 'pneumatic chemistry' making use of Hales' pneumatic trough. And as mentioned earlier, Priestley described in it a method to produce sparkling water, among an amalgam of other things. One of the crucial observations though was how a candle burns out under an inverted water-sealed jar, before running out of wax. This air under the jar became 'injured', so much so that a flame would die out and a mouse could not survive in it.

That candles will burn only a certain time, is a fact not better known, than it is that animals can live only a certain time, in a given quantity of air; but the cause of the death of the animal is not better known than that of the extinction of flame in the same circumstances; and when once any quantity of air has been rendered noxious by animals breathing in it as long as they could, I do not know that any methods have been discovered of rendering it fit for breathing again. It is evident, however, that there must be some provision in nature for this purpose, as well as for that of rendering the air fit for sustaining flame; for without it the whole mass of the atmosphere would, in time, become unfit for the purpose of animal life; and yet there is no reason to think that it is, at present, at all less fit for respiration than it has ever been. I flatter myself, however, that I have hit upon two methods employed by nature for this great purpose. How many others there may be, I cannot tell.

Priestley tried diverse methods to restore the bad air. Letting it stand over water, mixing it actively with water, letting it stand over earth. He tested which life forms can survive in putrefied air. Insects such as beetles and butterflies seem to do rather well. When he put in plants, he was surprised with the results he sometimes obtained.

When the air has been freshly and strongly tainted with putrefaction, so as to smell through the water, sprigs of mint have presently died, upon being put into it, their leaves turning black; but if they do not die presently, they thrive in a most surprising manner. In no other circumstances have I ever seen vegetation so vigorous as in this kind of air, which is immediately fatal to animal life. Though these plants have been crouded in jars filled with this air, every leave has been full of life; fresh shoots have branched out in various directions, and have grown much faster than other similar plants, growing in the same exposure in common air.

This observation lead him to conclude that plants, instead of affecting the air in the same manner as animals do with their respiration, reverse the effect of breathing and "tend to keep the atmosphere sweet and wholesome, when it has become noxious in consequence of animals breathing, or dying or putrefying it."

---

70   Priestley 1772

He then started a new experiment: he took a quantity of air in which mice had been breathing until it was too toxic to survive. He split the air in two parts and put one in a vial immersed in water and the other in a glass jar, standing in water, in which he put a sprig of mint.

> *This was about the beginning of August 1771, and after eight or nine days, I found that a mouse lived perfectly well in that part of air, in which the sprig of mint had grown, but died the moment into the other part of the same original quantity of air; and which I kept in the very same exposure, but without the plant growing in it.*

He not only used mice. On 17 August he put a sprig of mint into a quantity of air in which he burned a candle until it had burnt out. Ten days later, a candle burned perfectly well in it. He repeated this experiment no less than eight or ten times during the summer. But with the mice as well as with the candles, with mint as well as groundsel, balm or spinach, he did not obtain stable, reproducible results. He repeated his experiments several times under various conditions, sometimes putting a mouse back in after the air had been restored by the plant. "But this is so exceedingly various, that it is not easy to form any judgment from it..."

After a great many trials he found a possible way to restore this 'vitiated' air, but he still did not know what really caused this restoration. He was in doubt and could not draw any firm conclusion. "Towards the end of the year some experiments of this kind did not answer so well as they had done before, and I had instances of the relapsing of this restored air to its former noxious state. I therefore suspended my judgment concerning the efficacy of plants to restore this kind of noxious air." He resumed his experiments in the summer of 1772. He put mice in vials with air rendered noxious by mice breathing in it and that had been restored during a week's growth of mint. Franklin and Pringle saw the plant grew vigorously at about day three or four. He continued to grow plants in jars, testing the quality of it at regular intervals. He tested it against nitrous air, but did not describe exactly when and how much, nor did he give the figures. He clearly got uneven results and threw out the old plants, putting in fresh ones. He suspected that plants might have a contrary tendency to their restorative actions in some stages of their growth. He presumed that the restoration of the bad air depended upon the "vegetating state" of the plants, that is, upon them growing.

Still, he felt certain enough to conclude that the immense production of vegetables on earth is capable of counterbalancing the combined damage done by the respiration of animals and the putrefaction of organic matter. He happily cited Franklin who wrote him in a letter: "The strong thriving state of your mint in putrid air seems to shew that the air is mended by taking something from it, and not by adding to it." And he added:

I hope this will give some check to the rage of destroying trees that grow near houses, which has accompanied our late improvements in gardening, from an opinion of their being unwholesome. I am certain, from long observation, that there is nothing unhealthy in the air of woods; for we Americans have every where in our country habitations in the midst of woods, and no people on earth enjoy better health, or are more prolific.

Priestley was pleased with what he had found. He flattered himself that he had accidentally hit upon a method of restoring air which has been affected by the burning of candles, and that he has discovered at least one of the restoratives which nature employs for this purpose: it is vegetation, the growing of plants. And Franklin's idea enforced his hypothesis, that plants take something from the vitiated air rather than adding something to it.

That plants restore noxious air, by imbibing the phlogiston with which it is loaded, is very agreeable to the conjectures of Dr. Franklin, made years ago, and expressed in the following extract from the last edition of his Letters, p. 346.
"I have been inclined to think that the fluid fire, as well as the fluid air, is attracted by plants in their growth, and becomes consolidated with the other materials of which they are formed, and makes a great part of their substance."

Neither Franklin, nor Priestley had any idea of what exactly was happening inside the plants. They conjectured that part of the fire as well as part of the air become fluid again when the plants ferment.

Priestley hypothesized, backed by the suggestions of the famous Franklin, that plants improved the bad air by withdrawing something from it. While animal respiration destroyed the air by adding something to it, of which plants could benefit from by absorbing this aerial manure. This effluvium that would connect all living beings in a circle of interaction was phlogiston. Phlogiston was the key element in the chemical theory of that time and it fitted nicely within the scheme of things Priestley was drawing up. During 'phlogistic processes' such as animal respiration, combustion or putrefaction, the subtle fluid of phlogiston was given off to the air. As soon as a certain volume of air was saturated with phlogiston, it was unfit for use and could be called 'vitiated', in other words 'phlogisticated'. Only some process that could 'dephlogisticate' the air could restore it to its former healthy state. With the help of candle and mouse, Priestley in essence described these qualitative changes in the air.

During the following years Priestley worked on a great variety of different things and among them was the employment of an analytic method that would allow him to properly 'examine the state of the air': the nitrous air test. The air to be tested reacts with 'nitrous air' (nitric oxide), forms a red-brown vapour which dissolves in water, thereby diminishing the volume of air remaining. The measured diminution is a measure of the 'goodness of the air'. (This method was invented and developed by Landriani, the Italian 'father' of pneumatic medicine.[71]) Still, he did not succeed in obtaining consistent results, as well as being unable to recognise the right factors that play a role in the plant-air interactions. He got mixed results and had mixed feelings:

> *Upon the whole, I still think it probable that the vegetation of healthy plants, growing in situations natural to them, has a salutary effect on the air in which they grow. For one clear instance of melioration of air in these circumstances should weigh against a hundred cases in which the air is made worse by it, both on account of the many disadvantages under which all plants labour, in the circumstances in which these experiments must be made, as well as the great attention, and many precautions, that are requisite in conducting such a process. I know no experiments that require so much care. Particularly, everything tending to putrescence, every yellow or ill-looking leaf, &c. must be removed, before the air can be injured by it, and I did not at this time watch my plants with so very much attention as i did when I first made my experiments; though the method I now use examining the state of the air was much more exact than any that I was acquainted with at that early period of my observations of air.[72]*

Priestley is in doubt. The experiments don't produce reliable results. And he does not know what the reason(s) might be. His method with the nitrous air test was much more reliable than his earlier methods such as the time a candle burns or a mouse breathes. Of one thing he is convinced: phlogiston is the key element in the interaction that takes place when animals breath.

> *As I think I have sufficiently proved, that the fitness of air for respiration depends upon its capacity to receive phlogiston exhaled from the lungs, this species may not improperly be called dephlogisticated air. This species of air I produced from mercurius calcinatus per se, then from the red precipitate of mercury, and now from red lead. The two former substances yield it pure; ... that this air is of that exalted nature, I first found by means of nitrous air, which I constantly apply as a test of the fitness of any kind of air for respiration, and which I believe to be a most accurate and infallible test for that purpose.[73]*

The dephlogisticated air he produced himself is great stuff: candles burned better and longer in it, wood burned with great crackles throwing sparks in all

---

71   Beretta 2000
72   cited by Nash 1952
73   Priestley 1775

directions and a mouse lived for a whole hour, while it would not survive longer than a quarter of an hour in common air. But it did not help him clarify what plants do in ameliorating the air.

Things got even more complicated when he performed some more experiments in June 1778. He removed the air-producing plants from the vial, the insides of which covered with a green kind of matter which continued to yield air even when the plants were removed. This convinced him that the plants had not, as he had imagined, contributed to the production of the pure air. But he continued to put plants in vessels, put them sometimes near the stove, sometimes on the window sill, and sometimes they produced pure air, sometimes not. After describing several experiments with small vials that were harbouring this 'green matter', he remarked that "I have had some appearances, which, extraordinarily as it will seem, make it rather probable, that light is necessary to the formation of this substance; but many more observations, which I believe can only be made in the summer season, will be necessary to determine this."[74] Priestley got lost in a jungle of too many and contradictory facts. He was totally confused about what could be cause or effect or just incidental manifestations. In the end he started thinking the air came from the water, knowing that gases can be soluble in water. After all, he was the inventor of soda water.

As he described in the appendix to his book, he made some more experiments on 19 February 1779 by placing two jars of pump water in the same south window, one of them neatly covered by some brown paper. In about ten days the water in the uncovered jar had yielded about four ounces of air, and the covered jar only a few bubbles. He returned to the jar after a journey and on 2 April the uncovered jar had yielded ten ounces, "so pure that one measure of it and one of nitrous air, occupied the space of 0.84 measures; whereas the covered jar had very little more than one ounce measure, and with this the measures of the test were 1.55 measures; i.e. by no means so pure as the former." (The measurement of the pure air in which two measures were reduced to 0.84 indicates that the sample contained a lot of oxygen, probably around 40%.) Priestley concluded that light disposes water, containing enough substances to make a deposit of greenish or brownish matter and then to yield dephlogisticated air. But "... having no great attachment to any particular hypothesis, I am very willing that my reader should draw his own conclusions for himself."

As Nash describes the state of affairs in Priestley's laboratory: "The whole of Priestley's work on this (and other) subjects presents a mélange of the most ingenious and perceptive observation and experimentation combined with maddening failures in interpretation."[75]

74   cited by Nash 1952
75   Nash 1952

In the summer of 1779

This was the state of affairs when IngenHousz retreated to the countryside in June 1779 and started his series of experiments on plants and sunlight. He combined a wide-ranging knowledge, with great manual dexterity and experimental insight.

On 1 August he put some leaves from an apple tree in a jar and put it in the sun, between 10am and 5pm. At the same time he put leaves from the same tree in a green bottle. The next day he used leaves from nettles and from sage. Each day he noted the quality of the common air, and noted the weather and amount of sunshine. From each jar or bottle he made note of the colour and measured the quality of the air. That means: each time adding another measure of nitrous air - sometimes up to four times - and measuring the diminution of the air in the glass tube of his eudiometer.

He took notes of all the data meticulously. He put a jar in which the day earlier apple leaves were producing pure air out in the sun for a day. He put potato leafs in a jar during the night. He noted when clouds covered the sun for most of the day. Sometimes he measured the air from a jar a second time, after having made fresh nitrous air. He wrote down that he thought the data were not trustworthy when he was not careful enough and common air was able to enter the jar.

On 8 August he was on his way from London to his country house: "I filled a great number of jars with different leaves about 9 o clock in the evening, when almost no leave gave more any bubbles of air in the water except potato leaves" Under number 53 he noted: "in the morning (aug 9) at 6 o clock I found that all had given but a very small quantity of air (1/10 of in the sun shine)".

In experiment nr 65, he referred to Bonnet and used boiled water. For the next experiment he used pump water and saw how green matter formed at the bottom of the jar, which made him refer to Priestley.

In experiment 79 he bound a glass cylinder to a tree to be able to put leaves in it, without detaching them from the tree. When noting the results of an experiment he remarked that this must be "often repeated before we conclude anything".

He again referred explicitly to the work of Bonnet, his illustrious predecessor. He explicitly quoted Bonnet's work as the reason for performing these specific experiments.

On the night of 10 August he put a handful of sage leaves in a jar and found the air the next morning diminished and bad (it even extinguishes a flame). In the margin of his logbook[76] he noted:

> Which indicates that plants absorb air by night without correcting the remainder, but I think that they will however correct bad air or inflammable air or air breathed which must be tyed[?]

---

76   Breda Collection IngenHousz A11

Next, on 11th of August, he noted a little further down the page, in the margin:

NB: Latter experiments demonstrated me that plants mend common air in the sun

IngenHousz here made the step that had eluded Priestley. He focused on the link between the light of the sun and the production of dephlogisticated air. Still, he wasn't easily satisfied, he was unsure as to whether more factors could be involved in the process he was studying and he wanted to find out.

He started taking measurements before and after noon, suspecting that plants deliver more and better air in the afternoon.

By experiment numbered 333 he noted that "plants spoil air at night and mend it again during the day." He was making up his mind and writing it all down in the manuscript of his book, for the publishing of which he returns to London on 12 September.

The secrets of plants in print

While he was taking notes on his experiments, he had been developing a line of thought, building hypotheses into a framework that could throw light on the interaction between sunlight and vegetable life. He had not just been taking notes, at the same time he had been working hard at the manuscript of the book that appeared in print in October.[77]

EXPERIMENTS
UPON
VEGETABLES
Discovering their great Power of purifying the Common Air in the Sun-shine
and of
Injuring it in the Shade and at Night
to which is joined,
a new method of examining the accurate
Degree of Salubrity of the Atmosphere.

by John Ingen-Housz,
councellor of the court and Body Physician
to their Imperial and Royal Majesties,
F.R.S. &c. &c.
London:
printed for P. Elmsly, in the Strand;
and H. Payne, in Pall Mall, 1779

---

77   IngenHousz 1779

In line with the tradition of the era, IngenHousz formulated a title that says it all in one sentence. He described what he discovered as: the powerful capacity of plants to ameliorate the air that has been corrupted by breathing, a power that is being fuelled by the light of the sun. On top of that he described how plants reverse this beneficial action during the night or in the shade. As an added bonus he promised to introduce the reader to the new measurement technique he applies to quantify this atmospheric activity.

IngenHousz dedicated his book to Sir John Pringle, his great friend and fellow doctor and investigator. He thanked his benefactor for all he did to this foreigner. He recognised publicly how Pringle's recommendation to the Austrian "August Sovereigns" to inoculate their family opened the door to emoluments and honours wide open to him. He did not want to leave this country without leaving public testimony of his real sentiments towards Pringle. In the same sentence he admitted that it was the reason why he was in a hurry to get this work to the press, without having enough time to finish it as desired. He signs "Sir, your very much obliged and faithful friend and servant, J. INGEN-HOUSZ, London, October 12, 1779."

In the preface he sketched the state of affairs in the study of the airs and the atmosphere. He opened:

> *The common air, that element in which we live, that invisible fluid which surrounds the whole earth, has never been so much object of contemplation as it has in our days: it never engaged so much the attention of the learned as it has of late years. This fluid, the breath of life, deserves so much the more the attention and investigation of philosophers, as it is the only substance without which we can scarce subsist alive a single moment, and whose good or bad qualities have the greatest influence upon our constitution.*

He put the problem upfront in the next sentences by indicating that plants are capable of restoring what the breathing of animals damaged, but that plants are equally able to poison the air in certain circumstances and that therefore this needs urgent and more attention from philosophers, especially the attention of those whose profession it is to preserve health and to cure diseases. Being a philosopher and a physician, he found this a particularly pressing field of interest.

He referred without further ado to the works "of that excellent philosopher and inventive genius, the reverend Dr. Priestley, his important discovery, that plants wonderfully thrive in putrid air; and that the vegetation of a plant could correct air fouled by the burning of a candle, and restore it again to its former purity and fitness for supporting flame, and for the respiration of animals". He quoted from Pringle's address to the Royal Society in November 1773. All plants, from the fragrant rose to the deadly night-shade, are beneficial to mankind, as the winds convey our vitiated air towards them, for our benefit and for their nourishment. Or as Pringle had said:

> From these discoveries we are assured, that no vegetable grows in vain, but that from the oak of the forest to the grass of the field, every individual plant is serviceable to mankind; if not always distinguished by some private virtue, yet making a part of the whole which cleanses and purifies our atmosphere.

IngenHousz expressed his happiness that he had had some degree of success in investigating this important subject, as the following pages will show. But "I am far from thinking that I have discovered the whole of this salutary operation of the vegetable kingdom; but I cannot but flatter myself, that I have at least proceeded a step farther than others, and opened a new path for penetrating deeper into this mysterious labyrinth."

Next he summarized the state of knowledge in the science of airs. First he praised Priestley, who however despite his "candour and modesty" has underrated the value of his useful production of his inquiries and cites from "Experiments and observations relating to various branches of natural philosophy, with a continuation of the observations of air", page 269, how Priestley was disappointed by the practical applicability of the nitrous air test: "I have frequently taken the open air in the most exposed places in this country, at different times of the year, and in different states of the weather, &c.; but never found the difference so great as the inaccuracy, arising from the method of making the trial, might easily amount to or exceed." IngenHousz was thinking much more favourably about the test as he was now using the method introduced by Fontana. He felt one was now able to judge the degree of purity of common air as good as the degree of heat and cold by a good thermometer. When making ten observations of the same air, the difference would scarcely amount to 1/500th of the two airs employed in the experiment.

This improvement on Priestley's method would now make it possible to unveil in what manner the faculty of plants is induced to increase the salubrity of the air. He cited repeatedly from Priestley's work (from 1778) how the latter said to be disoriented by contradictory results. In his preface IngenHousz summed up Priestley's confused and confusing work. And he felt free to step in where Priestley opened the door for his readers to draw their conclusions for themselves. "Thus far this matter was carried on when I took it up in June last." He wanted to check two possible ways in which plants ameliorate the air. Do they absorb, by absorbing as part of their nourishment, the phlogistic matter, leaving the remainder as pure air; or do the plants possess some particular virtue hitherto unknown, by which they change bad air in good air? The former seemed to be Priestley's bed, he suspected the latter to be the case.

A prepared mind

IngenHousz did not come to his hypothesis and the experiments to test them by random trial and error. He stepped into a fertile field of investigation that had been explored by Priestley, Lavoisier, Fontana, Landriani, to name just of few of his contemporaries. The nitrous air test had been the subject of intense scientific and political debate for some time (as will be explored in chapter five). He did have some idea of what he was looking for. Discovery comes to the prepared mind, and the same was true for IngenHousz. This is how he described it in the preface to his book:

> I was not long engaged in this enquiry before I saw a most important scene opened to my view: I observed, that plants not only have a faculty to correc bad air in six or ten days, by growing in it, as the experiments of Dr. Priestley indicate, but that they perform this important office in a compleat manner in a few hours; that this wonderful operation is by no means owing to the vegetation of the plant, but to the influence of the light of the sun upon the plant. I found that plants have, moreover, a most surprising faculty elaborating the air which they contain, and undoubtedly absorb continually from the common atmosphere, into real and fine dephlogisticated air; that they pour out continually, if I may say so express myself, a shower of this depurated air, which, diffusing itself through the common mass of the atmosphere, contributes to render it more fit animal life; that this operation is far from being carried on constantly, but begins only after the sun has for some time made his appearance above the horizon, and had, by this influence, prepared the plants to begin anew their beneficial operation upon the air, and thus upon the animal creation, which was stopt during the darkness of the night; ... that this office is not performed by the whole plant, but only by the leaves and the green stalks that support them... that even the most acrid, ill-scented and even the most poisonous plants peer form this office.

It is hard to image that this was really a vision that suddenly cleared the mind of IngenHousz. He wrote this after or during months of concentrated work, totally submerged in this subject, while the ideas that fill this preface bubbled up in the margin of his lab logbook. He continued to summarise the characteristics of this botanical process. Young leaves did not seem to yield as much good air as fully-grown ones. All plants contaminated the surrounding air by night, although some were worse than others. Flowers did the same, equally by day as by night. The same appeared to be true for fruits and roots. Delicious as they may be, peaches could kill us, if we would be locked up in a room with enough of them. And finally, that the sun in itself had no power to mend the air without the presence of plants.

He warned the reader that in every experiment of this kind some difference in the result will be observed. Too many different circumstances were involved in producing the same quality consistently, such as more or less light, and leaves more or less crowded in the jar, shading each other.

He furthermore expressed his gratitude to Abbé Fontana for being able to use and expand on his method before he had made it public. He also thanked the August Sovereigns in Vienna for letting him spend the whole summer to pursue his research, allowing him to perform more than 500 experiments from June to the beginning of September. Moreover, he wished to apologise for possible inaccuracies in expressing himself in a language which was not his mother tongue.

He ended with a methodological reflection:

> I had this object in view some years ago; but, as I did not enjoy such a favourable disposition of mind and body as was necessary for a task, in which all possible steadiness, perseverance, and close attention were requisite, I deferred the undertaking till I should find myself more fit for it.
>
> Detached experiments may indeed be very useful when a sufficient number is collected to draw some conclusion from them; but, without pursuing methodically the same object, discoveries are to be expected only by mere chance, and are even sometimes overlooked. I owe to the example of my worthy friend, the Abbé Fontana, the thorough persuasion which I now entertain, that natural knowledge can make but a very slow progress in the hands of those who have not the patience and assiduity enough in pursuing one and the same object, till they discover some things undiscovered before; or till they find that the difficulty of the undertaking surpasses their abilities.

On the nature of plants

Before embarking on his main story, IngenHousz made an aside introducing his readers to the works of Dr. Priestley and the doctrine of air. He described the basic ingredients that make up this doctrine: nitrous air, phlogisticated air, dephlogisticated air, fixed air and inflammable air, and last but not least the eudiometer.

Then he introduced the reader to the nature of plants, and especially their leaves, by referring to the work of Charles Bonnet, a Swiss investigator from Geneva who published in that city in 1754, *Recherches sur l'usage des feuilles dans les plantes, et sur quelques autres sujets relatif à l'histoire de la vegetation.* Together with Priestley's work, this book must have been a major source of inspiration for IngenHousz as may have been clear from his remarks on Bonnet's work in his logbook quoted above.

This celebrated author has taken a great deal of notice of those air bubbles which cover the leaves when plunged under water. He says on p.26 that the leaves draw their bubbles from the water. He is the more persuaded that this is the case because he found these bubbles did not appear when the water had been boiled some time, and appeared more when the water is impregnated with air, by blowing in it. He had also observed, that they did not appear after sunset. Page 31, he explains his opinion farther upon his head: he says, that these air bubbles are produced by common air adhering to the external surfaces of the leaves, which swells up into bubbles by the heat of the sun; and that the cold of the night is the reason why these air bubbles do not make their appearance at that time.

IngenHousz considered the cause of these bubbles to be more important than Bonnet himself. And it gave him a crucial headstart in interpreting the experimental facts of his long methodical sequence of tests over Priestley's seemingly haphazard looking series of observations. What IngenHousz did may look simple or obvious, but to his fellow natural philosophers it didn't:

If the sun caused this air to ooze out of the leaves by rarifying the air in heating the water, it would follow that, if a leaf, warmed in the middle of the sunshine upon the tree, was immediately placed in water drawn directly from the pump, and thus being very cold, the air bubbles would not appear till, at least, some degree of warmth was communicated to the water. But quite the contrary happens. The leaves taken from trees or plants in the midst of a warm day, and plunged immediately into cold water, are remarkably quick in forming air bubbles, and yielding the best dephlogisticated air.
If it was in the warmth of the sun, and not its light, that produced this operation, it would follow, that, by warming the water near the fire about as much as it would have been in the sun, this very air would be produced. But this is far from being the case.
I placed some leaves in pump water, inverted the jar, and kept it as near the fire as was required to receive a moderate warmth, near as much as a similar jar, filled with leaves of the same plant, and placed in the open air, at the same time received from the sun. The result was, that the air obtained by the fire was very bad, and that obtained in the sun was dephlogisticated air.

IngenHousz eliminated a parameter which was not relevant for the phenomenon in question. He took good care to standardise the procedure, taking two samples of leaves from the same plant, while building in a control by putting similar jars in the light of the sun and in the warmth of the stove.

And that was what he did throughout his book, demonstrating with the results of his simple but effective experimental settings. First he looked for

the source of the dephlogisticated air. Did it come out of the leaves or out of
another part of the plant? Did it really come out of the leaves or is it coming
out of the water itself? Was it part of the substance of the leaves itself, or was it
secreted by the leaves after some transmutation has taken place? Did the plants
have to be alive, or would dry leaves do it just as well? Was it relevant that
the plants are immersed in water and not in their natural setting? Did various
species of plants differ in their capacity of producing good air? Or were other
factors at hand? Did this beneficial quality of plants depend on the fact that
they are growing (the act of vegetation)? Was there a difference between various
species of trees in their production of pure air? (So one could decide which ones
to plant best near our houses.) And was there a difference between young small
leaves, and big full-grown leaves?

And what did plants do when the light has gone? This was IngenHousz'
next logical step, after he recorded that light is crucial for the dephlogisticating
activity of plants. His measurements indicated that

> *The bad air which plants yield by night is so inconsiderable in comparison to*
> *the quantity of dephlogisticated air which they yield by the day-time, that it*
> *amounts to very little. By a rough calculation I found, that the poisonous air*
> *yielded during the whole night by any plant could not amount to onehundreth*
> *part of the dephlogisticated air which the same plant yielded in two hours time*
> *in a fair day.*

The evaporation of bad air in absence of sunlight was "far over-balanced" by
the beneficial operation of plants during the day. He conjectured that this foul
nightly production is dispersed in the greater atmosphere but could possibly
be harmful when concentrated in a small room, and recalled the stories that
are being told of people found dead in their bed in a room full of flowers. The
practice in hospitals to put the flowers out in the hallway at night is a silent
remnant of an old suspicion. Flowers, however nicely scented they may be, have
a malignant influence, in a small quantity, by day and by night. The same is true
of fruits and roots. He had to admit that he was quite surprised that even the
most beautiful fruits were aerially poisonous. He even tried green beans, saw
bubbles oozing out, only to find that it was bad air.

Plants also turned out to be better at improving air fouled by breathing or
burning, than in improving ordinary common atmospheric air. "As plants seem
to delight in foul air, because this air impregnated by phlogiston affords more
proper nourishment, viz. phlogiston to the plant."

He kept on trying new experimental set-ups to approach the problem from
ever differing angles. He put two plants in seperate glass jars and put one on the
window sill in the sun, the other in the room in a shaded place. The air in the
jar plus sunshine improved after a few hours in the light, the air in the shaded

jar became worse. Then he inverted the experiment, putting the jar near the window in the shade and vice versa. The air in the two jars was after a few hours reversed in quality.

On the way he elucidated some more botanical processes: "Vegetables seem to draw the most part of their juices from the earth, by their spreading roots; and their phlogistic matter chiefly from the atmosphere, from which they absorb the air as it exists." He conjectured here that plants get their water through their roots and the carbon dioxide through their leaves, the two pathways by which we now know they indeed get hold of the basic ingredients for running the photosynthetic process.

He was writing his book in great haste and had difficulty retracing the right data in the huge amount of notes he took while performing his long series of experiments during the summer. "I have no time to search in my notes all the particularities I have observed upon this subject. I can say in general from remembrance, that some water plants were very willing to absorb a good deal of air, principally when they were placed with roots and all in the air; and that they readily absorbed air fouled by breathing." He would have liked to say more about the capacity of plants to absorb other airs, but cut the treatment of this topic very short.

On air

After having established his main thesis on the interaction between plants, sun and atmosphere, in the final sections of the first part of his treatise, he concentrated on some methodological issues.

The plants he used were suspended in water. IngenHousz realised very well that this experimental construction was not the natural setting for the plants, but had chosen to do so because there was no other way to capture the produced air in a medium that was not hurtful to the plants. This didn't pose a problem whatsoever to the many water plants he used. He knew that water contains some air in soluble form and argued for the use of water that does not contain either too much or too little air. In the former it would rush into the bulk of the plants and distort the elaboration of dephlogisticated air, in the latter it would absorb too much air from the bodies plunged into it, leaving less air to be collected and measured. He tried it out by putting plants in water impregnated with fixed air (Priestley's famous seltzer water) and found that they produce less dephlogisticated air than in a control jar with pump water. The real explanation for this phenomenon, or for the rationale behind choosing a certain kind of water, remained unclear.

Another unsolved problem was the green matter that plagued Priestley. He wondered how this wonderful substance, which he considered a "green vegetable matter", was able to produce so much fine pure air continuously. Still, he found it no more or less amazing that the same transmutation which happens within the cow when grass is changed into fat, or the production of oil from water in an olive tree.

Finally he examined the reliability of the test that measured the salubrity of the air, the nitrous test. And he was forced to conclude that it was an undeniable test of the goodness of any air, but at the same time it had to be accepted that in certain types of airs this test would fail. That's because he mixed inflammable air in the jar with the plants in some of his experiments. The nitrous test showed an improved air quality. Yet, this air exploded with a bang on contact with a flame (not knowing that a mix of hydrogen and oxygen make for very explosive stuff). Yet when he put a chicken in that air, it died in less than a minute, showing that the air was not suitable for sustaining live. These contradictory results puzzled IngenHousz greatly, but not enough to make him give up the nitrous test. Not in the least because the test "holds up perfectly well in atmospheric air, which is the chief object of our observations".

This provided him the opportunity to sing the praise of the eudiometer, the existence of which we owe to the reverend Dr. Priestley. And to Abbé Fontana, who brought Priestley's truly great discovery to a considerable degree of accuracy. This method from then on allowed researchers to examine the quality of the atmosphere on different spots throughout the world, and study the changes the atmosphere undergoes at any one spot. And here the doctor painted a bright future in which we would be able to adjust the air we breath to cure or prevent various ailments. He equally sketched a wider perspective in which common air is no unchangeable fluid, but just one of the many substances that in an endless network of interactions between plants and animals, matter and fluids, heat and cold, are transmuted into each other. And that's why "in the course of three months, which I spent in my solitary retirement, I scarce found the degree of salubrity of the common air just the same during two days."

These fluctuating values were not, he stressed, the result of the inaccuracy of the testing method. He and Fontana did counter this argument by measuring samples of air after quite some time, getting the same results as on the day they were taken.

Experiments *in extenso*

The second part of the book is completely devoted to the detailed description of the "series of experiments made with leaves, flowers, fruits, stalks, and the roots of different plants, on purpose to examine the nature of the air they yield of themselves, and to trace their effects upon common air in different circumstances".

But first and foremost he started with a detailed exposition on the nitrous test, as performed with the instrument and the technique developed by Fontana. Fontana himself had not yet published his version of the test, IngenHousz explicitly asked him for permission to do so. He was even allowed to have some engravings made of the instrument, based on Fontana's drawings. The eudiometer and its scientific, technical, historical and cultural background will be discussed in a following chapter. Suffices to say for now that the basic principle is as follows. One takes one measure of the air to be tested. To this is added, over water, sometimes over mercury, one measure of nitrous air (nitrous oxide). Described in modern chemical terms this reaction looks like this:[78]

$$2NO + O_2 \longrightarrow 2NO_2$$
'nitrous air' plus oxygen yields red fumes that dissolve in water

The NO (nitric oxide) is a colourless, insoluble gas. NO_2 (nitrogen dioxide) forms a red gas which dissolves easily in water. Mixing nitric oxide with oxygen over water results almost instantaneously in the formation of red fumes that dissolve in the water and 'disappear' from the gaseous volume. If only these two gases would be mixed, at the right volumes, no gas would be left after the reaction, as the diminution would be complete. But as common air in the atmosphere is about 1/5 oxygen, there will always be a large amount of residual gas left over after mixing. The rest consists of nitrogen, a little carbon dioxide and a handful of other gases in minute amounts. Doing the nitrous air test with a sample of common air would result in something like this:

$$2NO + O_2 + (N_2 + CO_2) \longrightarrow 2NO_2 + (O_2 + N_2 + CO_2)$$

The nitrogen (N_2) and the carbon dioxide (CO_2) are the normal constituents of atmospheric air apart from oxygen and they are left as gases above the water. The more oxygen in the tested sample, the more of the volume will disappear. Very pure air (i.e. saturated with dephlogisticated air, as IngenHousz and his contemporaries would call it) results in a greater diminution than ordinary air. The greater the diminution, the more dephlogisticated air (oxygen) in the sample, the better the air.

While testing with ordinary air, Priestley established that a maximum contraction occured when mixing one volume of ordinary air with two volumes of nitrous air. After a few minutes only 1.8 volumes of gas were left. The diminution was as great as the whole volume of added air and about 20% more. Doing this with very bad air (with no oxygen in it) the final volume would be three volumes as no NO_2 would be formed. Intermediate values would indicate intermediate grades of goodness.

78   Conant 1950

Although Fontana and IngenHousz significantly improved on Priestley's method, the test is still not as easy as it sounds. IngenHousz spent more than twenty pages to point out twenty types of errors one can make while performing the test. This was not just a question of being scrupulous or being a fusspot, it was a central principle in performing reliable experiments. At the same time, it illustrates how anxious he was to explain the procedure of the test in such way as to make it reliably reproducible by others.

This set the scene for the complete reproduction of his long series of experiments. In all, he listed 124 experiments, falling very short of the 500 he mentioned elsewhere. This may make clear that he does not reproduce his lab log in that exact chronological and exact manner. He summarized and compiled them in neat little chapters.

For each experiment he provided information on the type of leaf of plant he used, the time of the day and the duration of the exposition to the sun, the results of the nitrous test, one following the Fontana method, another following the original Priestley approach. Each test asked for the subsequent addition of five measures of nitrous air. Doing this two times for each tested sample, added up to ten measurements for each experiment. If IngenHousz considered this as separate observations, they would easily add up to more than 500. Every experiment got a note in the margin of the "quantity of the two airs destroyed". If after adding five measures of nitrous air, the final volume left is 2.84, it meant that of the six volumes mixed together, 415 have been destroyed. This figure in the margin gave a clear and conveniently arranged indication of the quality of the air produced in that particular set-up.

In a first series of experiments he put leaves of different plants (willow, vine, *Becabunga*, *Lamium album*, grass,...) in the sun between 10 or 11am until 2 or 4pm. In a second series he collected all experiments in which he demonstrates the difference in the purity of the air with the leaves of the same plant at different times and with different durations. Shorter exposure delivered less dephlogisticated air. When dark or on rainy days, less was produced. Next he noted the moment of the experiments to see if the time of the day made a difference. In another series he assembled all experiments in which he could learn whether there was a connection between the number of plant mass and the amount of air produced. Another series investigated the difference between the air produced by day and at night. Leaves of oak, French beans, artichokes, potatoes, willow, sage, *Persicaria urens* (water-pepper), lime-tree, yew-tree, apple-tree, all produce either no air to put to the test, or air that was of very poor quality. Next he compared night time with day time in the shade. During the night the air in the jars turned bad, putting the jar afterwards in the sun, the air improved. Putting plants in jars in rooms away from the windows or in the shade of dark bushes, produced air of poorer quality than common air.

Then came a series of experiments that show how "the damage done to common air by the night is very inconsiderable compared to the benefit it receives in the day". Overnight the amount destroyed is about 300 or more. This he followed up with a series in which he started out with air damaged by breathing or burning, demonstrating how plants can restore the damage done. A mustard plant in an upturned jar gave off bad air after a night (117 destroyed), after a quarter of an hour it already improved (121), after 1.5 hour in the sun it was at 176 and after a full three hours it reached 180.

He also tries acrid stinking and poisonous plants (*Hyoscyamus*, *Laureoceratus*, deadly night-shade, tobacco and *Circuta virosa*, among others), who delivered dephlogisticated air just like as other plants. Then came the flowers that produce bad air. Flames were extinguished in it, mixing with one measure of nitrous air gives 1.54 or more. The same resulted from a series of tests on air produced by roots, and by similar experiments with fruits. Pears, apples, peaches, lemons, plumbs, mulberries and silberts (hazelnuts) passed the test. Following up on these findings, he tested green stalks, which turned out to be the only other plant parts that yield dephlogisticated air. A mulberry stem with grey bark didn't seem to do anything, giving off air of about the same quality of common air (190 destroyed). He kept on trying various approaches to measure ever different aspects of the phenomenon he was focussing on. He tested the plants in different kinds of water. He compared the purest air he could get from plants (from the mysterious green matter) with the one which is made artificially from red precipitate (652 vs. 750).

He spent considerable time on trying to understand what happens with inflammable air (hydrogen gas). After lots of contradictory results (of air measuring as good in the nitrous test, yet killing little chickens and exploding with a loud bang) he concluded: "This result convinced me that the nitrous test really fails in this kind of air; for though it gave all the appearance of good air, yet it exploded with a loud report; and a chicken placed in it grew immediately sick, and was ready to expire in six minutes, when I took it out quite motionless."

He tended to conclude that some plants (water plants seem to be rather good at it) are able to correct inflammable air. But he thought that it needed more time for it, probably a number of days to perform this transmutation, in contrast with phlogisticated air, which they could correct within hours. Still he thought that plants change the air into an as yet unknown kind of air, which couldn't be measured by the nitrous test. He proposed the hypothetical name of 'fulminating air', even more explosive than inflammable air not knowing what we know now that hydrogen mixed with oxygen yields a more explosive mixture than the hydrogen on its own.

He finally explained how he abridged the measuring method. As he had so many experiments to do and samples to test, he made his own variant on the methods of Fontana and Priestley. His pragmatic adjustment was that he vigorously shook the tube in which he brought the two measures of air together. He shook for exactly 30 seconds, after which he let it stand for one minute, pouring water over it, bringing it to the same temperature, as it had taken on warmth from his hands while shaking. To measure common air, mixing two measures is enough; to test dephlogisticated air, he needed to add more measures of nitrous air. He was able to perform the whole operation in three or four minutes. The accuracy was such that when measuring the same air in ten trials, the difference in the results did not amount to 1/200 of the bulk of both airs. The degree of salubrity of common air in general ranged between 103 and 109, meaning that the remaining column of air in the glass tube occupied between 103 and 109 subdivisions.

This method differred from the one Priestley used as Priestley mixed the two airs and let them stand before he put them into the large measuring tube, without shaking. "If this method is pursued accurately, and if the same interval of time is observed between joining the two airs and letting them up, the result will, however, be found different in different experiments, as Dr. Priestley makes no scruple to allow."

IngenHousz was more meticulous in fine tuning the measuring method. He compared tubes of different sizes, experimenting with different or no shaking times, with letting the mixed airs stand for longer or shorter times.

The work continued

The book was only a beginning. In an "advertisement" on page lxii, at the end of the preface, he writes: "As the author intends to publish a French translation of this work, he thinks it is his duty to give public notice of his intentions, that no one may give himself any unnecessary pains about it." He knew by then that he would publish editions in other languages. And the work continued also in other ways. In a postscript he described how he continued his experiments during the period the book was being printed, accumulating more data that would make a second volume necessary. He speculated cautiously about how the heat of the sun could corrupt the air, looking at the planes of Hungary or the low region around Rome, where this mischief could be remediated by planting a sufficient quantity of plants, principally trees, by draining marshes, preventing inundations and keeping the rivers within their dykes.

After the publication of his book, the author took the coach to Bath, to see the person to whom the book had been dedicated. Sir John Pringle was ill and taking the waters. Patron and protégé spent some intense days together. Journeying back to London, IngenHousz broke off his travel at Bowood, near Calne, the residence of Lord Shelburne. IngenHousz had known Lord Shelburne for years but this must have been his first visit to Bowood. This is where Priestley had been the resident chemist, teacher of the Shelburne children and tender of the library. IngenHousz and Priestley must have discussed the implications of that summer's experiments with excitement.[79]

It was just one stop on a long journey back to Vienna. While travelling, he continued taking air samples and evaluating the quality of the common air. In October he was back in London.

On 22 October his diary-cum-logbook[80] says, at the entry numbered 367: "Being returned to London I now and than grade some experiments". In the margin he wrote: "the air near my lodgings in Pall Mall proved to be better than ever I found it in the country Q. Is this constant? Q. is it owing to the vicinity of the trees in Carleton Gardens?" IngenHousz seemed to be continuously asking himself questions, eager to be able to explain all those phenomena he saw.

On 3 November he sailed in the Thames estuary between Sheerness and Margate, two or three miles from shore. He tested the air and found it better than any land air. In the middle of the sea the weather was too rough to perform experiments, but he took three bottles of sea air, to test them on land, when he arrived in Ostend on the 5th. On 7 November he arrived in Bruges, the 8th he spent in Gent, the 12th he arrived in Brussels. The following day he took a sample in the park: "l'air prise du bois de la ville a onze heures occupoit 106, l'air en haut de la ville etoit 104 1/2 l'air de ma chambre à coucher n'ayant pas renouvellé 108". The following day he travelled to Antwerp, taking a sample at Vilvoorde, because he forgot to take a last measurement in Brussels. From Antwerp he continued to Breda, to get to Rotterdam via the Moerdyck. On 4 December he was in Amsterdam. And all along the way he took samples and tested the common air against nitrous air.

From Amsterdam he returned to Antwerp, Brussels and went on to Mons, arriving in Perone [Péronne] on Christmas Day. By the 29th he was in Paris. He went through Passy, then a village near Paris, north of the river Seine opposite the Champ de Mars (where now the Eifel Tower stands). It was the place where the delegation from the American colonies was staying. One of the delegates was his great friend Benjamin Franklin. They must have been glad to see each other again and without any doubt had plenty of things to talk about. Chimneys and stoves, lightning rods, air balloons, gases, metals, gunpowder, heat conduction, electricity and the secret life of plants. No traces are left of their meeting.

Page 45 of his logbook opens on the year 1780 and the title "Continuation of experiments on plants and air". On 30 April he was back in Brussels, where he stayed until 23 May to return to the French capital on the 27th. By summer he was back in Vienna, where his old life resumed its pace, being reunited with his wife, with whom he only had lived for three years since they got married. On the basis of his correspondence[81] with a dear friend in Holland, Jacob Van Breda,

79 Beale 1999 p 22
80 Breda Collection IngenHousz B1
81 Haarlem Teyler's Museum 2243-2247

we are able to reconstruct parts of his life in Austria. Van Breda took care of the translation and production of the Dutch edition of *Proeven op plantgewassen*, which appeared in Delft, later that year. At the same time a French and a German edition saw the light, entitled respectively *Experiences sur les plantes* and *Versuche mit Pflanzen*. These editions were edited considerably and extended, based on the further experiments and measurements IngenHousz had been doing during his travels through Europe. No fundamentally new findings were presented, but IngenHousz was the prototype of a researcher who never stopped collecting new evidence for his claims.

Under a new sovereign

IngenHousz wrote Van Breda a final letter from Paris 5 June 1780, which mentions his most recent contributions to the *Philosophical Transactions* on electricity and powder, and the papers he submitted on the compass and the atmospheric conditions at sea. He also mentioned the fact that Priestley had referred to IngenHousz' work on inflammable air in his last book (part IV, 1779).

The next letter came from Vienna on 21 August, stating that he had been in the Austrian capital for ten days already but hadn't been able to do anything useful due to ceremonies and obligations of various sorts. On 6 October in a letter that opened for the first time with "My dear sir and friend", indicating that they knew each other well enough to become colloquial, he announced to Van Breda that the French edition had finally been published and that he had requested that a copy would be sent to him as soon as possible. On 1 November his letter began with a complaint about the slow delivery of the letters from Breda. He suggested sending the letters to his brother in Breda, who could send them on the Brussels where Prince Von Starkenberg would put them in the imperial courier service. Further he wrote:

> *Ik heb experimenten genoeg gereed om het tweede deel te componeren, dog ik heb myn gedagten nog niet kunnen bepaalen tot de egte invloed der planten op de gemeene lugt voor zo ver als het acidum aerium betreft, en daarom heb ik het article niet durven plaatzen in de Franse editie. ...*
> *Ik ben nu zeker dat planten ook ~~phlogist~~ vaste lugt uytwerpen en ik kan zelfs de hoeveelheit daarvan bepalen.*

He has performed sufficient experiments to complete a second volume, but he had to admit that he was still not wholly convinced of the real influence of plants on the common air as far as the aerial acid is concerned. That's why he had not put anything on this topic in the French edition. He was persuaded however that plants expel fixed air and that he could even quantify the amount.

He found it more difficult than expected to inhale and exhale this dephlogisticated air, so he needed to invent a better method than the one Fontana developed for patients to benefit from this air. And about the eudiometric method he noted:

> *Ik volg de methode van Fontana niet meer omdat die teveel tyd vergt. De differentie van 1/100 of 2/100 in verschyde experimenten is geen groot verschil. Men heeft reden om zig te verwonderen dat deze ontdekking tot zo een trap van volmaaktheid heeft kunnen gebragt worden.*
>
> *Door langdurige exercitie word men er zo handig in, dat men zelde meer als 1/100 verschil heeft. Het is nodig om verschil te hinderen, dat men altijd met een veder de bodem der plank daar de trechters ingeneden zijn, van de lugt bellen zuyvert die er altyd van zelfs onder hangen. Men moet ook de eudiometer altyd onder water laten, of als men hem er uyt doet, en laat opdrogen, hem eerst van binnen uytspoelen met zeepsop om te beletten dat het water er niet aan blyft hangen met droppen.*

He was improving on Fontana's eudiometric procedures. The accuracy he obtained was more than satisfying. He stressed the importance of certain

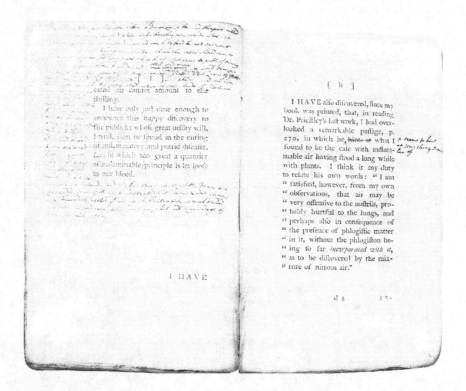

Page from *Experiments upon plants*, 1779, on which IngenHousz scribbled his notes for incorporation in future editions.

manipulations, such as using a feather or soap to prevent air bubbles from accumulating in the test tube. And he pleaded for measuring the goodness of the air all over Europe.

In November 1780 Maria Theresia died and her eldest son Joseph II came to the throne. Joseph was a proponent of enlightened absolutism, inspired by the writings of Voltaire and the Encyclopedists and the example of Frederick the Great. He saw reason as the ultimate guiding principle, to be executed by the state to the benefit of all. Many of his new rules and regulations were in accordance with the grand principles of the enlightenment: German as the compulsory language (replacing Latin - a controversial issue in a multilingual empire), spreading education (compulsory elementary education for boys and girls), reducing the power of the religious orders and the clergy while propagating religious tolerance, emancipating the peasantry and abolishing the death penalty. But introducing some modernising but austere reforms - ranging from banning corsets to demanding breast-feeding - made him less popular with some Viennese.[82]

However, all these innovations based on rational thinking created an environment in which IngenHousz and his scientific curiosity must have flourished. His brother-in-law was one of the foremost Linneans in Europe. The home of the Jacquins was the setting for a group of artists and scholars who met weekly for discussion, music and light entertainment. One of the regulars was Mozart, who would become acquainted with the Jacquins in 1783. He dedicated the Nocturnes K 436-9 to his close friend Gottfried Jacquin, the younger brother of Joseph. Jacquin's botanical theories, which sexualised nature in an unprecedented way, have influenced culture in more than one way.[83] IngenHousz probably never met Mozart, but the nearness of this little network suggests he was very close to the new culture of arts and knowledge that was arising in Europe in the second half of the 18th century.

In December he wrote to Franklin, as annotated in his letter register:[84]

> *dec a Passy B Franklin I continue to pursue the object of my book as to the influence of the vegetable kingdom upon the animal. I think to have discovered the ways by which nature changes a part of the atmosphere into nitrous acid. I think now that plants in the shade and at night emit fixed air and decompose common air partly unto fixed air by communicating to it their phlogiston.*
> *but I want more decisive facts before I publish it.*

A new year started without much ado. IngenHousz' mind was frantically active. He wrote to Pringle, as indicated in his letter register:

82   Hasquin 1978
83   Hunter 1997
84   Breda Collectie Ingen Housz A6

1781 jan 2 Pringle that I have made some discoveries that I am now convinced that plants do not only decompose common air but also trow out themselves a good quantity of fixed air that I will send in a few days to the society in Rotterdam a new and easy method of inspiring dephlogisticated air

IngenHousz waited for the Dutch translations that Van Breda had put in the post. This was one of main topics of many letters: translations, amendments, corrections, additions &c. to manuscripts of his books or articles, about to be published in various Dutch, French or German journals or books. Their correspondence was quite intensive, a letter to Delft departed from IngenHousz desk almost every fortnight.

On 29 August 1781, he wrote:

Hoe beter de gedephlogisticeerde lugt is, how harder den slag is. Ik denk egter niet dat dit phenomon enige betrekking met den donder heeft, derwyl men geen bewys heeft dat er tusschen wolken knallugt is, en derwyl de electrische vlam alleen genoeg is om die beweging of schudding in de lugt te veroorzaken waar in het geluyd van den donder bestaat.

Ingniting a mix of oxygen with hydrogen resulted in loud explosions, making him speculate about the origin of thunder. In his August letter IngenHousz further described how *Conserva rivularis*, the green organism growing in glass vessels, produced excellent air, measuring over 250 degrees. He started dreaming about installations that could capture this good air for the health of all. The air in Vienna was often under 90, while in Delft, where Van Breda lived and performed the same tests of the atmosphere, it was usually much better. IngenHousz wondered if that had something to do with the winds coming in from the sea. He asked Van Breda if he could find any difference with offshore winds. Later that month he reported on his invention of a new kind of air-pistol, fired with inflammable air. He further experimented with an air pump based on a new principle and gave a recipe for soap bubbles filled with inflammable air which exploded when coming in contact with a flame: "a very nice phenomenon". Yet he did not think that this had anything to do with thunder, as he reckoned there was no inflammable air in between the clouds. But it was not just chemistry that occupied IngenHousz:

De Engelsche natie heeft alle agting verloren door de barbaarsche handelwys op S. Eustacia. Men moest haar door represailles tragten te toonen dat onder beschaafde natien zulke ongehoorde wreedheid niet onstrafbaar moet of kan gedult worden – ik hoop dat onze regeerders noyt zo lafhartig zullen zyn

This last paragraph alludes to the fact that the island of Santa Eustacia in the Western Indies was sacked by British troops in 1780 largely because of its notorious role in the smuggling trade to America. It reveals IngenHousz' social concerns. He is against (state) violence and repression and does not refrain from criticising even the country in which he loves to live and where he has many close friends.

In October he formulated some doubts on the effect of the water over which he tested the goodness of the air. Until now, he had found no differences between sea or fresh water, warm or cold water, rain or well water. He seemed however to find now a difference between warm and cold water and indicated he planned to investigate this further.

His work on experiments, instruments and the editing of his articles was frequently interrupted by visitors. On 1 December he mentioned the visit of the Russian Tsar, his wife and the whole imperial entourage. They were clearly impressed with IngenHousz' demonstrations and the Tsar requested a reprint of the article on dephlogisticated air, so he asked Van Breda to send it as soon as possible via his brother in Breda to Vienna.

Ten days later he returned a letter full of corrections for a print proof of the French edition on dephlogisticated air. In an aside he complained about the Dutch printers. He found them slow and selfish (not in the least because French and English printers offered him free copies which the Dutch didn't).

Controversy

On Christmas day he wrote to Delft that he had received news from London that Priestley has criticised him in his new and fifth volume of his book. He hadn't seen the book yet, but relies on the August/September issue of the *Critical Review*. He felt that Priestley sacrificed their friendship, being envious about his discoveries. Priestley maintained that the air IngenHousz described and measured was obtained from the water and only attached itself to the leaves, just like the air bubbles that attach themselves to the skin when having a bath.

This is the first written expression of the disenchantment between these two great natural philosophers. IngenHousz must have been really incensed about Priestley's attitude. He spent four full pages of his letter on this topic. IngenHousz had to admit that he was wrong to insist against the advice of many of his friends to praise Priestley in the foreword of his book, especially when he now saw how Priestley denied that the vegetable kingdom influences the animal kingdom. Priestley now claimed, just like Cavallo, that water deposits its air bubbles on the leaves, which removes the phlogiston, so that the rising air is a little purer than common air. But why, asked IngenHousz, would there be a difference in the quality of the air when different species are used? Why would the bubbles appear only on one side of the leaves? Why is the air better with a plant that gives off more air? If Priestley and Cavallo were right, then the amount of air would depend on the amount of water, not on the surface area and on the type of plants used, as his experiments have shown. IngenHousz was indignant because he had done more experiments and therefore had more data than his critics. The contradictory and ill founded claims of Priestley showed

> dat men jalours van de uytvinding is, en dat men selfs met een dollen kop will over hoop werpen liever dan een ander de eer van de ontdekking toe te eygenen.

that one is jealous about the discovery and that one is even preferring to throw everything in chaos rather than to grant another the honour of the discovery.

He went into quite some detail on the experiments of Cavallo: three vessels, one with a branch of vine, one with a green cloth, a third with plain water. In the first dephlogisticated air was found, in the second a little bit, in the third nothing. IngenHousz analysed these results and referred to part five in his book, where he included some warnings to ensure that no air clung to the glass of the cups before starting to measure. And he suspected the cloth wasn't sufficiently wrung out under water to remove all air.

IngenHousz decided to tread carefully with people such as Priestley and Cavallo. They were the kind of people who thought that Divinity would be endangered without their help and that the heavens had to be protected against the cruelties of humans. He saw them as self-declared preachers of freedom and integrity, who presented themselves as lambs but were "doorgaans grammoedige vervolgers en niet zelden kwaadaardige dwepers": more often than not wrathful prosecutors and cruel fanatics. IngenHousz wanted to stay out of their way, and not to step on their toes.

A torrent of ideas erupted as a result of his enthusiasm. He ended with the thought that he had been entertaining for some time: if the water was giving off no air, the air produced by the plants would be purer. That could explain why the air is usually purer when more air is produced, as the contribution of the water itself is limited. He described a new experiment with an Agave with common air in a vessel over mercury for 45 minutes, producing a great amount of extra air with a high goodness, showing that the plant expelled more air than it absorbed.

Another priority claim

It was January 1782, whilst in Vienna, that the collection of articles went to print. In Milan and Paris the printers were ready to go to work to on the Italian and French editions. He sent Van Breda the figures (engravings) that had to be inserted in the Dutch edition. IngenHousz received more visits from the Tsar and his wife, who stayed sometimes for the whole day to watch him perform his experiments.

Not all went well with the experiments. The "niterlucht" or nitrous air seemed to be of variable quality, depending on the copper with which it was produced and the manner in which it was recollected. The type of bottles and retorts and their connections and seals were not as good as he would have wished.

In March, the French printer had not started yet, but the Viennese printer delivered 5 pages every week, which were sent to Van Breda for translation. Vienna saw also the visit of the Pope, ironically described as "een allerzeldzaamst phenomeen politico-religiosum", a very curious politico-religious phenomeen. In 1782, IngenHousz published another paper clarifying the role of the vegetable kingdom on the animal creation, illustrating that he started to grasp the essential role of plants in sustaining life on earth. After critique from fellow members of the Royal Society and from competitors from overseas, he invited some of his friends to assist in some decisive experiments, which he described in an exact account in "Some farther considerations on the influence of the vegetable kingdom on the animal creation", which was read at the Royal Society on June 13 1782, in which he concludes:

> If all what I have hitherto should not be thought sufficient to take away the prejudice which Mr. PRIESTLEY's fifth volume, and Mr. CAVALLO's book on Air, may have produced in the mind of some philosophers, I should advise them to be present, at least once, at the most beautiful scene which they will behold, when a leaf of an agave Americana, cut in two or three pieces, is immersed in a glass bell or jar of pump-water, inverted and exposed to the sun in a very fair day in the middle of the summer, when this plant is in its full vigour; and when they shall have seen those beautiful and continual streams of air, which rush from several parts of this vegetable, principally from the white internal substance of it, I will be answerable for their laying aside all farther doubt about the truth of my doctrine.
>
> After having now demonstrated, as I think, in the clearest manner, that vegetables diffuse through our atmosphere, in the sun-shine, a continual shower of this beneficial, this truly vital air; that plants immersed in water, far from robbing it of all air, impregnate it fully with a better and more salubrious air; let us not pass so wonderful, and hitherto not even so much as suspected,

an operation of nature, without admiring the designs of that infinite wisdom, who has employed, such wonderful, and at the same time such beneficial means to preserve from destruction the living beings which inhabit our earth.[85]

Note that he goes further than Pringle did in his laudatio to Priestley. They saw plants as being of service to mankind. They had the idea that man was the pinnacle of creation and plants were only there to keep that up. IngenHousz comes up with a wider perspective in which all living things are mutually interdependent. Every thing on earth has its own place and nothing seems to be any more important than any other. The only difference is that man has been given the capacity to think about it.

And thinking they did. At the time IngenHousz was developing his theory, another man claimed the priority of the discovery of the illusive botanical process. Van Barneveld, pharmacist at Amsterdam, had published an article in the transactions of The Utrecht Society on the effects of sunlight on the air production by plants. Van Barneveld behaved as if he was unaware of IngenHousz' book, while the first copies of it had already been circulating in Amsterdam only a few weeks after it had appeared in London. Van Barneveld's dissertation recounted a whole series of experiments just like IngenHousz' without mentioning IngenHousz even once. Curious was that Van Barneveld told that he wrote the article in 1778 and handed it in on 23 October, while the dissertation itself was dated 6 March 1780. IngenHousz asked Van Breda to investigate this case and to find out of the real date of the publication.

Having read Priestley's article, IngenHousz has turned milder in his opinion of him. He found no insults in this work but "zoo veel fyne, en aan geestelijke personnagiën veeltyds ygen, streken dat ik wel geen rede heb er hem qualyk over te bejegenen", so many fine, typically clerical, tricks of rhetoric's that he had no reason to treat him nastily. Priestley claimed to have said that plants regenerate the air, whereas he never actually said so, having doubts about, e.g. the nature of the green matter in his vessels. But Priestley kept strategically quiet about some of IngenHousz' experiments and findings, thereby favouring the idea that he, Priestley, was the real discoverer.

Molitor, who took care of the German edition of IngenHousz' book, was enraged and was about to write a critical introduction in the German edition. IngenHousz suggested Van Breda could translate this also in Dutch. IngenHousz himself was going to write a letter to the Royal Society to defend himself and offer new proof for his "system".

On the Dutch plagiarism front, if that was what it was, news spread. It was not yet completely clear to IngenHousz what exactly had happened, but Van Breda had revealed that the secretary of the Utrecht Society told him that the piece by Van Barneveld was indeed dated 23 October but that it had been

---

85   IngenHousz 1782

returned to the author with comments and demand for improvements. The date of the resubmission was unknown. No copy was kept at the society, so Van Barneveld must have been able, according to IngenHousz, to add whatever he liked, even using IngenHousz' terminology. So he could have sent it in later, after having read IngenHousz' book and whilst claiming to have written his piece at an earlier date.

Letters were going backwards and forwards between Vienna and Delft. A new publisher/printer had been found in Paris (a suggestion by Franklin). Copper engravings were recycled for new language versions and sent across Europe, others were corrected and remade. In the mean time, the experiments continued. IngenHousz was still trying to find out what exactly the green matter was that appeared in water that had been standing for a while. He dried it, grounded it and ... it still produced good air.

A new year in Vienna

On 5 February, IngenHousz reported to Van Breda to have received a letter from Priestley. While having a deep difference of opinion, they kept in touch. Priestley didn't only tell him about experiments on the relationship between metals and phlogiston:

> *Hy schrijft my verder dat het article van mynen brief wegens myne klagten over zyn gedrag, waarop deze een antwoord is, hem aandoening geeft. Hy zegt niets onvriendelyks voorgenomen te hebben, en dat, indien hy niet genoeg zyne gevoelens van vriendschap geuyt heeft jegens my in het vyfde deel, hy zulks zal verbeteren in het zesde deel. Ik begryp niet waarom hy dit schryft dewyl ik in myne brief aan hem niet geklaagt heb over het niet uytdrukken van zyne vriendschap, maar dat het my wonder voorquam van zo ene jeuking voor critiseren te betoonen aan een man waarvoor hy betuygt vriendshap te hebben.*

Priestley wrote that he was touched by IngenHousz' complaints about his behaviour. He claimed that he hadn't meant to harm him, and if he had not properly expressed his friendly feelings in part five, he promised to put things right in part six. IngenHousz did not understand why he wrote such things while he never complained about friendship but about the way he criticised somebody whom he claimed to be his friend.
His work drew a lot of attention:

> *Ik heb onlangs drie vrydagen achtereen byna een paar honderd heren en damens meest van den eersten rang by my gehad savonds, als ik al merenmale*

gedaan heb om te beletten dat zy my afzonderlyk niet zoveel tyd ontnemen. Voor ik de experimenten begon, heb ik een redevoering voor gelezen. In de tweede redevoering heb ik bewezen dat ons gansche lichaam, mogelyk geen 1/20000 uyt gezonderd, niets is als geconsolideerde lucht, en na de dood wederom niet in aarde, maar in dezelfde lugt zal veranderen. Dat den dampkring het generale magazyn is van alle dieren en gewassen en ook van veele andere lighamen. Deze dry redevoeringen maakten veel indruk, en men solliciteerde my sterk om ze uyt te geven.

He gave public demonstrations, sometimes to hundreds of high society men and women. He invited them all at once in order to save time. He loved to speculate with them that air has a nutritious function and told them that the human body consists of fixed air (carbon dioxde) and that after death we don't change into earth, but into air. The atmosphere was the general store house for all plants and animals and other organisms. Although many urged him to publish this story, he did not yet dare to do so.

A third contender

In his letter of 26 February 1783, he wrote Van Breda that he did receive Jean Senebier's book *Mémoires physico-chimique sur l'influence de la lumiere solaire pour modifier les êtres des trois règnes de la nature et surtout ceux du regne végétal*, published in Geneva the year before. Here was a new contender to claim the priority of the discovery. Senebier claimed to have had the same ideas as IngenHousz and to have written letters about it in May or the summer of 1779. IngenHousz would have loved to see the letters as the proof of this claim.

Ik heb het eerste deel nog maar doorlopen, waar in ik myn boek byna uytgeschreven vind. Met enige niewe aanmerkingen. Den autheur drayt ook rondom een center met Priestley en Barneveld. Hy heeft iets diergelyks <u>analogui</u> gedagt over my. En heeft het zelfs geschreven in may 1779 aan de Heer Van swinden, of ten minste in die zomer.

He had only leaved through the first part and it was clear to him that Senebier had copied IngenHousz' experiments and added some remarks in places. Senebier seems to be playing the same tricks as Priestley and Van Barneveld. He now pretends that he had been thinking something analogous about this subject in May 1779 and that he wrote about it to Van Swinden. Jan Hendrik van Swinden is one of Senebier's correspondents, mathematician in Franeker who had won together with Charles Coulomb the international prize of the Académie Toyale des Sciences in Paris in 1776.

IngenHousz found Senebier's eudiometric method unreliable as he was using vessels that were too big. Probably the reason why Senebier did not find that plants produced air in the shade or why plants in fixed air gave more and better dephlogisticated air compared with plants in just water.

In another letter in March he summed up more inaccuracies and mistakes that could be found in Senebier's book. Air was dissolved in water and bubbles were set free even without plant leaves. Senebier asserted that this was only fixed air. According to IngenHousz, water saturated with fixed air formed bubbles on the submerged leaves but caused the death of the plant. Senebier said that his book repeated many parts of IngenHousz' book but that there were many differences. IngenHousz deplored that Senebier didn't make clear what these differences were. He assumed Senebier did so in order to make comparison between the two works difficult thus avoiding critical comments. Now, according to IngenHousz, one had to leaf through the book page by page to find the differences. Compared to this, Priestley at least was much more honest: he focused on the deviating facts and opinions, making discussion easier. Senebier played dirty tricks to be able to flaunt with another's discoveries, as nobody would take the time to compare the two works in depth, tactically evading the chance that someone would see his work as an echo of IngenHousz'.

Still, Senebier's work stimulated IngenHousz to investigate what kinds of air could be dissolved in different kinds of water. And the work on the mysterious green matter continued too. A piece of meat in pumped water was after a few days filled with a green crawling mass, IngenHousz called it "round and oval insects". After a while they fixed themselves to the bottom and showed no more sign of life. After a few months this changed in a crust, a moss-like mass with white filaments. Fontana called it "planta-animalis". Was it a plant, or an animal? Or was it an insect that changes into a plant, as IngenHousz sometimes thought. Nobody was sure what kind of organism this really was. He sent a sample of the green mat-like structure, looking like *Conserva rivularis*, to Van Breda. It can still be found in between sheets of paper in the Haarlem archive of the Teyler's Museum.

In the period between these experiments and speculations, a detective story unfolded over the question who had been the first to identify the sunlight driven power of plants.(More about this in chapter six.) Senebier still claimed he developed similar ideas to IngenHousz' in March 1779, but did not publish it because he wanted to publish it as part of a greater volume. The note in which he pretended to have explicated this idea could however nowhere to be found. IngenHousz has asked many people in his intellectual network to unearth it. Simultaneously IngenHousz seemed to be annoyed that all this hassle takes up so much time, he ended his letter with "this letter is long winding and that's why I cut it short here."

During the summer one of his guests was Jean-Henry Hassenfratz, a young and ambitious natural philosopher and close collaborator of Lavoisier. Hassenfratz was using the eudiometer for his own meteorological observations to measure the goodness of the air, in his garrison in Falaise. IngenHousz received Hassenfratz in his own laboratory, demonstrated his experiments and worked with the eudiometer together with his eager guest.[86] Hassenfratz told the whole story in a very long letter which he sent almost immediately to Lavoisier on 18 August 1783. It is the first direct proof of the close relationship between the young correspondent of the Societé Royale de Médicine and the famous chemist. Lavoisier had a great interest in the experimental work of IngenHousz (and Senebier) on the release of oxygen by plants. Hassenfratz wrote him:

> *Sachant que vous vous proposiez de repeter les experiences de MM Ingen-Housz et Senebier, soit pour vous assurer par vous-même des résultats obtenus, ou pour vous décider l'espèce de contradiction qu'ils ont dans quelques expériences, je m'empresse de vous faire part de celles de M Ingen-Housz a bien voulu avoir la bonté de faire pour nous et que nous avons manipulées en partie.*

He was glad to recount the eudiometric method of IngenHousz, in order for Lavoisier to be able to reproduce these experiments. Then followed the description of the experimental set-up, how plants submerged in a glass vessel completely filled with water, produced air in the sunlight, which air had not been, so they were able to prove, first dissolved in the water. Then they could test the air with the eudiometer and showed it was oxygen. They repeated this under various conditions over five consecutive days. The length and detail of the letter suggests these two men had already been discussing the eudiometer extensively. Hassenfratz had criticised IngenHousz' working methods with the eudiometer earlier in a letter to the Societé Royale, but after having extensively worked with him, he had to admit that IngenHousz had very good reasons for his methods. He asked Lavoisier:

> *Je vous prie, Monsieur, d'avoir la bonté de mettre sous les yeux de la Societé le procédé de M Ingen-Housz pour le comparer au mien. Le progrès des sciences étant le seul moteur qui m'amine, je serai trop heureux, si le sien est préféré, d'avoir contribué à faire suivre un genre d'observations dont on reconnaîtra d'autant plus l'utilité que l'on en aura receuilli d'avantage.*

He asked Lavoisier to bring the topic of the eudiometric method once more to the attention of the assembled Society and to tell them explicitly that IngenHousz' method was to be preferred over his own. He had no problems admitting he was wrong, as that was the only way science would make progress.

---

86   Grison 1996

And progress was also what mattered most to IngenHousz. On 14 August, he reported to Van Breda that people started to have more confidence in the newest and useful inventions. Graf Von Rosenberg had told him the lightning rod on the mountain of Lurcyory functioned properly and the underlying village had been spared from being hit by lightning. In the time span that the church was normally struck 25 times, with horrible consequences for the inhabitants, now the lightning had only struck twice. The lightning rod was a spin-off from Franklin's work on electricity. Papers on the preventive invention had been published in London on 1751 and in Paris in 1752. Attracting the devastating energy from lightning would enable the prevention of damage to churches, high houses and ships. On both sides of the ocean, experiments with lightning rods were set up and reports of the new invention appeared everywhere. Debates raged over the exact form the rods were deemed to have, pointed or rounded, and slowly the idea took hold that the rod should be grounded to be effective. But the public debate over the desirability of putting up the rods proved even hotter. Religion saw thunder and lightning traditionally as the hand of God, punishing men for evil thoughts or deeds. It would be blasphemous to try to evade this heavenly judgement. But Joseph II was the epitome of an enlightened monarch. His anticlerical politics interlinked in 1783 with the abolition of the 'Gewitterleuten', the ringing of the church bells to chase away thunderstorms.

Previous attempts of introducing the lightning rod dating back to 1755 had all failed because of religious opposition.[87] The lightning rod had become a symbol of enlightened politics. IngenHousz, being a friend of Franklin and interested in all things electric, was involved in the installation of the first lightning rods to protect the Viennese powderhouse, as can be understood from his letter to Franklin in May 1780:

> If you could communicate to me some short hints, which may occur to you about the most convenient manner of constructing gun powder magazines, the manner of preserving the powder from moisture and securing the building in the best manner from the effects of lightning, you would oblige me.[88]

He described the installation of lightning rods on the imperial powder warehouses and on the Viennese Hofburg in the first volume of his *Vermische Schriften*. Another invention is mentioned almost casually, a little further on in IngenHousz' letter on 14 August:

> om het te schielyk uyt drogen van een drop water op een plat glas te verhinderen leg ik een stukje talk of fyn muscovisch glas er op of nog beter een stukje fynglas waar van hier nevens een beetje.... Die membrana vitrea vind men overal in glasfabriken op de grond liggen.

87   Hochadel 2002
88   Labaree vol 32 p 341

IngenHousz described what was probably the first application of a cover glass under a microscope: to prevent the fast drying of the drop of water on the glass plate he put a fine sliver of glass on top, a better practice than using a slice of talcum or muscovic glass. This 'membrana vitrea' could be found as waste in any glass factory. Moreover, a little fragment of the glass plate is still enfolded in the letter. IngenHousz' pragmatism led him exactly to do this kind of thing and without pump or circumstance circulate this technical improvement to anybody who wanted to know. It also led him to the description of the erratic and random movement of particles in fluid, now known as the Brownian motion. More about this later.

It was also encouraging that he received a letter from Senebier, via Landriani, with the request to describe the experiments that prove there is an 'influxus noctarum' (the nightly outflow of fixed air). Something about which Senebier's publicly had expressed his doubts. If these would convince him, he promised to revoke his earlier statements in public. IngenHousz loved to show his experiments to Landriani, but not to Senebier. He did not trust this man who executes trifling tests. But he thought that this was a sign that Senebier might start recognising his mistakes.

This influxus nocturnus was a crucial argument, according to IngenHousz, to conclude that nobody before him had any idea about the influence of plants and the light of the sun on common air. How could Priestley, Senebier or others, possibly have seen that plants improve the air in the sun and not in the shadow, if they had not investigated what happens in the absence of sunlight? To know this, one had to investigate it. And when running investigations, one would see that they have a opposite effect in the dark. As nobody had ever mentioned this, it was more than likely that nobody had properly investigated it, neither in the sun nor in the shade. IngenHousz was still a doctor first and foremost and he ventilated his opinions on the pest, which he thought to be an infectious disease. That's why he wondered why the Emperor was giving the orders not to disinfect the letters from Constantinople anymore, where the Black Death had errupted. One contagious letter could easily spread the pest to other countries. He must have been receiving advice from stupid doctors, such as Stoll who thought the pest was caused by undigested food in the stomach. However, he asked Van Breda not to mention his opinions at the Society in Holland, as it would be dangerous to oppose the will of the mighty.

Ignorance, doubts and caution

On 8 October he wrote to Van Breda that he appreciated his translation of the article on the electrophore very much. He thought there are only few people who really understand Franklin's theory on electricity. He doubted that even Volta really understood it. He also was in doubt about the similarities that many people saw between magnetism and electricity. He warned that little was known about the reality of the 'magnetic fluidum', if it existed at all, "sodat wy die twee kragten niet in alle verschybselen kunnen van elkander scyden", so we can't properly distinguish both forces from each other in phenomena.

The dispute with Senebier kept popping up.

> Dewyl de zaak my aangaat, moet ik zoveel mogelyk weten wat daaromtrent geschiet is. Als men publiek schrijft moet men voor de vuyst te werk gaan en niets mysterieus agter houden. Met wat regt kan Senebier zeggen dat hij die analogue ideeën had <u>eer ik er op dagt.</u> als hy de voorrede van myn boek met aandagt gelezen had, zou hy gezien hebben dat ik er wel ver van af was, van op dat voorwerp niet gedagt te hebben en ik tot het maken van de experimenten omtrent myne ideeën overging. ik was zeer vol van differente ideeën, zonder te kunnen voorzien wat my de experimenten zelf zoude aanduyden. Hoe zoude ik hebben kunnen op het denkbeeld vallen van planten in de schaduw en in het ligt te stellen zonder enige idee te hebben dat ik er iets onderschydens uyt zou hebben konnen vinden. indien ik geen ideeën gehad had dat wortelen vrugten bloemen andere werkingen hadden op de lugt als bladeren, hoe zou het my ingevallen zyn om in zo een korte tyd zo veel differente ontdekkingen daar omtrent gemaakt te hebben?

He had to admit that this case was important and he wanted to know what had really happened. If one made his findings public, one had to be bold and not keep anything secret. He asked what gave Senebier the right to say that he had similar ideas to those of IngenHousz, <u>before he had thought them up?</u> If he would have read IngenHousz' introduction to his book with any attention, he would have noticed that he was still far from his final conclusions and that that was exactly the reason why he had started his series of experiments. He was himself full of conflicting and varying ideas, so he could not foresee what they would lead to. How would he have been able to formulate his hypothesis about plants in the shade and in the light without some idea that he would find something significantly different? Hadn't he had the idea that fruits, roots or flowers had different effects on the air, he would not have investigated them and would not have made so many different discoveries in such a short time.

These thoughts had been praying on his mind for a long time. He had some ideas about the influence of plants on the atmosphere in 1773, but it seemed meaningless to publish them without any experiments to prove it. Senebier insinuates that Mr IngenHousz had not the least ideas on these matters, and should have asked him first. This kind of tricks sounded to IngenHousz rather 'jezuit'.

Indien ik geen idee gehad had dat wortelen vrugten bloemen andere werkingen hadden op de lugt als baderen, hoe zou het my ingevalen zyn om in zo en kote tijd zo veel differente ontdekingen daar omtrent gemaakt te hebben? in 't kort ik was zedert 1773 vol van allersoorten van ideën omtrent den invloed van der planten op den dampkring, maar wat had het publiek te doen met voorgegaane ideën? het komt my slegt voor van zig daarop publiek te willen beroepen en er dan nog by te voegen dan den H IngenHousz geen de minste ideën daaromtrent gehad heeft zonder het eerst aan hem te vragen of zulks waar is of niet. Zulke loopjes komen mij wat jesuitisch voor.

Van Swinden, in the mean time, admitted not having received anything from Senebier, in contrast to what Senebier himself wrote on page three of his book, as quoted by IngenHousz in his letter of 8 October:

> *... qu'il avoit eu des idées analogues avant que Mr. IngenHousz y pensat; qu'il avoit communiqué ces idées à mr. Bonnet, qu'il avoit aussi envoyé un appercus sur ce sujet à mr. van Swinden et Volta.*

Senebier invoked on the same page an illness which, in 1780, had prevented him from performing the right experiments. In this respect Van Barneveld got more appreciation from IngenHousz, as he did at least did perform the necessary experiments with his own hands.

He on the other hand continued his experiments, sometimes in the Botanical Garden in front of an audience of hundreds of people. He suspected that Senebier would be swiftly informed of all his experimental progress. That's also the reason why he refrained of mentioning things to Van Swinden, and asked Van Breda to do the same, as Van Swinden would more than likely reveal it immediately to Senebier. Van Swinden was for that matter very critical about the variable results IngenHousz obtained with his eudiometer. That was no surprise, according to IngenHousz, as Van Swinden did not really understand the eudiometric principles. Different plants on different days in different vessels would give different results. But the same air from the same bottle measured twice should give the same results. But taking one measure of air to test, or two, would give different results as the contact surface between the two airs would be greater in the latter case. This was an observation done by Van Breda, for which IngenHousz is very thankful.

Still, in all this polemic turmoil, IngenHousz asked in a postscript for the recipe of cumin cheese. Typically Dutch and unavailable in Vienna, IngenHousz missed some of the culinary roots:

> *bezorge my een beschryving van hoe men de beste komynekaas maakt, hoe men*
> *de melk traiteerd, en hoeveel kruidnagelen en komynzaad men by een zeker*
> *gewigt van kaas doet &c.*

IngenHousz letters sometimes are like a puzzle. He writes in bouts, at moments when he finds the time. He regularly excuses himself to his correspondent when he has to stop because somebody had arrived for a visit, often high placed people. Letters are written over the course of several days, which results in topics that change abruptly from one line to another, while the thickness of pen and quality of the ink change as well. He excuses himself for his mixed-up stories, in which he repeats himself, as he writes what comes up in his mind, and because he does not keep copies of his letters:

> *ik schryf nu mogelijk dingen die ik reeds geschreven heb en dit moet gy my*
> *verschonen, dewyl ik geen copy van de brieven hou, en maar schryf wat my in*
> *de kop komt als ik den pen in handen heb*

On 22 November 1783 he announced that he sent three articles to the *Journal de Physique* in Paris, two on acidic waters, one on the green matter. He told Van Breda that he did not make them polemical, although he could not avoid mentioning Senebier, as the latter described different results. IngenHousz expanded his thoughts on what might be necessary for gaining right and trustworthy knowledge about the nature of things. A natural philosopher philosophises:

> *ik tragte die dry stukken zo te schikken dat ze niet polemisch zyn, maar voor*
> *een bewys streken dat myne ontdekkingen wezenlyk zyn. ik heb dog niet*
> *agtergelaten om mr Senebier te noemen daar ik van hem verschil. ik begryp*
> *niet waarom hy in de voorrede in een vage wys zegt dat hy van my verschilt,*
> *dog dat hy uyt discretie my in het boek niet noemt. dus legt hy al den last op*
> *den lezer om byde onze werken met alkander te vergelyken. en wie zal zig die*
> *moyte geven?*
> *dus pronkt hy met myne ontdekkingen zonder het te zeggen, en doet geloven dat*
> *wy verschillen daar wy het eens zyn. Dewyl het, wel ver van iets onaangenaam*
> *te zyn, is van in denkwyze van een ander te verschillen, het integendeel zeer*
> *voordelig is voor de wetenschappen, kan niemand kwalyk nemen van zinder*
> *de minste agterhoudenheid rond uyt voor de vuyst te zeggen, waar in men van*
> *elkander in denkbeeld verschilt, en op wat gronden dit anders denken steunt. in*
> *metaphysische disputen kan men niet we aan den ander geschaakt worden en*

exclamatien zig somtyds bedienen, dog in natuurkundige stellingen moet men niet zeggen dat men <u>calumnieerd</u> als men aan planten beschryft een invloed op de lugt die by nagt opgesloten zynde dezelve schadelyk maakt voor ademhalende dieren, en dat maar op ze een losse voet dat men zig de tyd of moeyte niet geeft het te proberen. zou den H. Senebier ook ymand als calumnierende doen doorgaan, die opentlyk beweerd dat er vergiftigde adders, dolle honden, arsenicum &c in de wereld zyn, alleen omdat het misschiens beter voor ons zoude zyn als ze alle van den aardbodem verbanne waren. wy begrypen alle de relatie niet die den schepper tusschen alle lighamen gemaakt heeft en het past ons niet op de werken van de natuur te vitten. men moet ze nagaen.

He deplored that Senebier said in his introduction he was of a different opinion to IngenHousz, but failed to say precisely in what respect he differred. For reasons of discretion, so he pretends. In doing so, however, he put the burden on the reader who had to find out for himself. The reader would have had to compare the two works painstakingly, and who would have had the time and energy to do that? IngenHousz ironically asked.

But in doing so, Senebier walked away with my discoveries, continued IngenHousz, pretending that they agreed on several points. This was conterproductive for the sciences because it thrives criticism and contradiction. But that implied that scientists had to say openly what they thought, where they differed in opinion from each other and what grounds these opinions were based on. In metaphysical discussions one can use just exclamations, but in physical questions it is no argument to say one is <u>blasphemous</u> when one writes that plants produce something harmful to animals during the night, especially so if one does not even try to investigate it. Would Senebier find someone blasphemous who says that adders are poisonous, dogs can be rabid or arsenic deadly &c.? Even if you could admit that the world would be better off without these threats. None of us, so continued IngenHousz, understands the true relationships between all bodies that were made by the creator, and it is not to us to criticise them. We just have to investigate them.

rewind

The quest for the secret of the sunlight 1779-1780

One of the recurring themes throughout the 1780's in IngenHousz life was the priority question, which leads into fulminant discussions with Priestley, Senebier and Van Barneveld. The priority question is one that pops up now and again in histories of science. It seems to be in some ways unavoidable, while it usually proves difficult to solve them completely.

IngenHousz wrote to Franklin about the problems he was having with Priestley and asked him for advice. Franklin abhorred polemics and verbal controversies as much as IngenHousz. Franklin writes to his friend it was always better to spend what time he "could spare from public business in making new experiments than in disputing about those already made."[89]
They have to conclude that it seems to be part of the game. Man is a social animal, and in socially organised groups there are forms of hierarchy, some of which are based on the fact if you are first, fastest, strongest, richest or cleverest. Luckily, being happy does not solely depend of these kind of picking orders and both Franklin and IngenHousz are free to go their own way.

Some other theme's that make an entry in this episode (I will come back to them later) are the relationship or possibly conflict between reason and religion.
Another theme, very much in the realm of the philosophy of science, is the theory ladeness of discoveries. IngenHousz suspected how light had something to do with the plants purifying the air and worked towards it in his series of experiments, meandering along various possible variables, which he eliminated one after the other. So when he finally struck gold and identified sunlight as the motor behind the photosynthetic machine, his prepared mind found it easy to recognise as the crucial factor.

89   Cohen 1952

Letter from 12 January 1782 to Van Breda, including drawing of bottle with valves and tubes (Fig V) destined for *Nouvelles experiences et observations sur divers objects de physique*, which would only appear in 1785.

A life in letters,
the story continues
1780-1799

What exactly plants did with the light they received from the sun was far from clear. But to IngenHousz and a handful of natural philosophers in the 1780s it was clear that it was another example of the wonders of nature that deserved further research. To IngenHousz it was just one of the many subjects that formed an interlocking web of topics, unhindered by any distinction between scientific disciplines that did not exists at that time. He was involved in a constant exchange of information. His logbook of outgoing letters, in which he registered the letters he wrote to whom and what they were about (leaving a wide margin to note what was said in the response - clearly a nice idea but not often executed), shows how he really was a 'man of letters'. On some days six or seven letters are entered, and there are not many days when nothing is written. A typical day might look like this:

> 1774 feb 24
> aan de heren Wolfert <u>Van Hemert en Gerard Backus</u>
> aan <u>H Collond</u> dat ik ontsteld was wegens het wedervaren van myn vrint Franklin, en met ongedult zyn defensie verwagt. ik zend een volmagt om bij den H Roberston in de Royal Society het volume LXI en volgende indien het gedrukt is af te halen en het te zenden met den eersten courier aan Mylord stormond te Parys
> a <u>Mr Philippe Baron</u> de Toussaint a Brin en Moravie
> to dr <u>Benjamin Franklin</u> Craven street London I am in the utmost consternation about his alarming circonstances and beg to be informed of them
> to Sir <u>John Pringle</u> Baronet president of the R S Collin will soon publish a treatise upon the virtues of the roots and flowers of cercinain hemorraghii uteri
> a mrs <u>Tourton & Baur</u> a Paris je lui envoye un certificat de vie de l'ambassadeur de france

It would take a lot more research to identify all those he corresponded with and to find out what each letter was about. They go to London, Brno or Paris, often asking for more information on experiments, theories or recently published books or articles, but equally about the money business, administrative obligations and personal relationships. Each entry is in the language of the correspondent, painting a polyglot canvas of IngenHousz' communicative life. It was winter now in Vienna, 18 February 1784, and IngenHousz wrote another letter to his friend in Holland.

Another winter in Vienna

Together with the mysterious kind of green matter, which was a question that bothered many philosophers all over Europe, from Priestley to Fontana, another returning theme in IngenHousz' letters from the beginning of the 1780s was the hot air balloon. All over Europe hot air balloons were going up, following the first demonstration of the Montgolfière brothers in Paris. IngenHousz wondered about the best air to fill the balloon. Montgolfier used inflammable air, but how could they get the necessary amounts in an affordable way? Franklin wrote about leather and horn that was burnt, but IngenHousz doubted this could be workable. He was the president of a club of ballooning enthusiasts who wanted to get a big balloon in the air in Vienna also. They collected 3000 ducats to finance their project. If they didn't assemble enough funds everybody would get their money back, if they collected more, the surplus would go to the poor. But in the end they were not able to collect enough money. It was a hard winter in Vienna. It was bitterly cold.

> *de koude is wederom toegenomen. heden morgen was myne thermometer van fahrenheit op 6 ½ graden, dat onder het friesen op 25 ½ uytkomt. den groten arm van des Donaus ligt by de stad niet gans toe maar in hongarien ligt hy wel zo uren lang vast of meer zo dat indien den dag erst aan het boven end van de rivier begint, de ganschen stroom zal gestopt zyn en dan zullen er ysselyke over stromingen geschieden.*

The thermometer only rose to 6.5 degrees F. The Donau was frozen solid in Hungary. The fear was great that when it started thawing, huge inundations would occur. Luckily his house was in the higher parts of town. But in March these fears came true. Massive chunks of ice started to break up and drift, taking two bridges in Vienna with them. His house was not damaged, but greater parts of the lower town were flooded, the pump water was spoiled and the post couldn't get through to the rest of Europe. His report testifies this was not just an ordinary winter. The winter of 1783-84 was the most severe witnessed for years. On average the temperatures in Europe were about 2°C below the norm of the second half of the eighteenth century. This was the result of a gigantic dust cloud spewed out by the Laki vulcano on Iceland, which started with a blast on 8 June 1783, lasting till February of the following year.[90] A "dry fog" stinking of sulphur and accompanied by a scorching heat covered large parts of Europe during summer, killing people (an estimated 30,000 extra deaths that summer in Europe), damaging harvests and creating waves of panic. During winter, the cloud reflected the sun's light away. Benjamin Franklin wrote to the Literary and Philosophical Society of Manchester that the sun's "... effect of heating the earth was exceedingly diminished. Hence the first snows remained unmelted. Hence the air was more chilled."[91]

In the midst of all this catastrophic news IngenHousz asked for more information on the making of cumin cheese, as he had found an Hungarian who was going to make it for him.

On 20 March IngenHousz received Senebier's new book *Récherches sur l'influence de la lumière solaire pour métamorphoser l'air fixe en air pure par la vegetation avec des expériments et des considérations propres à faire connaître la nature des substances aëriformes*, published in Geneva in 1783. He was pleased to see that Senebier was at last performing some decent experiments himself, using better equipment, and formulating his own thoughts instead of rephrasing chemical doctrines from others. Still, he thought the man from Geneva gave pointless explanations for his bad results in the past. And he still saw no reason to accept Senebier's hypothesis that plants "fixe lugt aan de son elaboreren", that they process fixed air in sunlight.

In the mean time it was spring 1784 and the long awaited French edition had still not been printed. The German edition was already being in its second print. A second volume in German was ready to go to the press. He felt like he would be overtaken, at least in the French language, by Senebier and had sent, in November 1783, three articles to the *Journal de Physique* in Paris in order to get his experiments and theory in the public eye. Unfortunately, they had already so many of his articles, that they would not be published in time. Some of Senebier's mistaken ideas which Ingenhousz would have liked to correct, had been put right by Senebier himself in his own book by now.

These concerns didn't stop IngenHousz from following up on the developments in the world of chemistry. In his 16 June letter he expressed his doubts about the experiment of Priestley and tended to put his trust in Lavoisier (who he valued as a chemist) who put steam through hot steel pipes. He had an apparatus like that under construction himself.

90   Pain 2005
91   Thordarson 1995

A critical mind

In the letter of 30 June another aspect of IngenHousz character and personality emerged. He talked about Anton Mesmer and described his experiments as 'crazy'.

> dat de Prysenaars aan de overzotte experimenten van Mesmer geloof slaan, verwonder ik me niet omdat de regering de dwaasheid ygen aan het mensdom niet tegen gaat. … het is een soort van toovery onder een niewe benaming. … Mesmer zeide dat terwyl er maar een god is, een religie, een natuur, een zon, een maan, een aarde &c, volgt het ook natuurlyk dat er maar een ziekte is, en dus ook maar een remedie, en dat is de magnetismus animalis.
>
> En de mens heeft 7 polen omdat er 7 planten zijn &c. Met één woord dit is wederom een van die verschynselen die bewyzen dat het geen Boileau Despean dat zo aardig zegt van alle dieren die op aarde, in de lugt en in het water sweven, het sotsten van al de mensch is.
>
> Die guychelaars zoals mesmer doen de gekste dingen, wekken doden tot leven en maken bijgelovige zwakke geesten onzinnig. de broer van de graaf van saxen is devoot en melacholiek geworden, de graaf van klam heeft zich de hals over gesneden.

He was not surprised that the Parisians were going along with the super silly treatments of this quack. Animal magnetism was not more than some magic with a new name. He dissected Mesmer's rhetorics: as there was only one god, one religion, one nature, one sun, one moon and one earth, there surely was only one illness with one remedy, which is animal magnetism. He quoted approvingly Boileau who said that of all animals on earth, the silliest is man. He gave some warnings for the negative effects of these seemingly innocent charlatans: the brother of the Duke of Saxen had become pious and melancholic, the Duke of Klam had cut his throat. He described how he went to see another charlatan at work, Father Gasner who pretended to be able to exorcise bad spirits. Mesmer had been debunked by the scientists, but still attracted huge crowds.

IngenHousz knew what he was talking about. Around 1775, Mesmer had been doing very good business in Vienna with his magnetic-magical treatment. He sent statements of his ideas on animal magnetism to many academies of science in Europe. The only reply he received came from the Berlin Academy, arguing that: 1) Mesmer's claims that magnetic effects could be communicated to materials other than iron and concentrated in bottles contradicted all previous experiments, 2) Mesmer's evidence, based on the sensations of one person, was inadequate and even inappropriate to prove his ambitious claims, 3) the abscence of effects in healthy persons made his claims very suspect and 4) other explanations could easily account for the phenomena Mesmer described. Mesmer's attempts around this time to demonstrate the effects of animal magnetism to IngenHouz were even more negative and publicly humiliating.[92] While Mesmer demonstrated the magnetism of a single teacup in a group and elicited convulsions by pointing a magnet toward a relapsed Miss Öterlin, one of his notorious hysterical patients, IngenHousz surreptitiously tested the effects of strong magnets which he had concealed. He found that the patient reacted only to objects which she believed were magnets or that were connected with Mesmer. IngenHousz therefore denounced Mesmer as a fraud. In response, Mesmer publicly attacked IngenHousz' scientific ability and demanded a court-ordered commission to establish the facts concerning his treatments. He was ultimately observed during eight days by a local hospital physician who couldn't find any evidence for a genuine therapeutic effect. IngenHousz' denouncement had undermined Mesmer's reputation and thus prevented him from gaining the recognition from doctors, scientists, academies and even the public. Nevertheless he continued to try and treat the blind daughter of the secretary to the Empress Maria-Theresia. That ended in an ignominious public failure, forcing him to leave Vienna in January 1778. He went to Paris, where he set up shop and had initially lots of success in this fashionable city where people were easily carried away by sensational reports of novelties, inventions and medical marvels. That's what IngenHousz had heard about and it worried him because the government was not speaking out against these misleading stories.

He recounted this story as he had been informed that Mesmer and his therapies had been put to the test in Paris by a royal investigative commission, presided by Franklin, and counting other illustrious members such as Lavoisier, Guillotin and De Jussieu.[93] He had been corresponding with his friend in Paris about this quack from the time he had driven him out of Vienna in 1778: "I hear that the Vienna conjuror Dr. Mesmer is at Paris ... that he still pretends a magnetic effluvium streams from his fingers and enters the body of any person without being obstructed by walls or any other obstacles, and that such stuff, too insipid to get belief by any old woman, is believed..."[94] Franklin and Mesmer had met in Paris as the magnetic magician was using the glass armonica, a Franklin invention, as musical background at his séances.[95]

During summer, IngenHousz was visited by the Emperor himself, by Volta and many other travellers, among whom the Duke of Tuscany. These high ranking visitors might have been pleasing to the ego, people like Volta offered the occasion to continue the scientific debate. (On the other hand, all these travelling savants and noblemen kept him from his work, a nuisance he sometimes complained about.) Volta decided against Senebier and agreed with IngenHousz. In a letter dated 9 June and which became public, Senebier had

92 Lanska 2007
93 Best 2003
94 Lopez 1994
95 Gallo 2000

tried to put the blame on IngenHousz, using more "Jesuit tricks". IngenHousz was not pleased with this behaviour and was planning to put things right in the second volume of the German edition and to publish the letter and his response in a polemical booklet. And asked Van Breda:

> ik denk dat u het met my eens is van zulke soort van luyden met geen lafhartige vrees onder d'ogen te zien, en niet al te veel venerati voor een predicantenmantel te tonen als hy niet gedragen word door een opregt, openhartig man.

> Won't you agree not to be afraid of this kind of cowardly people and not to have too much reverence for a minister's cap if that is not worn by a sincere and candid man.

He was unhappy with the rhetorical tactics of Senebier and held a plea for basing discussions on facts and experimental data, not on personal reflexions and clever tricks and sophismata.

> Laat ydereen die will myne ontdekkingen bevegten met experimenten, maar men moet zig onthouden van personnele reflexiën van slimme streken en sophismata.

To IngenHousz, honesty was something of a second nature. In his letter of 21 September, he told Van Breda that his German publisher had printed the year 1785 on the title page instead of 1784, so to be able to sell the book as new at the Leipsiger Buchmesse. IngenHousz loathed the idea and had asked to destroy all these front pages. If he had not seen it in time, all his discoveries would have been postponed by a year and people like Senebier would not have refrained from using that to their own advantage. In the mean time, Senebier's letter had been published in *Journal de Physiques*. IngenHousz was in dubio if he would publish his answer in the same journal, as he thought it was for publishing discoveries, nor for polemics.

In November 1784, IngenHousz was forced to send his letters and parcels to Van Breda via Paris, as the relations between Austria and the Dutch Republic were troubled, prohibiting direct postal traffic. He would send it on as 'petite post' with the ambassador, but they would only sign with their initials, he could not address it to Delft nor mention anything else but physics topics, no politics. On 15 December, he received a letter from Landriani who asked him to reconcile with Senebier and forget about the whole quarrel. IngenHousz wrote to Van Breda on the 18th to tell him he wouldn't mind cancelling his polemic booklet, if Senebier would also stop the discussion. He realized that he would bring a greater sacrifice than Senebier, because his reply had only been published in a German journal, while Senebier's story had gone around the world already.

Landriani wrote that Senebier told him in a letter that his book appeared in July 1783, based on experiments he did in the summer of 1782 and therefore could not have known about IngenHousz experiments. IngenHousz remarked finely that he did not tell anything about Lamberthengi who had visited IngenHousz in the beginning of '83 and whom he had told all about the many mistakes in Senebier's book. Lamberthengi communicated everything to Senebier, who in turn repeated his experiments and must have found his mistakes. Part 1 of his book was printed in April '82, too early to report on experiments performed in the summer. Parts 2 and 3 were printed in the beginning of '83; in which volumes he could have published the results.

Both Volta and Landriani claimed they believed in the nightly production of bad air. Landriani however did not believe just like Senebier that plants produce air under water. IngenHousz sent him some extra information.

In his first letter of the year 1785, IngenHousz touched on one of his other topics of interest, reappearing regularly in his correspondence:

> a propos van vrijmetselaars! u zult door de couranten haar wedervarendheid hier vernomen hebben. Weet u niet, of men in ons land ergens overgezet en gedrukt heeft een zeker vrymetselaarsjournaal, dat hier te wenen sedert Dec 1784 alle 3 maanden uyt quam, onder den titel _journal für Freymaurer, bearbeitet von den brüderen der zur wahren eintracht im orient von wien, als manuskript für freymaurer_ Uw stuk bestaat uyt 18 bladeren en is verdeeld in vier delen. het eerste vervat verhandelingen over de mysterien alle volken over de ygentlyke essentie der gehyme der vrymetselaars. het tweede redevoeringen over vrymetselaars en natuurkundige onderwerpen. het derde zyn gedigten. het vierde bestaat uyt vrymetselaars niews van andere plaatzen. men heeft van dit journal een ongelooflyk debiet gemaakt door gansch Duytsland. en derwyl die societyd in nauwer verbondschap staat met vreemde logien komt het my niet onwaarschynlyk voor dat d'een of ander harer broederen het zoude overgezet hebben in 't Frans of hollands tot gebruyk en stigting der nederlandsche logien. indien dit zo mogt zyn , kan u het wel weten van een uwer vrienden die de graad van _meester_ in een logie bezit. van geen ander mensch kan u er iets van weten, derwyl geen exemplaar mag verkogt worden als aan meesters, die er dog geen van kunnen bekomen als door de hand van de grootmeester der logie. Het is het enigste egt boek over die geheime zaak. Daar word er nog een uytgegeven ergens in Duytsland onder de naam van _Vrymaurer bibliothek_ en een ander genaamd _Archiv fur Freymaurer und Rosenkreutzer_ zo er enige dier werken mogten vergezet en gedrukt zyn in ons land, verzoek ik het te mogen weten. het journal der Freymaurer kost hier 6 duytse guldens jaars by subscriptie. Dog het drygende onweer zal mogelijk de verdere uytgaaf verhinderen.
>
> ik verzoek my te willen onderrigten van den inhout des artikels van de wetten

van het genootschap te uytregt waarin van het terugzenden der teorie gesproken
word. Deze wetten vind ik niet in de drie delen die ik heb. maakt dit een boekje
op zig zelf uyt?

The freemasons intrigued the rational thinking and indepent IngenHousz. It
was a secret organisation and he tried to get hold of their publications in order
to understand what exactly they were doing. Since the end of the previous year
the freemasons in Vienna published a journal and he asked Van Breda if he had
the right connections (preferably the 'grootmeester' or grandmaster) to get hold
of this or other publications from the lodge. In his next letter, some weeks later,
his verdict was hard and sharp, based on his inside knowledge of this secret club,
which he learned through his freemason friends in England. He thought their
ceremonies weird and mysterious conjuring tricks. He got to read their secret
journals and they were only worth the pain to see how they show rash zealotry.

De vrymetselaars hebben hier den shop gekregen, de Leydse franse courantier
heeft het edict niet goed overgezet omdat hy waarschynlyk ook een v.m. is.
Den k segt niet prestiges maar guychelary, en met veel regt, derwyl alle hare
ceremonien en gebaarden by de receptien niet als belachelyke guychelaryen zyn.
niemand weet beter als ik wat de gansche orde is, derwyl ik by de voornaamste
v.m. in Engeland voor een broeder gehouden wierd. Zy hebben geen geheime
als hieroglyphen, waaromtrent zy zelfs niets verstaan. De ene houdende ze
voor de overblyfselen van de braminen van oude tyden, d'ander voor symbolen
van de christenen, andere voor symbolen van de Egyptische priesteren. Zy
hebben geen systema, weten niet wat zy voor insigt hebben. zy spreken niets als
van bouwen, den tempel van salomon, niet van steen maar een mystike tempel,
den tempel van menschlievendheid, van wysheid, van liefde, weldadigheid
&c. Daar zyn reeds 6 deele van hare journalen hier uytgekomen, die extra
gehym gehouden worden. ik heb nogtans middel gevonden om ze alle te lezen.
ze zyn de pyne waard te lezen om de onbezonnen dwepery des menschen te
kennen anders zyn ze niet waard om er den a. aan te vegen. Nu ze beginnen
te schryven moet zo een dwepery vanzelfs vervallen.

He sees them as zealots raving over a mystic temple of brotherhood, wisdom
and love in a weird mixture of old symbols and languages. Their texts are not
worth reading except for getting to know their rash zealotry. The paper is
only good to wipe your a(rse). It was the same fanaticism that IngenHousz so
deplored in Senebier's attitude. From letters from Landriani it was clear to him
that Senebier still did not want to accept that plants had a negative effect on air
during the night. He didn't seem to be able to accept the results of the numerous
experiments. He seemed to admit to Landriani that he indeed realized the
mistakes he made after IngenHousz pointed them out in the beginning of 1783.

Contrary to IngenHousz' and Fontana's opinions, Senebier stuck also to the idea that the green matter was vegetative and not animal-like. They remained of this opinion because they saw it move. The fact that IngenHousz kept on studying the strange phenomenon demonstrates that he still was not convinced about the real nature of the thing. It couldn't be a real plant, because it did not die, nor did it grown from seeds of roots.

Whatever his almost institutionalised doubt, from to time to time he drew rather strange conclusions:

> *de lucht in wenen is 3 of 4 graden beter dan in Engeland of Holland. dat kan verklaren waarom mensen van die van daar komen meer spyzen tot zig nemen. Dit heb ik ook bestendig in myn yge geobserveerd. de meeste menschen eten hier viermaal daags. teweten smorgens een ontbyd met sterke koffie en room, ten een of 2 uren het middagmaal, hebbende tenminste 4 differente spyzen, 2 of 3 uren na het middagmaal eten de meesten wederom wat brood vrugten chocolaat koffy met melk of iets anders, dan nog een avondmaal. Ik leef zoals ik in Engeland gewoon was, nemende een ontbyd en ten 2 a 3 uren het middagmaal. dog me dunkt dat ik zonder half zo veel beweging te doen als ik in Engeland pleeg te doen, omtrent dubbel zo veel spys in quantityd gebruyk. savonds gebruyk ik niets. Dus schynt het my zeer apparent dat de lugt hier fyner si even als op zee.*

The air in Vienna was slightly better than in England or Holland, and that explained probably why people there ate more. Most people in Vienna ate four times a day. A breakfast with strong coffee and cream, lunch consisting of at least four dishes. Two or three hours later some bread with fruits, chocolate and coffee with milk and later dinner. He lived as he was used to in England: a big breakfast and a meal at 2 or 3 in the afternoon and no evening meal.

IngenHousz knew what he wanted. His independent mind and lack of ambition explain why he was not a society-man. He was member of the Royal Society, and that seemed more than enough to satisfy him. He turned down the many requests to become member of other societies. Duke Arenberg wanted him for the Imperial Academy in Brussels, the Elector of Pfalz wanted him for the Mannheimer Academie. Mr Le Roy asked him to become member of the Academie Royal des Sciences, but he wouldn't oblige because then "one has always to answer all those questions one is asked".

In the continuing story about the priority question, IngenHousz tried to reconstruct the whole sequence of events again in his letter of 11 May. He remarked that Senebier wasn't the first to tamper with the dates of letters or publications. Fontana did and Priestley was another heavy weight example: when he met Franklin the previous time in Paris, his friend received a letter from Priestley in which he communicated some discoveries, while the letter was dated five months before the letter was received. He gives these examples to illustrate that probably Senebier picked up something about IngenHousz' discovery. There are many possible routes. Through Pringle or Magellan of the Royal Society. Magellan was a Portuguese expat in London, a close friend to Priestley, an unpaid confidential agent of the French governement and one of the main channels of communication between the English pneumatic scientistst and the French chemists[96]. Another possible leak could have been via Lamberthengi who might have intercepted IngenHousz letter to Graf von Rosenberg in which he asked him for permission to remain in London a little longer to finish his experiments. Whatever the fact of the matter, Senebier did not have the guts to call upon anything but ideas and words, as he could not reproduce any facts from experiments.

> *Men heeft er fontana al publiek van beschuldigt zyne uytvindingen te datummen op tyden wanneer hy er waarschynlyk nog niet op dagt, en van ontdekkingen aan te kondigen die hy niet heeft gemaakt. Ik heb reden om te denken dat priestley er ook niet vry van is, terwyl de H. Franklin, de laatste keer als ik in Parys was, een brief bequam van Priestley, ontdekkingen bevattende, welke brief omtrent 5 maanden voor den ontfangst gedatumd was. Ik wil hierdoor maar alleen beweren dat het niet onmogelyk is dat Senebier al vroegtydig de wind van myn ontdekking, misschien wat verward, heeft gevat. Hoe dit ook zy, hy had de couragie niet van in zyn boek zig op iets anders als <u>ideen</u> te beroepen., zig waarschynlyk niet betrouwende van enige proeven te spreken, omdat zyne vrienden hem zoude schuldig konnen kennen van onwaarheden te schryven.*

IngenHousz really was tired of this unproductive bickering and excused himself for this long-winded letter.

On 2 June he mentioned that Franklin has asked him to accompany him to America. He would have done so if Maria Theresia would still have been alive. Other big and small news travelled up and down between Delft and Vienna. IngenHousz made some new and breakthrough discoveries but did not publish them as he needed more proof first. Landriani wrote him to say the discussion with Senebier was now closed as the latter had written a brochure. IngenHousz had not seen it yet, so couldn't say anything about it. Half of the German edition (500 of 1000) had already been sold. He read the dissertatio inauguralis by Ricardo Lubbock from Edinburgh 1784, written in Latin, in which the existence of phlogiston as entity was refuted. Scheele had described

---

96   Mason 1991

a new eudiometer, but IngenHousz did not approve of the model as it took 24 hours to read the results. He had finished the corrections in the Greek edition of his book. He received three parts of Memories of the Utrecht Society, but many parts were missing and the sequence was clearly thrown in disarray. "Den binder moet een sat lap zyn", the binder must have been drunk.

It was an occasion to reopen the debate on Barneveld's claim to priority, based on fiddling with the dates of submitting his dissertation. He strongly urged to put things right with the help of the secretary of the society and held a plea for a more sophisticated and accurate administration of what was submitted, read and published on which day and by whom. Van Barneveld must have picked up some of IngenHousz' ideas when he talked about them at the beginning of December 1779 in Amsterdam, while presenting the first copies of his book in the Netherlands. He must have taken some essential thoughts and used them in his dissertation. He said that plants turned bad air into good air in the light and made good air bad in the dark. But he did not say that plants made bad air worse in the dark. Van Barneveld even claimed that air did not deteriorate after days and even weeks in the dark. According to IngenHousz this proved that Van Barneveld had not done the experiments but only guessed the results. If he had done them he would have seen the air worsening very quickly.

> Ik verneem door Landriani dat zyne afgod <u>Senebier</u> nog al gestadig bezig is om argumenten uyt te vinden die bewysen dat het tegen Gods wysheid stryd dat de planten by nagt de lugt besmetten. Barneveld schynt ook als een tweede atlas de hemel op zyn schouders te willen dragen, uyt vrees dat zo een klyn mannetje als ik ben, het wys gebouw van het geheel al overhoop smytte.
>
> ...
>
> Met zo een kinderagtige vooringenomenheid zullen zy liever geloven blind te zyn, als zy het effect voor ogen hebben, dan te bekennen, dat de planten by nagt de lugt mephitiseren.

Both Senebier and Van Barneveld kept denying that plants spoil the air during the night. Senebier was, according to another letter from Landriani, still looking for arguments to prove that it was at odds with God's wisdom that plants contaminate the air in the dark. Also Van Barneveld seemed to want to carry the heavens on his shoulders like a second Atlas, "from fear that a little lad like me would be able to destroy this wise building of the universe. With such naive prejudice, they will rather be blind with the phenomena before their eyes, than admit that plants mephitise the air at night."

Almost robbed of the light of the sun

In his letter of 22 October 1785, IngenHousz needed another three full pages to analyse the way in which Priestley, Senebier and Van Barneveld had each tried to claim to be the first in discovering the sunlight driven air production of plants.

> *Barneveld heeft ook zorgvuldig Priestley van de uytvinding uytgesloten met zig te <u>verwonderen dat hy dit niet opgemerkt had</u>. Hy kon Priestley met zo veel voordeel noemen, als den H. Ing. met overhoopstoting van zijn gansche pretensie. Hy moest zelfs tot den 8 sept. 1780 niets gehoord hebben van myn book, schoon toen reeds de franse editie al wereld kundig was, en er in verschyde journalen al van de ontdekking gewag gemaakt hadden.*

Van Barneveld tried to eliminate Priestley by questioning that he had not seen the mephitic effect of plants during the night. At the same time, he pretended not to have heard anything anout IngenHousz' book until 8 September 1780 (when the first volume of the French edition was already circulating) which makes this a very dubious story.

> *Priestley, Cavallo en andere reeds voor het ynde van 1779 de uytvinding voor al te gewigtig aanziende (als mede senebier die waarschynlyk een exemplaar uit Engelland gehad heeft of tenminsten de zaak reeds vernomen had door d'een of d'ander van zyn veelvuldige correspondenten) om niet afgunstig er over te zyn, hadden al begonnen grimmig er op te worden en wilde gaarn den <u>influxus nocturnus</u> uyt den weg ruymen (want dit standhoudende zagen zy dat alle pogingen vrugteloos moesten aflopen) maar de proeven genomen zynde durvde Priestley er niet mee voor de dag komen. Cavallo was stoutmoediger en kon nog Priestley niet overhalen om my langs die kant de ontdekking uyt de hand te wringen. Hy vond er een ander middel op uyt en was eventwel nog genoodzaakt van mynen naam een diep stilzwygen te houden, tot hy den lezer ver genoeg in zyne belangen overgehaalt had. Nu geloove ik, dat Senebier en Priestly de zaak genoegzaam opgeven of ten minsten er niet verder zullen op aandringen en er een speld by steken. Barneveld alleen is nu nog te bevegten.*

Priestley, Cavallo and others must have seen how important IngenHousz findings were, even before the end of 1779. The same was probably true of Senebier, who must have heard about them via some correspondent in England. These men had tried to eliminate the <u>influxus nocturnus</u> or nightly emanations, because when they accepted this, all other attempts would be futile. The experiments done, Priestley did not dare to come out with them, Cavallo could not convince Priestley to wrestle the discovery from IngenHousz' hands, but finally decided to keep silent about IngenHousz in the hope that everybody would forget. IngenHousz believed that Priestley and Senebier had given up, and that only Van Barneveld needed some riposte.

> U zult er [in een nieuwe dissertatie voor het journal de physique] denkelyk nieuwe denkbeelden in gewaarworden en oude in weerlegt vinden. Ik heb nog niet al de experimenten, meer als duysend, kunnen compareren of volmaakte conclusies er uyt te trekken. U zult daar uyt ook kunnen oordelen dat ik my van het sonneligt niet kan of mag laten beroven? De gevolgen van de kenniss des nagt en dag invloed der planten moet met der tyd onyndige gevolgen omtrent het ontstaan van de wetten van de natuur. Deze dissertatie en die onder den naam van den H. _Schwankhard_ gaat zal het volumineus werk van den abt Bertholon _De l'Electricité des végétaux_ byna overhoop gooien. De _ideén_ van van franse schryvers zijn gemenelyk extra vrugtbaar. en als by ongeluk de _moederideé_ vals is vervallen alle daar uyt gesproten ideén. Senebier heeft het zo gemaakt met zijn _idee_ dat de fixe lucht by na al het werk des wasdom der plante bewerkt, en daarom bevond hy by ondervinding (geconcipieerd in zyn cabinet of in zyn bed onder het dromen) dat er in de planten in alle regen in de lugt &c vaste lugt in overvloed was. hy werkt nog sterk om die fatale _influxus nocturnus_ uyt de wereld te verbannen.

IngenHousz asks Van Breda rhetorically if he would allow, having collected all these experimental data, for him to "be robbed of the sunlight". He speculated that learning more about the influence of plants by day and night would deliver new and astonishing insights in the laws of nature. He would publish these novel thoughts in a dissertation under the alias of Schwankhard, in which he would refute the ideas of Bertholon on electricity and plants. No matter how interesting they were, but if the basic idea was wrong all its derivatives are wrong too. The same is true of Senebier's idea that fixed air is necessary for the growth of plants (and here actually, IngenHousz was wrong and Senebier right) and that's why he found this by experience (IngenHousz ironically wrote: "by experience, conceived in his cabinet or in his dreams").

Apart from that, IngenHousz also unveiled the way Lamberthengi learned about his discoveries: he saw Van Barneveld's dissertation in the Memories of the Utrecht Genootschap in Milan.

On the 2 November he asked Van Breda more information on the new Teyler's Natuurkundig Genootschap, Society for Physics, in Haarlem. (The location by the way, where IngenHousz letters to Van Breda have been kept.) He had read Priestley's dissertation on phlogiston in _Phil Trans_. Vol 75 and agreed with him that phlogiston is an essential part of metals. At the same time he agreed with Lavoisier that by burning dephlogisticated air and inflammable air, one obtains water.

Another quack got a dusting by IngenHousz:

> *Lavater, predikant uit Zuric, die altijd al een dweper, een quast een welsprekende zot geweest is, is nu een grote mesmerist geworden. Hij somnabuliseerd zijn wijf die dan prophetesse wordt, door een muur heen leest, yder een zyn gedacht ekent &c &c. Te Parys had men die farze reeds by na vergeten na dat mesmer verdwenen is naar london.. De Switzers, die zeer veel dwepers zyn, en de luyden in Duytsland, die zijn physionomie werk gekogt hebben, en hem voor wat wonderd aanzien, lopen nu gevaar voor de helft van de tyd gesomnabuliseerd te worden. Moye meysjes van een zekere soort worden in parys meest het voorwerp van de somnabulisation. men gaf 20 louis d'or om de formula somnabulisandis wel te leren en men deed de proef. men behoefde maar by zichzelf te mompelen hoe men begeerde dat het meysje zich zoude buygen, zitten, liggen &c en als men er al mede gedaan had dat men gewild had, wist zy zelfs niet wat er gebeurd was.*

Lavater, a pastor from Zurich, in IngenHousz' words, had always been a zealot and an eloquent madman and by then had become a mesmerist. He somnambulised his wife who became a prophetic and could read through walls and read others' minds. In Paris they might have almost forgotten these farces after Mesmer left for London, but in Germany and Switzerland, many people seemed to believe it. In Paris, beautiful girls of a particular occupation, were somnambulated, so that one could do whatever one liked with them. And afterwards they didn't even remember.

Lavater was widely known as a physiognomist, who could "read" faces and was holding the doctrine that one should focus on the permanent anatomical features of the face such as jaw structure, nose and ear size, brow angle and so on, over which the individual had no manipulative control as they were created by God and Nature as a legible public language. He inspired novelists and painters to portray the good, noble and heroic as beautiful and fair of countenance, and villains as dark and disfigured.[97]

And there were more things that triggered IngenHousz' curiosity. The freemasons seemed to be having a lot of success in Vienna, which made IngenHousz wonder. The Emperor thought they were magicians that might bring chaos to society. He himself had been asked several times by very good friends to join the club, but he had refused. Not only was he not a society-man, as he had shown on other occasions, he had never heard anything interesting or useful resulting from it, and he feared that membership would lead to commitements and liabilities. He suspected that so many people joined them in search of intercession and protection.

---

97 Porter 2003b

> *U zult dor de nieuwspapieren zien dat de vrymetselaars hier gevaar lopen
> by den vorst voor guychelaars niet zonder gevaar van wanorde te stichten,
> gehouden worden. Ik heb zo veel van myne intieme vrienden daar onder
> gehad, die my alles gezegt hebben dat ik het noyt de pyne waard geagt heb my
> onder haar te doen opnemen om niet genoodzaakt te zyn van met luyden om
> te gaan wiens gezelschap my ten minste onnuttig was, en die vragen & konnen
> lastig zyn. Ze vermenigvuldigen hier zo zeer dat menigte er zig by voegden
> enkel om voorstand en protectie.*

On 3 May he learned from the *London Chronicle* that a new book by Priestley
had appeared. The third volume of *Experiments and observations relating to
various branches of Natural Philosophy with a continuation of the observations on
air*. He couldn't get hold of the book in Vienna, so he asked Van Breda if he
could pick it up and send it as quick as possible to Vienna. He was very curious
because Priestley had promised him two years earlier that he would mention
him in a next edition, so hopefully he had put things right in this new book.
By the time of the 6 June letter it had become clear that Priestley said nothing
about IngenHousz in his new book, not a word about the intricate way in which
he had appropiated the light of the sun.

> *omtrent het geen wy reeds hadden opgemerkt omtrent zyne ingewikkelde
> manier om sig het sonneligt op planten toe te eygenen. hy moet met de zaak
> verlegen geweest zyn en niet wel geweten hebben hoe sig te zuyveren. Dit denk
> ik te meer omdat hy volgens zyn brief aan my, zeer wel weet dat wy hem als een
> afgunstig physicus hebben doen voorkomen, zeggende zelfs, dat myne vrienden
> hem hielden <u>als den sultan die geen competiteur tot zynen troon oner zyn ogen
> dulden kan</u>.*

Priestley kept silent about what already had been remarked about the ways he
used to appropriate the light of the sun. He must have felt a little ashamed
about the whole situation, not being able to clear himself from the blame,
according to IngenHousz. The more so because Priestley, according to his
letter to IngenHousz, knew very well that he had been perceived as a envious
physician and that his friends even described him as <u>a sultan who did not
tolerate a competitor to his throne</u>.

To IngenHousz it was clear that Priestley, by keeping a solemn silence on
the subject and not mentioning IngenHousz at all, was hoping that everybody
would think he discovered the phenomenon. IngenHousz thought it was proper
to mention something about this in the introduction to the second volume of
his new French book. The problem was that nobody had written about this in
the English language, so the fact of the matter would easily be forgotten.

All this hassle did not prohibit IngenHousz from continuing to perform experiments. He measured the amount of electricity in the air and investigated whether electricity would be able to enhance the growth of plants, as some famous natural philosophers claimed. His results indicated that there was no influence whatsoever.

On 24 November he expressed some philosophical thoughts on experimental work:

> *Men kan niet te voorzigtig zyn in experimenten. Ik denk nog al vry gelukkig geweest te zyn in het punt van voorzigtigheid omtrent het uytgeven van experimenten. Ik weet nog niet dat Priestley, Senebier, of ymand anders iets tegen myne experimenten met fundament heeft in kunnen brengen. Indien u iets diergelyks zoud gewaarworden hebben in 't lezen of experimenteren, zou het my genoege zyn er van onderrigt te worden. Al dat men heeft tegen geworpen tegen enige artikelen in myn boek was al door den abt Fontana my opgegeven als waar, en ik heb hem ook overal voor den opgever, als billyk was geciteerd. ik zie nauwelyks ooyt een paragraaf van Senebier in, of ik vind er mistastingen in.*

He tried to be very cautious in revealing experimental results. He only published them when he was sure about the data. That's why Priestley, nor Senebier or anybody else had been able to formulate fundamental critique on his claims. If Van Breda would hear or read any fundamental criticism or experience anything contrary to his findings, IngenHousz would really appreciate were he to communicate that immediately. Fontana had indicated some mistakes to him and he was only too pleased to be able to correct these and to mention him as the source. He did not see, on the other hand, a single paragraph by Senebier without a mistake.

Two days earlier in the evening Archduke Ferdinand came along and stayed for more than three hours. They talked about these and other scientific topics. He would definitely come back to continue their conversation.

In the beginning of 1787 he told Van Breda that soon the son of his colleague Jacquin would leave Vienna for a trip through Europe. Jacquin jr. was a brilliant investigator and did his exams in mathematics when he was 12. He was 20 then and lectured about airs an other topics to big audiences.

IngenHousz also received a letter from Franklin from Philadelphia. It contained two volumes of the *Transactions of the American Philosophical Society* which opened with a letter from IngenHousz about the theory of fire and smoke in connection to the workings of chimneys (36 pages and two plates). He quoted Franklin closing his letter with:

> *Hij sluyt dezen brief met deze woorden:* I have great pleasure in having thus complied with your request, and in the reflection, that the friendship you honour me with, and in which I have even been so happy, has continued so

many years without the smallest interruption. Our distance from each other is now augmented, and nature must soon put an end to the possibility of my continuing our correspondence: but if consciousness and memory remain in a future state, my esteem and respect for you, my friend, will be ever lasting.

Franklin was 71 at the time. He felt he was getting older and weaker and he was realistic enough to recognise that they wouldn't see each other again.

It is January 1787 when Spallanzani came to visit IngenHousz, on the way from Constantinople to Pavia, where he was professor.

Enige dagen geleden had ik het genoegen den beroemde apt Spallanzani by my te hebben op een rys van Constantinopel teruggaande na het Milaneesch, daar hy professor te Pavia is. Hy is een extra groote vriend van Senebier, dog denkt hem een man van gering oordeel te zyn, en zegde dat Bonnet er ook vry geringe gedagte van had en hem zelf meer dan eens vermaand had om meer oplettende te zyn in zyne geschriften.... Hy zeide my zyne ongenoegen aan Senebier te hebben doen bekent maken wegens de onbezonnenheid van in de overzetting van zyn boek een invectief tegen my geplaatst te hebben... Hy bekende ook dat het hem al vry klaar voorquam dat de intentie van Senebier direct geweest was om myn werk te verdelgen, zig zelf met de een van de voornaamset in myn werk bevat te bekronen , en zelfs my voor een <u>homme de mauvaise foy</u> te doen passeeren.

Spallanzani was a good friend of Senebier, so he knew him very well. He was well placed to say that Senebier was a man of little judgement. He had told him more than once to be more attentive in his writings. He said he had expressed his dissatisfaction with Senebier's thoughtlessness in attacking IngenHousz in his book. He admitted that it was clear to him that Senebier had wanted to destroy IngenHousz' work and crown himself with the most important findings, and even to ga as far as depicting IngenHousz as <u>a man of bad faith</u>.

Spallanzani admitted that IngenHousz was right to think that Senebier had been fighting a psychological war, portraying IngenHousz as untrustworthy and unreliable, and that his book contained the real and correct theory. It was improper to claim that his book was better than IngenHousz'.

By March his dissertation on the influence of electricity on plants had been widely read and not many comments came forward. De La Métherie urged him to publish more and bundle everything in a book, as he had performed some experiments that confirmed IngenHousz' ideas. He too complained about colleagues who had remarks and critique but never performed any experiments. One of the many visitors of IngenHousz ("yesterday Weilburg the Prince of Nassau came along") who wanted to see him at work and observe his experiments, was a Venetian nobleman who told him that a tree with an isolated conductor would produce leaves and flowers more quickly - Bertholon claims something similar. He wrote to Van Breda that this proposition was not based on experimental evidence but on a theory which they didn't even seem to understand themselves. He had connected some chestnut and lime trees to a conductor and was curious to see any results: "Hoe meer boomen, hoe gewisser de conclusie zal zyn", the more trees, the more convincing the conclusion would be. In May he reported that all these conductors didn't give any difference. Temperature and humidity were the real growth-influencing factors.

A new year begins with unrest

The first weeks of 1788 were coloured by the unrest in the Dutch republic. A civil war had broken out and the King Of Prussia had come to the rescue of his sister, married to Stadholder Willem V to halt the revolt of the patriots. On 17 January he wrote Van Breda how this turmoil did upset him.

> dat men als quantzhuys plunderingen en alle soorten van insolenties sterk verbiet en flauw tegengaat.
> het insulteren en benau maken van honderde eerlyke familien die in dat wel vry land vry dagten en zig vry uytten, zou men de aller vigoureuste maatregelen moeten nemen om het quaat tegen te gaan, en de eerste die men op het fyt kon apprehenderen aan den deurstyl van het huys daarby geweldenaryen gepleegt had op staande voet moeten ophangen, het welk veel beter is als door het schietgeweer de schuldige en onschuldige te verdelgen.
> ...
> ook onlusten in Breda, waar zijn broer en diens familie leeft. Zij zijn vorlopig gebleven, in de hoop dat het onhyl op zyn hoogst is geweest. IngH vreest dat het geld, samen met zoveel families het land zal verlaten, en waar gaan dan al die loonwerkers, van Yndhoven, Tilburg, te Leyden &c dan hun brood mee verdienen? Ze kunnen toch niet leven van plunderen en rooven?

He thought it was wrong to forbid this kind of plundering for appearance's sake. One needed to take vigorous measures to stop the insults and threats to innocent families, in a country where one could think and express oneself freely. Chain the looters to the door styles of the houses they plundered, and anyone committing violence against the innocent should be hanged on the spot (better than shooting indiscriminately) ... IngenHousz clearly was quite upset, also because there had been revolts in Breda and he feared for his brother and his family. They had stayed, hoping that things would get better. But IngenHousz was afraid that many people, as well as their capital, would leave the country. And what would the workmen then earn their income with? They surely couldn't make a living by plundering and robbing?

He apologised for so much emotion, that his pen deviated from their philosophical dissertation, but thinking about all these disasters brought about by greed and revenge, moved him deeply.

> *Vergeef me dat myne pen van onze philosophische onderhoudingen afwykt voor deze maal, het geschiet, zonder dat ik het byna gewaarword door de volheid van myn gemoed en het overdenken van de onheilen door hebszugt en wraakzugt begonnen en nu volbracht wordende.*

And there was not only political unrest: also philosophically speaking something was on his mind. He had read about the new chemical system by Lavoisier. He was not pleased with it and hoped that people would resist it. Everybody seemed to be against it, in France and in Austria. He did not make a secret of his dislike of Hassenfratz, the main inventor of these new hieroglyphs that made up this new language or cabala, who had send him a copy of his new book about which he asked IngenHousz' opinion on behalf of his colleagues.

> *U zult misschien de nieuwe <u>nomenclatura chimica</u> van Lavoisier <u>cum suis</u> reeds gezien hebben. Ze behaagt my in geene opzichten, en ik hoop dat er een generale opstand tegen zo een dolle onderneming plaatz zal nemen. ik heb daarover myn gevoel reeds aan de schryver van de Journal de Physique mede gedeelt. hy wenste dat hy het met myn verlof kon uytgeven, maar daar ben ik tegen. in Franrijk is men er in 't generaal tegen, hier ook. ik heb myn oordeel daaromtrent zelfs niet willen ontvynzen aan den voornaamste bewerker van deze nieuwe taal <u>Hassenfratz</u> den uitvinder van de nieuwe hieroglyfen die een voornaam gedeelte van de nieuwe spraak of van de nieuwe cabala uytmaakt, in een brief van dankzegging voor het toezenden van een exemplaar van dat singulier boek, over welkers inhoud hy aan my in naam van zyne collegaas myne advise vraagde.*

In his letter of 20 February IngenHousz demonstrated his particular wilfulness and independent mind. He had been offered to succeed Van Swieten as the president of all universities and scholars of the whole empire - but he declined. He was most happy when he had no power over others and others not over him. He loved being honoured without being envied. He was able to move among the powerful of this earth, but preferred not to use these occasions to court their favours and instigate the jealousy of many. The Empress had offered him a higher salary, noble titles and public decorations. But he told her that he wanted nothing more than a continuation of what he already received. He knew now from 20 years experience that he was infinitely more quite and happy than if he would have been led by power and pride. Sometimes he stayed at home for three or four days, where he lived in a nice house that costed him only 1200 gilders a year, his salary was brought to him every three months and he had no enemies.

denkt het gelukkigst te zijn als men geen macht (bewint) heeft over anderen en zijn ook niet over jou, dat men agting kan verwerven zonder wangunst te baren. Hij bevind zich tussen de groote vorsten der wereld, maar prefereert van die gelegenheden zo min mogelijk gebruik te maken en alle gunsten te declineren die hem bewint over anderen en de afgunst van velen moest verschaffen.

My werd aangeboden B. [Baron] Van Swieten op te volgen en dus het opperbewint aller universityten en alle geleerde van een magtig en wyd uytgestrekt ryk te voeren. De Keyzerin bood hem niet alleen een grooter inkomen maar ook adelyke titels en publike decoratien. Ik sloeg dit alles af zeggende dat ik niets meer verlangde dan de continuaties van wat ik reeds ontfangen had ; en ik heb door een nu 20 jarigen ondervinding bevonden dat ik onyndig gelukkiger en geruster geleefd heb dan ik zou geleefd hebben indien ik door heerszugt en hoogmoed my had laten verlyden.

Ik ga somtyds in drie of 4 dagen niet uyt den huysch. Ik bewoon een pragtig quartier dat 1200 hollandsche guldens 't jaar kost waarin ik alle commodityten heb. men brengt my alle drie maanden myn inkomen thuys. ik ken geen vyanden van myn persoon.

He was happy in Vienna, but he was happy while travelling too. In July he left for Paris, after many weeks of delay because the right carriage could not be delivered. One of the reasons he wanted to go to Paris was to supervise personally the printing of his books and dissertations that had been delayed for too long already. He had some sixty articles ready for printing. He kept on making notes on so many new and exciting topics. Nature was just too wonderful and stimulated his curiosity like nothing else. Investigating nature and thinking about this great system was his source of happiness and contentment. Politics however, certainly in his home country but even in Austria, didn't make him happy. Too much freedom was suppressed.

From then on, Van Breda was to send his letters to Paris: "nu brieven sturen naar mijn adres au soin de Mrs Tourton & Ravel à Paris. Zal daar begin juny zijn, denkende hier te vertrekken 22 of 23 may." He planned to leave Vienna on 22 or 23 May.

He picked up his travel diary again (the same as before, in which the last entry was from 1779), turned it around and started filling it from the other side:

18 juilet arrivé a Paris. je suis aller loger au grand hotel de Toulouse rue du jardinet en face de la rue de l'eperon a raison de 1 ½ louis d'or par semaine. apres la premiere semaine je payera 4 louis par mois.

He arrived in Paris on 16 July, and automatically switched to writing in French. He stayed at the Grand Hotel de Toulouse opposite Rue du Jardinet, on the left bank around the corner from the university on Boulevard Saint-Germain. He paid 1.5 Louis for a week, after the first week only 4 Louis per month. He went out to buy clothes:

> J'ai fait faire un habit veste et culotte de drap de louvier double de poux de soye blanc (qu'on appelle a vienne gros de tour) J'ai en 3 ¾ d'aulne de drap a 30 L l'anne
> J'ai fait faire aussi veste et culotte de prunelle de soye noir 5 anne a 15 L l'anne f. 75 L. une veste d'ete de velour cannelé 3 annes a 14 L l'anne f. 42L.

He also bought a new map of Paris and an umbrella of taffeta. IngenHousz had a busy programme, visiting instrument makers in their workshops, attending lectures by famous professors, meeting people from the 'Republic of Letters', discussing anything from electricity to chemistry and beyond, over coffee or dinner. His diary gives an impression on what his days looked like:

> le 24 aout les ambassadeurs indiens etant venu voir les experiments physiques au college royal. j'ai vu les experiences de Mr Lavoisier de changer l'air en eau et l'eau en air. Les experiences sont vraiment interessant. On fait un vide dans une grande boule de verre. on y fait entrer de l'air dephlogistiqué. on fait ensuite entrer un petit jet d'air inflammable par un tube, dont l'embouchure s'ouvre au beau milieu du globe. on allume ce jet d'air inflammable par une etincelle electrique, il en resulte une petite flamme au beau milieu du globe, qui s'y contient par le jet continué d'air inflammable jusqu'à tout l'air inflammable soit changé en méphitisme incapable de soutenir la flamme. on trouve alors une grande quantité d'eau ramassé au fond de l'eau.

On 24 August he attended some experiments at the College Royale, together with the Indian Ambassador. He had also seen the experiments by Lavoisier in which air was changed into water and vice versa. He found these experiments really very interesting and carefully notes how Lavoisier did it. He pumped all air out of a glass bowl and filled it with oxygen (the old term dephlogisticated air was still used by IngenHousz). Through a small tube some inflammable air (hydrogen gas) was injected into the bowl. This was ignited with an electric spark which formed a little flame in the middle of the bowl. This flame burned until the last inflammable air had been changed in air that was too bad to sustain the flame. Then one found a big quantity of water on the bottom of the bowl. His next letter to Van Breda came from Paris, 16 September, mentions he arrived in the French capital by the end of July. He was happy to inform Van Breda that by now the printing of the second volume of the French edition of *Experiences sur les plantes* was in full swing. Already 8 or 9 pages had been printed. He wrote he hoped to be able to travel to London in November.

He went to Passy (where the American delegation stayed and where Franklin lived) to see their telescopes. He visited Abbé Rochon and his lens maker Mr Larossier. He met with Mr Berthollet, Mr Fourcroy, l'Abbé Quenard and Mrs Qeynard and Coulomb. Some of these meetings were coincidence, so he reported. These were probably unplanned encounters, as these things go, meeting friends of friends in an ever expanding network of acquaintances.

> *fin novembre chez Mr Coulomb, qui habite en hiver N 15 rue d'Argeteuil*
> *exp met electrophore*
> *le 9 dec Mr Berthollet fit devant moi le sel nouveau le muriate oxygène de potasse*

Paris was a giant play-garden for a curious kid like IngenHousz. Together with Coulomb he performed experiments with the electrophore; from Berthollet he learned how to make potassium chlorate.

On 22 December he met Comte de Carburi who told him that the Doge of Venice was suffering from gravel. IngenHousz noted the advice to take care while performing a lithotomy, removing stones from the bladder, to let the wound bleed profusely in order to prevent infection.

He had discussions with Réaumur and Bossier, makers of thermometers; he debunked another magnetiser Abbé Le Noble and learned to make exotic gasses from Mr Charles. He followed up on how Cassini calculated the distance between the earth and the sun and claimed that the light of the sun takes 7 minutes to get here, while Halley had calculated 8 minutes and 13 seconds. The speed of light was supposed to be 980,809,933 2/3 feet per second.

His plans were to travel to London in autumn but his medical condition prevented him from carrying on. The next letter to Van Breda was dated 19 March 1789, Paris. Bladder stones undermined his health.

> *toen ik uw laatste brief van 28 oct 1788 ontving was ik deerniswaardig, hebbende dan de gedugte toevallen van graveel en stenen. ik heb yndelyk op het ynd van nov van zelf zonder pijn een steen door de urethra gelost over de 6 linien lang en 2 breed, franse maad. Aderlatingen en andere middelen in 't werk gesteld voor de crisis hebben myn sterke constitutie zeer verzwakt en my in myne geleerde arbyde zeer vertraagt. Egter zal nu binnen 14 dagen het tweede franse deel der proeven met planten uytkomen.*

When he received Van Breda's last letter in October, he was in a bad state, having painful attacks of 'gravel and stones'. He got rid of one stone the natural way, by peeing it out, but the bleeding and other medical measures had weakened his condition and slowed down his scientific work. However, two weeks later the second volume of his French book would finally be ready.

His illness did not prohibit him continuing his exploration of the wonderful living database which is the French capital. He met Dr Guillotin:

> *Le docteur Guillotin, très bon médecin à Paris, me disoit qu'il a guéri plusieurs fois les fièvres intermittentes en faisant boire au malade un quart d'heure avant le paroxysme une tasse de caffée fort avec egales quantité de jus de citron. il y ajoutoit, qu'il a appris d'autres médicins, qui ont reussi de même.*

Dr Guillotin was a very good doctor, who told him he treated intermittent fevers with a strong cup of coffee mixed with an equal amount of lemon juice 15 minutes before the paroxysm.

Many of the discussions and experiments took place on the blurred borderline between chemistry and electricity. Lavoisier and other chemists who followed the new system thought electricity was a form of combustion which required the presence of oxygen or vital air (IngenHousz was using the old and new terminology intermittently). But he thought they were mistaken because the spark shines as well in fixed, phlogisticated, inflammable or vital air.

> *Mr Lavoisier et les autres chymistes qui suivent son systeme maintiennent que l'electricité est une combustion comme les autres combustions, qui requiere la presence de l'oxygène ou l'air vital pour avoir lieu. mais ils se sont trompé beaucoup car l'etincellee electrique brille egalement dans un air fixe, phlogistique, inflammable que dans l'air vital.*

And from the kitchen to the laboratory was only a small step, cooking is chemistry after all.

> *le 24 juin en dinnant chez mr Pelletier, Mr Bayen disoit qu'on peut conserver des feves de marai et de petits poids verts jusqu'au hivers en les en remplissant une bouteille a sec on bouche la bouteille bien avec un bouchon de liege. on la met dans un chadron rempli de l'eau froide. on fait bouillir cette eau pendant un quart d'heure. ensuite on met appart la bouteille sans le deboucher. on l'ouvre en hivers lorsqu'on veut s'en servir*

On 24 June 1789 he had dinner at Mr Pelletier's and one of table companions was Mr Bayen who told him how you could conserve broad beans and green peas until winter. One had to put them in a dry bottle and close it tightly with a cork stopper. Put the bottle in a big pot with cold water and heat it to boiling and let it boil for a quarter of an hour. Afterwards one could put the bottle aside without opening it. That's what you did in winter when you want to serve the beans or peas.

On 8 July he learned from Hassenfratz how one could make ivory soft in nitric acid.

> *Mr Hassenfratz me disoit (8 jul.) qu'en mettant pendant quelques heures un rameau d'yvoir dans l'acide nitreux delayé, il devient flexible comme cartilage. L'acide en ote la terre calcaire et la parti cartilagineuse reste mais alors ce n'a plus nature d'yvoir.*

In his letters to Van Breda he reported that physics and chemistry made great progress in Paris and that he had learned a lot already about the new system that still had a lot of opponents. (Sounding as if he was not opposed to it anymore.)

And not only the chemical sciences saw a revolution unfold. The summer of 1789 was hot in Paris and France. On 14 July the mob erupted after months of protests against famine and high food prizes. The Enlightenment philosophers had questioned intellectually the authority of the monarchy and the clergy, the lower classes would bring this critical attitude to the streets. They stormed the Bastille prison, dragged the governor out, tortured and beheaded him, only the start of a country-wide uproar against all things aristocratic. The king and queen with their palaces and sumptuous way of life were seen as a symbol for the luxurious and extravagant life style of the higher classes, and were an obvious target for the revolutionaries. The queen, Marie Antoinette, was a sister of the Austrian emperor. IngenHousz was travelling under the Austrian flag, in a private carriage. This made him conspicuous considered being wealthy if not aristocratic, and he left Paris on the 15th, fleeing North to Breda, being shot at in Belgium by revolutionaries. In Breda he was sad to learn that his brother had recently died in a traffic accident. He stayed in Breda to console his sister-in-law, his nephews and nieces. On 5 September he wrote his first preserved letter to Marinus Van Marum[98] from The Hague. He sent him the message that he would be in Haarlem the next evening, arriving by boat in Leyden and that he would be pleased to meet. He hoped to use some of his precious time during the day after, when he needed to be in Amsterdam by evening.

> *La Haye*
> *Dimanche 5 sept 1789*
> …
> *C'est à votre celebrité et à vos lumieres connues que vous devez attribuer mon importunité, autant qu'au désir qui m'anime toujours de rendre hommage aux savants et de profiter de leurs lumieres.*
> …
> *Je voudroi bien voir la fameuze machine Teilerienne.*

---

98   Haarlem Noord Hollands Archief

[on the outside:]
> *franco*
> *afgezonden met de schuyt van 12 uren van den Haag zo haast mogelyk te bestellen*

He would be honoured if they could meet. It is due to Van Marum's celebrity and intelligence that IngenHousz was so eager to meet him, even if this was to disturb him. He wanted also very much to see the famous "Teylerian Machine", the big electrical machine that he had built in Teyler's Museum. He also wrote that he was coming straight from Paris and that he was bringing some "literary news". As well as a letter from Landriani, who was in Vienna at the time. He sent the letter by the 12 o'clock boat from The Hague, marking on the envelope that it was to be delivered as soon as possible. The post went by water in Holland.

On 15 September he writes another letter to Van Marum, also in French which seemed to be their lingua franca, even when they were both Dutch-speaking. With the letter he sent him a copy of the first volume of the second French edition of *Expériences sur les végétaux*. He also attached five volumes of the "amalgame de Baron Kienmayer et un fil de platine". He told Van Marum that he did send simultaneously the second volume of his French book to Paetz Van Troostwijk. And he adds in a postcriptum: "Je compte de partir dans trois jours pour Londres."

Because the French Revolution was slowly but surely creeping Northwards, in the Netherlands too the sentiments were turning against anybody with royal affiliations. IngenHousz decided to carry on, left most of his luggage and his carriage in Breda and fled the continent, which was turning into a smouldering ruin where reason and honesty seemed to be ruled out by violence and revenge.

Back in England

His diary picks up upon arrival in London, 27 September. The sentence below shows he is back in English speaking territory. And his own medical condition resonnates through in his concern about the state of the art on the treatment of bladder stones.

> *Le 27 sept. arrivé à Londres.*
> *Dr george Fordyce told me, that he had relived more than 200 patients from gravelous complaints by giving them twice a day 30 drops of tart. purdeliquiam in water*

In October 1789 he writes to Van Marum that he received his letter two days earlier (the day before yesterday, "voorgister") and replies that the critique about the supposed flaws of Herschel's mirrors is ill founded.

Wat u van het gebrek der Herschelische holle spiegelen gehoort heb, is zo ik meen zonder grond, dewyl ik dit gebreck door d'een of ander wel zou vernomen hebben. De overige math. Instrument makers zyn al te jalours van de ongemene bequaamheden van Herschel om hem niet alle afbreuk te doen die mogelyk is.

He supposed that these rumours had been spread by other makers of mathemathical instruments, who were jealous and tried to show Herschel in a bad light whenever they could. As it happens, the mirrors were sensitive to fluctuations of temperature and that's why Herschel covered them at night with a tin disc with a leather brim, so no humidity was able to condense on it. But the best thing to do would be to come over to London and see for himself. Travelling to London was easier than going to Antwerp or Brussels. If he were to come, he needed to plan his voyage in such a way that he would be here on the third day of the moon to watch the heavenly bodies with the big Herschelian telescopes. Miss Herschel just discovered a new comet that would occupy and entertain all visitors. (Either IngenHousz' date on this letter is wrong, or the comet was not as newly discoverd as he claimed, as Caroline Herschel discovered her first comet in the summer of 1786 and her second just before Christmas 1788, her third in January 1790.)[99] He would be pleased to introduce him to the Herschels as they were very good friends.

He regretted there was no chimney at Teyler's Museum. When coming to London he would be able to see the newest types of stoves that were excellent to keep rooms warm and dry in winter. The maker is "Jackson, brasier in Berwick Street, Soho, London". And he mentioned that he spoke to Adams about the making of an eudiometer, which Van Marum was interested in ordering in London.

His next letter to Van Marum left on 7 December. Due to his ill health, the vastness of this capital and the consequences of "the fall of Magellan", he could not respond quicker to the letter he had received from Van Marum in Breda.

Ik heb enige dagen te Slough by Windzor doorgebragt met den befaamden Herschel. Ik heb alle de 7 satellites van Saturnus gezien, ik heb zelfs de 2 niew ontdekte op den ring gezien.
Nu druckt men een lyst van 2000 niew nebulae of stella nebulosae door hem ontdekt.

He spent a few days at Slough near Windsor with the famous Herschel. He observed all seven satellites of Saturn, even the two new ones on the ring. The new list of 2000 new nebulae was in print by now. And he added that even with the stongest telescope some nebulae could not be seen as seperate stars.

---

99   Holmes 2008

Onze denking kragt is te zwak om zo een uytgebreydheid van het geheel al te bezeffen.
Deze verwonderingswaardige man staat meer ontdekkingen in de astronomie
te maken dan alle zyne voorgangers tot nog toe gemaakt hebben.

Our intellectual power is not strong enough to comprehend such a vastness in its totality.
This remarkable man is capable of making more discoveries in astronomy than all his predecessors made until now.

And he reported that James Keir published his new book on chemistry *The first part of a dictionary of chemistry Etc.* Keir was not sympathetic towards the new system and even less towards the new nomenclature. It seems as if Prof. Black had not decided yet, he was now studying the system and most thought he is inclined to opt for the new system.

Schoon ik veele myner oude vrienden door de dood verloren heb, heb ik er egter
nog zoo veel in leven bevonden en niewe bekomen, dat ik tot nu toe nauwelyks
de helft heb kunnen zien of er my mede te onderhouden. Een enkele visiet
neemt gemeenlyk een gansche dag weg.

Stuur je antwoord naar den Heer Fagel by den Heer Baron van Nagel, extr.
envoyé van H.H. hoogmogende aan het hof van London.

I lost already many dear friends through death, but I still have so many alive or am getting to meet so many new ones, that I'm hardly able to meet half of them properly. Each visit takes up a whole day.

From now on he was to send his letters to Baron van Nagel, the secretary of Dutch Amabassador at the court of London.

In his next letter to Van Breda, sometime in the beginning of 1790, he reported he was back in England and left the tumultuous continent behind. In this undated letter he reported on his health. On the road, passing though Harwich, he suffered another nephritic attack and on arrival in London he peed two stones. After that, his pisspot was full of fine sand for a few days. He was using a new remedy (aqua mephitica alcalina) and he hadn't suffered since.

Ik agt my verplicht u enig verslag van myne zelve te geven. Onderwegen
te harwich voelde ik wederom een begin van een paroxysmus nephriticus.
Nauwelyks was ik te London of ik loste 2 stenen: enige dagen daarna was myn
waterpot bezt met een zand bezetzel. Sedert ik gebruykt heb de niew remedie te
weten het <u>aqua mephitica alcalina</u> heb ik niets meer gewaar geworden.

Het experiment van den H. van Troostwyk in het Journal de Physique van
nov. laatst bevat schynt my decisief voor het niewe systema van Lavoisier te
zyn. Dog laat ons horen van Priestley en Keir er van zeggen zullen.
Brieven aan myne adres net den enveloppe aan den Heer fagel by den H. Baron
van Nagel Envoyé Extr. & Plepip. van HH. MH. NN. te London en ter hand
gesteld aan de heer Tinnen commissaris der uytheemsche deputy in 's Hagen.

Van Troostwyck's experiments were reported on in the November issue of the
Journal de Physique and these seemed to IngenHousz to be the proof for the
new system of Lavoisier. He was curious about what Priestley and Keir would
have to say about it. Letters from then on had to be addressed to the Dutch
ambassador in London.

Apart from his health problems he was concerned about his investments all
over Europe which seemed to be in danger in this period of faltering economy
and wide spread confiscations of properties. Another source of grief was the
death of his dear friend Franklin on 17 April. This put an abrupt end to the
smouldering plans to emigrate to America. With spring in the air he considered
the idea of returning to Austria, but the political situation on the continent was
not conducive to travel. He was afraid that a contra-revolution would break
out any moment and he did not like the idea to get caught in the fire. France
was caught in the revolutionary terror; war was raging in Germany, Italy and
Austria.

On 31 May he sent a letter to Van Marum, telling him that Adams was
finally ready with the eudiometer and that the instrument will be shipped to
Rotterdam on the first ship sailing in that direction. He made some suggestions
on where Van Marum could stay when in London. IngenHousz was clearly
preparing everything for his friend to have a smooth stay in England. He
suggested the Bates' Hotel in The Strand for the first couple of days (2 shillings,
half a croon for one night).

Het beste is eerst in London 2 of 3 dagen in een hotel te logeren, dat kost vor een
kamer met bed 2 sch. (een halve croon). Het beste is Bates' Hotel in The Strand.
Anders gaat men in een burgershuis logeren, 2 kamers voor 1 of 1 ½ ginje, de
dienstmyd beryd thee waarvoor ze een fooi verwacht van 4 sch. Per maand.
In verscheidene van die logeerhuizen kan men met de familie eten voor 2 sch. Daags.
In gaarkeukens of taverns kost het 2 a 3 sch., in de beste 5 sch. Per hoofd.

Renting a room is another option, at two rooms for 1 or 1.5 guinies, with a maid
bringing the tea for 4 shillings a month. There were many places where one
could stay with a family for 2 shillings a day. In soup kitchens or taverns a meal
cost 2 or 3 shillings, in the best places 5 shillings.

In his letter of 8 July to Van Breda he told him it was no problem to buy a thermometer at Wedswood [Wedgewood] - costing three pounds in a box including all accessories. Van Marum would be coming to London to buy scientific instruments, so he could take the thermometer back to Holland. In the mean time Van Marum had arrived in London. One of the aims of the trip was to try to get the much covetted fellowship of the Royal Society[100]. On his second day in London he attended the Royal Society lunch which was open to foreign philosophers and which was chaired by Sir Joseph Banks. But past quarrels between Van Marum and men such as Cavendish (a close friend of Banks) kept the doors of the RS firmly closed to him.[101] Even in normal circumstances it was not easy to become foreign 'FRS'. IngenHousz introduced his friend to many of his friends and takes him to many instrument makers. Together with IngenHousz he visited Nicholson and admired his electrical machine with glass plates which Van Marum thought to be the best of its kind.[102]

IngenHousz spent the autumn of 1790 in Calne, at Bowood, as the guest and scientist-in-residence of the First Marquess of Lansdowne, Lord Shelburn, by then an elderly statesman after having served as Prime Minister and having helped to settle the American war of independence. Thomas Young gave a nice description of IngenHousz state of mind at the time in one of his biographies of men of science:[103]

> ... notwithstanding his dislike to the chilliness of the climate, it was always his favourite residence, and "where he enjoyed during many years", to use his own words, " that felicity which a free and independent man finds in the pursuit of knowledge and wisdom, in the society and friendly intercourse of those who have distinguished themselves by their learning."
>
> Dr. Ingenhousz was cheerful in his disposition, and often playful in his conversation. Though his pursuits were chiefly scientific, he was not destitute of taste for literature and poetry. He had a particular predilection for Lucan, and for the Cardinal de Polignac, and would frequently recite passages from their poems with great energy, and with a strong German accent. Nor did he disdain the comforts of commercial opulence, and he was often a visitor at the magnificent villa of the late Mr. Rucker of Roehampton. He had been introduced there by his friend Dr. Brocklesby, who was in many respects of a perfectly congenial disposition, and who had great pleasure in prevailing on him to partake occasionally of his own hospitality, when his table would otherwise have been solitary.

---

100 Wiechmann 1987
101 Levere 1970
102 Hackmann 1971
103 Young 1855

And by 2 January 1791 he was still there, although he had made a little trip to the Herschels in Slough where he saw the telescopes that impressed him hugely. His diary reads:

> I passed at the end of Nov 1789 some days at Slough, two miles from Windsor, to see the telescopes of Dr Hershell.
> I saw 7 manen (sattelites) of Saturn and the dark part of the moon
> I saw the moon thro a 7 feet telescope magnified 1200 times. I never saw any thing so delightfull.

William and Caroline Herschel had become close friends. He wrote to Van Breda: "Dit alles is het wonderwaardigste machien dat door menschen hande ooyt gemaakt is. Ik hoop dat er binnen weynige tyd in Haarlem een 20 voet Herscheliaansche tubus zal zyn." These must be some of the most wonderful machines ever made by man. He hoped that before too long a 20 feet Herschelian tube would be installed in Haarlem.

By February, he was back in London. Medical matters featured highly in his letters, such as the use of different forms of kina-extracts, imported from Latin America, to treat fevers. It had less side-effects and could be used at higher doses. The notes in his diary end somewhere here, but the correspondence with Delft continued.

In the letter of 21 June he described a technological innovation: the first copying machine. A few alternatives to copying by hand were invented between the mid-seventeenth century and the late eighteenth century, but none had a significant impact. In 1780, steam engine inventor James Watt obtained a British patent for letter copying presses, which James Watt & Co. produced early in that year. In the process used on Watt's copying machine the letter-to-be-copied was written with a special copying ink on a sheet of good quality paper and placed, when dry, in contact with a water-dampened tissue-paper. The two were held together for a few minutes in a type of mangle or screw press. The writing which offset on to the tissue gave an impression in reverse, but as the tissue was very thin it was simple to read the writing from the other side where it appeared the correct way. It must have been the first time IngenHousz encountered this machine and he was enthusiastic, and sent some of the copies he made.

The next day, 22 June, he wrote to Van Marum to report on his progress on his election as foreign member of the Royal Society. Van Marum had to be patient and wait for the right occasion. It might help to return to London to get personally in to contact with people such as Banks, Cavendish and Blagden. The latter already knew him and was very sympathetic towards his foreign membership. IngenHousz himself had by now been elected to the Council of the Royal Society and felt that his support for Van Marum's candidacy might now carry more weight. Van Marums letter on his electrical machine was read to the Society and IngenHousz had made a cardboard model to provide an illustration of it. But so many wanted to become a member and the criteria had been sharpened:

> *niemand durft zich opdoen, die niet een liberale functie heeft bekleed, een gentleman is en op wiens caracter niets te zeggen is. Een simple med. Doctor die buyten zyn functie geen natuurkundige wetenschappen bezit, zou nu zeer ligt kunnen verworpen worden.*

Never before has the honour of becoming a member been the object of such aspiration as it has been now. Being a gentleman is no longer enough to guarantee election. A mere med. Doctor with no scientific accomplishment outside his profession, may well be rejected. They are becoming very strict about foreign members.

But this was no time for travelling. Neither for IngenHousz and not for Van Marum. The French army threatened at the borders and the country was sufffering from considerable political instability and from a chronic shortage of funds. But Van Marum did what he could; he dedicated a paper to IngenHousz[104] and ensured his selection as member of the Hollandsche Maatschappij der Wetenschappen. IngenHousz thanked him for this intervention in his letter of 22 June.

IngenHousz spent July 1791 at Chevening, the estate of Lord Stanhope, at Sevenoaks in Kent, as we can learn from a short entry in another of his notebooks.[105] He studied Lord Stanhope's method of preventing the rotting of trees.

On 25 July he wrote to Van Breda about the latest unrest in England. The dissenters had attempted on the 14th to throw over church and state. Priestley was one of the revolutionaries; he had predicted this revolution in his writings against Edmund Burke. Priestley did not dare to join a party of the revolutionaries in Birmingham. Van Breda probably already heard about how Priestley had nearly escaped a certain death. All his writings and instruments were destroyed, which was a terrible loss for everybody. Still, it was a pity that such a great scientist was so cursed by fanaticism, which made him into an object of hatred in the eyes of the minds of peaceful men. So, he took pride in his prosecution and preached his doctrines. Some even claim that the drive behind his fanaticism was his lust for fame, and not his religious believes.

---

104 Van Marum 1791
105 Breda Collection IngenHousz B1

de Dissenters hebben op den 14 dezer een kansje willen wagen een oproer en daardoor een omwenteling van Kerk & Staat te weeg te brengen.

Priestley die een der eerste dezer revolutionairen was, en die zo een revolutie in zyn book tegen den H. Burke reeds geprophetiseerd had, wierd daags te voren wat benauwt en durfde zelfzyn persoon niet vertrouwen aan het gezelschap dat tot Birmingham op die dag een feest vierde. U zult al vernomen heben in wat vreselyk levensgevaar hy geweest is. Al zyn geschriften en instrumenten zyn vernietigt, dat een publiek verlies is. Het is jammer dat die geleerde het ongeluk heeft met een fanatismus behept te zyn die hem in de gedagte van vreedzame bedaarde lieden gehaat maakt. …glorieert hy in de vervolging, en preekt even sterk de unitarianismus, socimianismus en materialismus. De meeste menen dat de lust van faam de grond van zyn fanatismus is en dat hy niets geloofd omtrent de godsdienst.

All summer IngenHousz travelled from one friend to another, staying at their country houses, such as with Dimsdale in Hertford. On 13 September he received a letter from Van Breda who was informing his friend that his wife had died in childbirth.

Gisteren ontving ik uw brief van de 5 dezer bevattende zeer drukkende reden van verdriet. Ik neem in uw verlies een vriendelijk aandeel, te meer derwyl ik my herinner uw waarde gemalin gezien en gesproken te hebben. Reflexie, tyd en bezigheden moeten diergelyke nauurlyke voorvallen doen vergeten. Durum sed levius fit patientia, quidquid corrigere est nefas. Ik hoop dat het verlies niet veroorzaakt is door de hand van de vroedmeester derwyl ik die kunst … zedert 40 jaren… voor een moordkunst aanzie waardoor meer mensen elendig omgebragt zyn dan door alle oorlogen die sederd die tyd gevoert geweest zyn.

IngenHousz expressed his deepest sympathy, remembering the times when he enjoyed her presence. Reflexion, time and activities would help forget these sad events. He could only hope that this loss was not caused by the hands of the nurse, as he had witnessed the art of midwifery for 40 years as a deathly discipline, by which more people had been killed than in all the wars that had raged all these years.

His gravel was better, and he suspected that the medication he was taking also cured his gout, as he hadn't had a single relapse. He suspected that gout, arthritis and gravel might have the same cause. On 27 October he was back in Beauwood [sic] were he was planning to stay until the beginning of '92. In nearby Bath, he spent 1 shilling on a demonstration by Dr Graham who used to sell his therapy in a "musico-magnetico-electrical bed" and now claimed that his mud baths could make young women beautiful and old women young. "Deze klugt gaat alle dagen door." This farce was performed every day. He was stupefied by the thousands of people that visited Bath, to do nothing but play games, walk around, take treatments and visit shows of applied charlatanism.

In the letter of 14 December he described his continuing struggle with his mother tongue. Being away for so long and not speaking it regularly at all, he found it sometimes difficult to find the right words. He hoped Van Breda would correct his style and replace a word with a synonym if he used it twice in short space.

> *...derwyl u de stukken altyd overschryft voor den druk zal u den stiel wel willen verbeteren, de repetitie van het zelfde woord te digt by elkander staande, ramplaceren met een ander synonimisch woord &c. ik heb te lang myn vaderland afwezend geweest om zuyver nederduytz te schryven. ...*

Over the next year he took turns staying in London and roaming the country, visiting friends. In September he visited the Duke of Athol in the Scottish Highlands. Bowood however was the one place he really felt at home and he became part of the 'Bowood Set', a rotating group of more and less famous individuals visiting Lord Lansdowne and enjoying his hospitality. Doctor IngenHousz was very popular with this informal gathering of scholars, merchants, aristocrats and politicians. He was an entertaining companion, who could recount many anecdotes about far away places and famous persons, and could teach anybody a thing or two about science, medicine and astronomy. He often performed one of his many chemical experiments in which he burned pure metals in phials of oxygen which produced brightly coloured flames.

In his 9 avril 1792 letter to Van Marum he asked him whether he could bring some of his shirts and his waistcoat, which he had left in Breda, when he was coming to London in spring. The messages he got from Vienna however were not really promising. He thought he was going to stay a little longer on "cet isle heureuse". He was clearly having a good time in a country full of inspiring science and warm friendships.

He kept corresponding with Van Breda about chemistry and politics. On 5 November he wrote from Bowood Park:

> *Wat is de beste bestuursvorm? Is afhankelijk van land tot land.*
> ...
> *Ik ben al lang een cosmopolite en niet voor ingenomen zynde, denk ik met meer bedaardheid dar omtrent te kunnen oordelen als andere. egter weet ik wel, dat tot nog toe geen regering yder een heeft gelukkig gemaakt, niets is volmaakt, zelfs niet de wetten van de natuur. Wat reden, bij voorbeeld, kan een gezond oordeel uyt denken waarom onnosele schapen zo wel als menschen aan groote smarten onderhevig zyn voor den tyd van een natuurlyke dood? wy begrypen het niet, en moeten het met onderdanigheid aan nemen.*

het menschdom is een wild beest dat ten lange laasten zyn yge vreedheid en
onredelykheid gekent hebbende zig zelve een ketting heeft aangelegd, welke nu in
Vrankryk gebroken zynde dat wild dier wederom zyn natuurlyke woede aanneemt.

He wondered what the best form of government could be. He had to admit he
did not know. It depended from country to country. He had been a cosmopolitan
for so long and had lost all prejudice. He thought he was able to judge with
more equanimity than many others. What he was able to say for certain was
that until then no government had ever succeeded in making everybody happy.
Nothing was perfect, not even the laws of nature. What for example could be an
explanation for the fact that poor sheep just as well as humans were suffering
terribly at the time of their natural death? We do not understand, and we will
have to accept it humbly. ... Mankind is a wild beast that has recognised finally its
own savagery and irrationality and therefore put itself in chains. That chain has
been broken now in France so that the wild animal has regained his natural rage.

In a postscript to this letter he held a plea for always referring to the sources
of quotes and claims, in order to be able to trace back the letters or publications
and get back at the original.

Politics kept him alert and anxious and filled many of his letters. He wondered
how people still didn't seem to see the results of four years of the revolution in
France. Nobody was able to trust another and freedom meant that murderers
could run around freely. Those philosophers were bad politicians. And even
in England people defended the principles of this run-away democracy, he
wouldn't be surprised if a revolution would break out there too. He still planned
to travel to America, were this calamity to happen. And he dreamed of going
back to Vienna to join his wife, if the situation would allow it. Moreover, the
troubles in Europe made communicating by post difficult. Only four letters
reach Van Breda in 1793. Letters were sometimes opened by the letter carriers
in the hope of finding paper money. They then burned the letter. Some of these
thieves had already been hanged.

In his letter of 1 April 1794 he wrote that the new (chemical) system was
really the best, although it still was not capable to explain everything. But he
thought philosophically that probably no system would ever do so. But oxygen
had become a household word. IngenHousz had read some of the publications
of Dr. Beddoes. (He spelled this name in three different ways in just one letter,
none of them the right one.) Beddoes treated patients with nitrous oxide.
IngenHousz found Beddoes' reasoning behind this treatment a rather strange
theory in which too much or not enough oxygen would offer an explanation for
all diseases. A reprint from Beddoes' article dedicated to IngenHousz and dated
"Bristol 29th Sept 1794" describes a proposal towards the improvement of
medicine, expanding on the promising applications of the pneumatic medicine
and about the creation of an institute for pneumatic medicine.

IngenHousz' own books were translated, already three volumes had been published in German. If he were to stay in England for another year he would love to publish his works in English. The only publicly available printed testimony of his ideas in English, apart from the letters in the Philosophical *Transactions*, was the book from 1779 and that was rather strange. Moreover, there was a huge demand for it and if you could somehow still find a copy, they asked unreasonable prizes for it.

Writing was his main occupation, as he had still huge amounts of data that had not been used and ideas that had not been worked out. In May he bought a little house in the countryside, some 12 miles from London. He wanted to get out of town, away from the restlessness. On 8 May the guillotine took the life of his friend and colleague Lavoisier in Paris.

He sent the latest book by "state enemy no 1" Priestley (the book is "prul" or a waste of paper) and all necessary materials for making copies (a wetting book, a drying book and the right kind of ink) to Van Breda. It would be taken to Holland by Cuthbertson, a famous English instrument maker who lived partly in London, partly in Amsterdam.

After a description of what happened in Birmingham on Bastille Day 1792 when Priestley's house and belongings went up in flames, he sketched the characters of two of the world's biggest chemists he knew so well:

> *Zyn franse vrienden hebben hem altyd erkent voor een atheist. Hier passende hy de meeste voor een deist, en een man wiens gemoed vol hoogmoed en lust voor faam was. Hy verwagte, even als jean jacques Rousseau, een der befaamste menschen en wetgever te worden en een generale reform in 't kristendom te bewerken ...*
>
> *ik geloof dat niemand die man's caracter zo wel kent als ik, en den ommegang die ik met hem zelfs en de zyne gehad heb bevestigen my in dit oordeel.*
>
> *ik betreur myn vriend Lavoisier en verschyde andere vrienden daar ik intime kennis med had. Hoe kan ymand, zonder van oordeel beroofd te zyn, all deze vreedheden horen, zonder een afgrysen van democratien op te vatten.*
>
> *ik ga heden na myn buyte plaatsje terug en zal misschienin verschyde maanen niet meer in London komen.*

His French friends had always known him as an atheist, but here in England Priestley passed for a deist, a man full of pride and lust for fame. He had hoped to become like JJ Rousseau, one of the most famous men and legislators and to fundamentally reform Christianity. IngenHousz thought he knew his character well enough, through so many contacts with him and his friends, to be able to confirm that judgement. He regretted his friend Lavoisier and many other intimate friends. How could anybody, without having lost all his senses, hear about these atrocities without becoming sceptic about democracies, he asked himself. That day he was going back to his little country house and would probably not visit London for some months.

IngenHousz was back in London when he wrote a letter to Van Breda on 21 November 1794. It was a voluminous letter, it filled eight long pages with various notes on intestinal parasites and their treatment, on the use of different airs to treat several diseases (referring to the work of Dr Biddows [sic] of which he was thinking less critically than a year before) and a memory about a new way of curing cancers, dedicated to his nephew Henricus Ferdinandus, who had become field-physician.

> er moet nog veel verder van onderzogt worden, maar het zy my genoeg u in het ruwe een denkbeeld gegeven te hebben van een geheel niewe manier van een der schrikkelykste qualen te genezen, die tot nu toe niet dan door het mes en dat altyd met groot levensgevaar, kon uytgeroyd worden.

He saw the promise of the numerous new treatments that were being developed, such as treating breast cancer or the stumps of amputations with fixed air (carbon dioxide) but they had to be further tested. But the preliminary results were sufficient to give his correspondent an idea about the completely new ways to treat these terrible diseases, that could until now be attacked only with the knife and to big fatal risk. Water saturated with fixed air would also be a good method to treat worm infections in poor people, as it was easy to make and to administer. But there were still more questions than answers. Were the worms in our intestines the same as those living in the soil? Were all the experiments on the latter, in various sorts of juices and concoctions relevant to the treatment of the human parasites?

In the mean time he complained about the growing problem of insecurity in the streets. He had been attacked three times in a year. During the last of these attacks he was robbed by a gang of six or eight crooks in St James Street. And on top of that, often they were not even sentenced - if they were caught at all - because just one nitwit in a jury of twelve was enough to set one free.

On 16 janvier 1795 he wrote to Van Marum about his financial situation.

> Je dois me retracir autant que possible dans mes depences, car j'ai perdu tous mes revenus en espagne, en France et en hollande, et la plus part de ceux que j'avois en allemagne, avec danger de perdre le tout

He spent much of the summer in the country, where he would have liked to translate his works in English, but he failed to do so due to a lack of concentration and courage, troubled by the events on the continent. Instead he read a lot, mainly political works, such as on the situation in Geneva. He had a big sympathy for the new doctrine of the human rights, égalité &c. He was happy to hear that Van Marum is in safety, but warned him that their friend "Lavoisier se flattait long tems avec cette esperance, et ce n'evadoit pas pour cette raison". He did not really know whether he had to be glad that he did not go to America, as four provinces there are revolting. "Il n'y a point de tyrannie plus cruelle que celle du peuple."

Books and plants

IngenHousz read a lot thoughout his life. Books were his staple diet. It was more than leisurely reading. He kept notes on books he read or extensively copied passages, for later reference or use in his own texts. Some of these notes have been kept and give a view into his bookshelves and into his mind.

He read De Saussures work *Voyage dans les alpes* from which he learnt that the electricity is on average stronger on mountain tops than in the valleys below. (This is Horace-Benedict de Saussure, father of Nicolas de Saussure, the future famous plant physiologist from Geneva.) From Mr Lavoisier's 1793 *Traité elementaire de chimie* he made notes about the theory of detonnations, as an explanation of what happens with all the gases involved. He made notes on Hutton's *Natural Philosophy*. From a book by David Davies, Rector of Barkham, published in Bath in 1795 he learns that "...on the design of poor laws to provide for the employment of the able and industrious, for the correction of the idle and vicious and for the maintenance of the aged and impotent...". He picked up ideas from an old 1707 book *On the curiosities of nature and art in husbandry and gardening*. Further he read the book by Brisson on the "pesanteur specifique des corps", the 1779 book by Crawford on observations on animal heat, Francis Home's 1759 *The principles of agriculture*, as well as *De l'astronomie* by Lalande and De La Métherie's 1797 *Le theorie de la terre* (who seems to be still resisting the new nomenclature in chemistry). In his notes, one finds extracts from the *History of the Life of Marcus Tullius Cicero* in two volumes, by Conyers Middelton, principal library-keeper at the University of Cambridge, from *The history of the decline and fall of the Roman Empire* by Edward Gibbon, from the work of Soame Jenyns (member of parliament from 1748 to 1780, who "was a man of a very honest and deep sighted mind") and from Adam Smith's *An inquiry into the nature and cause of the wealth of nations*. Apart from these kinds of works, the medical literature got his full attention. Such as *Table of cases in which factitious airs have been employed* from the pneumatic clinic in Bristol in which Beddoes lists 79 patients with various diseases, from amaurosis and ascites over asthma and dyspepsia to palsy, typhus and cancerous ulcers, treated with hydrocarbonate azotic air or hydrogen, in varying doses and regimes.

By the beginning of 1795, IngenHousz reported to Van Breda once more on the hype about the air-therapy by Beddoes, which had become the rage. He still had doubts, but loved that he was kept informed by Beddoes how things went with his second cancer patient in treatment. He had the chance to invest a number of pounds in Beddoes' new clinic to be built in which all rooms would contain modified airs. IngenHousz observed it with a critical eye as the fantastic story of Beddoes unfolded (wonderfully reconstructed by Mike Jay).[106]

This seems to be the only letter to Delft from 1795, and the first of 1796, dated 16 February, might explain why four of the five letters which IngenHousz sent to his nephews in Breda never arrived. They disappeared through the "natural effect of equality and freedom". The postal services had become completely unreliable, it was better to give letters with travellers, preferably travelling via Hamburg. But there was also another reason why IngenHousz may have written fewer letters. He was busy with experiments and compiling his thoughts on strategies and methods to improve the yields of agriculture. This was no idle occupation.

At the beginning of the 18th century the population of England and Wales was estimated (no census was taken) to have reached some five and a half million. The birth-rate had risen only slowly in the first part of the century because killer diseases such as smallpox, dysentery, consumption, and typhus had taken their toll. Shortage of food, inadequate housing conditions and excessive drinking of cheap alcohol had disastrous effects on the poorer classes, while the rich fell hardly less to disease because of to a general lack of hygiene. But throughout the eighteenth century living conditions had been improving, and by the early nineteenth century the population had almost doubled to over nine million,[107] while simultaneously obesity was becoming a common phenomenon among the affluent.[108] But the population was growing faster than the food supply. In 1798 Robert Malthus would write his *Essay on the principle of population*, in which these growing concerns were articulated.[109] This was not a typical English problem. The growth of the population was taking Europe by storm. In the 1770's some 5000 families from the Southern Netherlands emigrated to Central Europe to escape from the growing pressure of the exploding population.[110] The Board of Agriculture's president John Sinclair asked IngenHousz for advice:

> Whitehall 19 July 1795
> ... it is impossible however to effect so desirable an object, without the union of philosophy and practice, and if you would have the goodness to direct your attention to this subject, I am persuaded that it would materially tend to promote the object of the board of Agriculture, as there is no department which it would be more desirable to bring to perfection than the Doctrine of Manures.[111]

106 Jay 2009
107 Black 1996
108 Wilson 2007
109 Jefferies 2005
110 Hasquin 1987
111 Breda, Collection IngenHousz A13

These must have been the first steps on the road to a full membership of the Board as an official document testifies that IngenHousz had become honorary member of the Board of Agriculture in the beginning of 1796.

> Board of Agriculture
> Whitehall, London, February 23 1796
> The which day Dr John Ingenhousz Body physician to His Imperial Majesty, was admitted a Foreign Honorary Member of the Board
> signed by order of the board
> John Sinclair President[112]

Sinclair asked him to write an essay on the food of plants and the renovation of soils, based on the 'memories' he had written for that institution in the past and which had been extremely well received by the board. IngenHousz reported about it to Van Breda.

> *Den President van den raad des landbouws, board of agriculture, heeft my sterk aangezet om iets tot die niewe nationale instelling toe te brengen. in gevolge heb ik eenige memorien omtrend aan hun gezonden, die alle van dien raad goedgekeurd zyn en waar voor ik verschyde zeer honorabele dankzeggingen hen ontfangen. Nu heeft men besloten die memorien te doen drukken op gemeene kosten, en in de werken van die vergadering in te lyven, onder den tytel van essay on the food of plants and the renovation of soils: het bevat onder andere een gansch niew project om in een enkele dag aan het uytgeputte land te geven al dat het, by braak liggen, niet dan in een gansch yaar kan bekomen; zo dat ik, in dit respekt, byna zou kunnen zeggen perargro loca nullius ante trita solo.*

The essay would contain the results of a new project in which IngenHousz had investigated how one could restore exhausted soil in one day better than letting it lay fallow for one year. The archives contained quite a few loose notes and records about experiments he had been performing on the environmental circumstances that influence the germination of seeds, the growth of plants and their formation of seeds. Below a small fragment to give the flavour of this work - in a later chapter, this essay will feature prominently:

> *1796*
> *april 20*
> *Experiments made in Dr Morton's garden at the British Museum with seeds steeped in various liquids, the ground was divided in squares, measuring four feet by four. The soil is not dunged but dug up 6 weeks ago*
> *april 21*

---

112 Breda, Collection IngenHousz A38

exp with grains steeped in oxygenous muriatic water, during 38 h
and with grains steeped in vinegar during 14 hours
or in vitriolic acid so mmuch diluted as to resemble vinegar
20 april
experiments with sprinkling soil with various liquids after putting in the seeds:
oxygeneted muriatic water, bitter salt, glauber salt, salt of tartar, diluted vitriol
and a control square without any liquid
22 april experiments with putting seeds at different depths in sowing[113]

The resulting essay was printed by May and enthusiastically received by many; not in the least the King himself who wanted IngenHousz to perform some experiments in Kew Gardens. These gardens were the work of Joseph Banks, a British botanical hero who had traveled around the world with James Cook in 1769 and who had created the botanical gardens at Kew with the full support of the science-minded King George II. In spite of all these high-society connections IngenHousz felt depressed by all the political adversities. Many of the manuscripts he sent to the continent got lost; books were published without the appropriate corrections (such as a German bundling of articles) and thus contained many mistakes. Letters kept being lost, scientific communication in the European networks in which IngenHousz felt so at home were breaking down. That's probably the reason why only two letters from IngenHousz to Van Breda from 1796 still exist.

He also had less cheerful things on his mind. A note testifies that his age and physical condition were making him prepare for the worst.

vor Mr A J Ingenhousz, to be send under a cover after the demise of Dr
Ingenhousz
Waarde Neve
London den 10 juny 1796
Ik schryf dese nog in volle gesondheyd seynde, maar wetende dat de tyd van
sterven natuurlyk is, heb ik by tyds alle myne wereldlse zaken in order gebragt
...
[uit vrees voor een revolutie in England heeft hij zijn fondsen in een keizerlijke
negociatie in Rotterdam belegd, de coupons zijn in handen van de heer Vander
Hoop]... de rest van myn geld is in handen van de heren Drummond, Charing
Cross, London ...
verder recommandere ik myne Ziel in uwe alle gebeden, en wensch U alle een
goed gedrag, en onder malkanderen een broederlyke eendragt
adieu,
J Ingen Housz

---

113 Breda, Collection IngenHousz A11

He informed his nephews in Breda that he was in good health at the moment but that "the time of death is all too natural". He had put all his earthly businesses in order. Fearing a revolution in England he had invested all his funds in an imperial bond in Rotterdam, his coupons were taken care of by Mr Vander Hoop and the rest of his money was in the hands of Mrs Drummond at Charing Cross in London. He wished his nephews "a good behaviour and brotherly concord.

In May 1797 he wrote to Van Breda about one of the books that had made a deep impression on him:

> Dr Darwin heeft zijn 2de deel uitgegeven <u>Zoonomia or laws of organic life</u>. het is minst ingenieus als extravagant en new. De niewmodische atheismus glanst er sterk in voornamelyk in het eerste deel. het article over de koortse en de materia medica zijn extra ordinaar ingenieus.

Erasmus Darwin published his *Zoonomia*. It was at least as ingenious as extravagant and new. His fashionable atheism was mainly very strong in the first part. The chapter on fevers and medical matters (Darwin was a doctor) was extraordinary inventive. Less fortunate was that IngenHousz no longer received a stipend from the court in Austria and as such "verdient geen stuyver", earned not a penny. Still not all was bad news, on 25 June 1797 he was able to inform Van Marum in a short note that the rules of the Royal Society had changed and foreigners could be nominated without being known in person, but through relying on their scientific accomplishments only. The next election was to take place at Easter of the following year.

And this was not the only way IngenHousz remained well reasonably connected with the world of science. He got mail with queries and interesting questions from the front line of medical research, this time from Beddoes in Bristol:

> from Dr Beddoes 29 jan 1798
> Dear Doctor
> I have a favour to ask of you – or rather (as you delight in doing good) & obliging your friends an obligation to confer upon you. Lord Lansdown may have told you that we are going to have a course of chemistry in Bristol – you who are so great ar devising beautiful exps will I hope tend your assistance in this way.[114]

It is unknown if he replied, but it is a fact that he did not go all the way west to contribute to Beddoes' scientific endeavour. Lack of time and failing health must have been preventing him from doing so. IngenHousz was back in London from another retreat at his country estate on 25 March. He found in the post the

---

114 Breda, Collection IngenHousz A13

Dutch translation by Van Breda of his essay on agriculture. He replied to him that he was very pleased about the translation which he found very faithful. He was now very curious about the effects of its publication. The art of agriculture was a very conservative one and new ideas didn't take root easily. As the essay had only come out in July, it was too late to see many tests of his hypothesis (he uses the word 'object') in the remainder of the year.

The steady stream of correspondence between IngenHousz and Van Breda became a trickle. IngenHousz had hundreds of notes and many urged him to record them into books, but the upheaval in Europe and England made him anxious and indecisive. He wrote the last preserved letter to Van Breda on 2 January 1798. It contained some thoughts on the treatment using muriate oxygené de muriate (potassium chlorate) for ulcera venera. And enthusiastic reports on two books he had recently read: Abbé Barruel's three volume *Memoires pour servir à l'histoire de jacobinisme* and Robinson's *Prooves of a conspiracy against all religions and all governements, carryed on in the meetings of the freemasons, illuminés and reading societies*. According to IngenHousz, these books should have received more attention. They should have been published 12 years earlier.

On 30 April 1798 he wrote to Van Marum that he returned to London especially to vote in the council of the Royal Society on Van Marum's membership. He was happy to tell him that he had been elected, without a single vote against him.

Return of the pox

IngenHousz international career had begun many years earlier as a result of his experience with smallpox variolation. In the 1760's he was one of the prominent inoculators in Europe, speaking up for this effective method to improve public health while demonstrating its effectiveness by inoculating the masses and the aristocracy. Thirty years later, a publication from an obscure surgeon from Gloucestershire brought this old speciality back to his attention. Edward Jenner published *An Inquiry into the causes and effects of the variolae vaccinae* in London in September 1798, based on his 'vaccination' of the eight-year old James Phipps with cowpox serum from the milkmaid Mary Nelmes on 14 May 1796. This has since become one of the icons of daring and breakthrough medical research. In this pamphlet Jenner defended and demonstrated his innovative technique. While variolation or inoculation inserted some (dried or live) smallpox pus through a scratch or a small incision into the skin, Jenner's vaccination used dried cowpox serum - hence the name derived from the Latin for 'smallpox of the cow' *variolae vaccinae*. Jenner was the first to consolidate the long-established folklore about cowpox providing protection against smallpox.

When IngenHousz read Jenner's *Inquiry* he must have experienced this as a threat to the - by then widespread - method of variolation. This method had significantly brought down the number of smallpox victims and he must have wondered if this new method could offer any advantages. As a clinician he must have been both excited about a new method and concerned about its possible unwanted effects.

IngenHousz puts pen to paper and writes to Jenner. On 12 October he firstly reported the case noted by a local Calne surgeon about a farmer who had contracted fatal smallpox from a freshly inoculated son, despite a past history of cowpox affecting the whole family. In a second letter he described a similar case of smallpox after cowpox infection. The ensuing correspondence between IngenHousz and Jenner has been described as ill-tempered and IngenHousz as being aggressive. This was understood as a furious quarrel between someone who felt his fame undermined by a young runner-up and a brash innovative pioneer. As Norman and Elaine Beale aptly demonstrate in their meticulously reconstructed account of what really happened,[115] IngenHousz' main concern was that Jenner proposed his new treatment on very little evidence. It would be dangerous to implement this new method without further large-scale trials.

His finding of several cases that contradicted Jenner's "doctrine" presented enough reason to IngenHousz to be cautious and not to promote the new technique with too much enthusiasm. He ended his letter of 12 October[116] with the sentence:

> *I will make no further observations, as it is far from my wish or my intentions to enter into any controversy with a man of whom I have conceived a very high opinion - Let it suffice, to have communicated to you in a friendly way, a fact which may awaken your attention.*

How more concerned and polite can one be? The polished and elegant style is partly due to IngenHousz character, but also partly because he must have been helped by Lady Catherine Fox, niece of Charles James Fox, unmarried and homeless but relative through his second marriage to the first Marquis of Lansdowne who had taken her in, together with her cousin Lady Elisabeth Vernon, in the late 1780's. Lady Fox was IngenHousz' amanuensis.[117]

In a second letter to Jenner, IngenHousz cited a new case that "invalidates the infallibility of your doctrine". These cases "seem to cast, as it were, a shade on the supposed popular opinion". This growing popularity was even more of a reason for IngenHousz to inform Jenner, in person and not on the public forum, of the possible weakness of his method.

---

115 Beale & Beale 2000
116 IngenHousz 1798b, as reproduced by Van der Pas 1964
117 Beale 1999

In his reply Jenner wrote politely and strategically to this high-ranking eminent opinion leader in vaccination matters. He admitted not everything was known yet and still quite a few things needed testing. Most probably, so he suggested, could the cases cited by IngenHousz be explained by imperfect infection by cowpox, offering not the proper defence against subsequent smallpox. In a second letter, only partly preserved but analysed by the Beales, it becomes clear that Jenner takes IngenHousz' critique to heart and realizes the critical importance of correct cowpox diagnosis. A third letter from Jenner to IngenHousz shows clearly how he traced back one of the rogue cases and he was forced to admit that it had been communicated wrongly. It concerned a case of smallpox infection before a infection by cowpox, definitely proving nothing with regard to the problem and certainly no basis for the criticism by IngenHousz. But IngenHousz did not give up and contacted many of his fellow doctors at the Royal Society in order to collect more cases that would support his critique on vaccination. Cowpox may have rendered patients immune to smallpox, but this could not be said with certainty nor could it be claimed in all cases.

In a last letter to Jenner IngenHousz went to great pains to put things straight and explained his position. His aim was the greater good of the public, not his own fame or name. He claimed to have abandoned the wrong case as irrelevant and the letter about it as not having been written. IngenHousz had no problem in admitting mistakes. But he urged Jenner to think twice before he published more on his method. He wished him success in his "laudable endeavour to promote useful knowledge" and ended this critical letter in a very positive and friendly manner suggesting Jenner should not feel obliged to send another reply, thus calling a sweet halt to this correspondence and giving a last bit of advice from an older colleague to "make haste slowly" (cited in Greek).

This correspondence illustrates how scientific insights are the result of a process in which criticism is just as necessary as encouragements. IngenHousz had been proven wrong in his refutation of vaccination, but his critical attitude had stimulated Jenner to delineate his clinical parameters more accurately. Jenner might since have become known as the founder of modern immunology, but IngenHousz, in his letters, was a necessary catalyst to make this happen. It is worthwhile reading the Beales' paper to understand how the classical account of this so-called bitter discussion between an "old dog" and a "young Turk" has been an inaccurate and biased interpretation based on limited source materials.

This episode again brings to light IngenHousz' attitude towards scientific knowledge and its application in the real world. He demonstrated how a better world with more people in better health was far more important to him than personal glory and power. Being older, and maybe a little wiser, he was mild-mannered in his social discourse, but could combine that with persistence in scientific matters if he thought it necessary.

Alone in England

The wars on the continent kept IngenHousz locked in England. He loved the country and had many friends there, but his boss and his wife were in Vienna. The situation was becoming even more complicated because Austria was involved in warfare too, needing money spending huge budgets on the war machine and cutting the budgets for the arts and the sciences. War taxes were levied on all salaries.

IngenHousz had always been travelling by government order and had been receiving his wages on regular basis through the banking system of the times. Messages started to arrive from the empire's capital that he had to return in order not to lose his salary. He had been away for almost ten years now and to the Austrian administration that was becoming too liberal an extension of his original travel permit. On top of that the financial crisis hit Europe. IngenHousz had invested large sums of money in bonds and enterprises all over Europe, from silver mines in Bohemia to canals in Spain. But huge parts of this capital had evaporated in the social turmoil of the last decade. When the local tax office in England wanted to subject him to tax as a permanent resident of the kingdom, IngenHousz saw the moment of real financial troubles looming. He wrote an extensive letter to the local tax authorities to explain the situation in which he was caught due to circumstances beyond his control. This is the transcript of a draft or a copy of that letter. It not only explains IngenHousz situation in his own words, but also summarises his autobiographical history of the late 1790's.

I undersigned humbly presents to the Committy of aproval the following consideration regarding the tax on income, with which the Parish offices or collectors of taxes of St James parish intend to charge me, being in the erroneous supposition that I remained such a long while in this country .. but few pose to gain money by practising physic. My case is this. I obtained leave from the Emp. Joseph to go to paris and London on purpose to publish several books on philosophical & medical subjects in French, Latin & English languises. two of those books being publish'd i found my self attacqued with a complicated disease (viz gravel and stone in the bladder obstructions in the liver, jaundice, gout) of which my medical friends thought i could not recover. i was obliged to stop my further literary pursuits. in the mean time the horrid revolution (of which the consequences put my own live in great danger, as being body physician to the Queens brother) made it prudent for me to fly from the bloody scene to my own country Holland, from which i went England where i arrived at the later end of october 1789, in a sickly condition so as to be pitied by all my friends. i took advise from sir george baker and other of my medical friends and begun a course of medicine, by which i was slowly, in the course of 3 or 4 years, so far recovered as to hope i should soon be able to undertake the journey to vienna, when i was ordered to return if possible. being much reduced in strenght of mind and body i found my self unable even to republish some works i formerly publish'd in London. though some booksellers offered me some hundered pounds for the reprint.

Considering that by being in this expensiv country i am deprived of the comfort of a wife and family, comfortable house, coach and horses, which the court of Vienna has always furnished me which, i ought rather to be considered as a down right madman, to live here in a scanty lodging of 52 pound a year, without been properly served, in an advanced age ... if my constitution was such as to allow me to undertake the long and tedious in this time of general confusion .. and dangerous journey to Vienna, besides my traveling equipage and luggage is still in Holland whither it is forbidden to me to go

As i reside here under continual orders of returning, and in a continual desire to follow these orders, the laws of nature, which i can not command, keep me still here, and possibly my keep me still stronger, though i risk to loose all my salary of the court, which is nominally about 500 p. but in reality 450 pound, of which i have ceded voluntary for contributing to this necessary war about 2/10 and after wars 1/10 more. the Emperor charged all salaries with a war tax with about 1/10 (i believe 1/5) i reduced my housekeeping at Vienna in proportion as i did my expenses at home. from the beginning of this year the counsel has suppressed the whole of my salary and threathen to stop it for the future if i do not return.

at the same time the Emperor exhausted by the continuance of the war ordered a forced loan of 30 percent on the capital of the bank, to furnish which i was obliged to order all my silverware, jewels of my wife and all superfluity to be sold and even to send to vienna the small sum of money which i had economised here, and was in the hands of mr Drummond at Charing Cross. i have now lately received new intimation to come back under the treat that my salary will be stopt for all ways if i doe not comeback, and it seems more than probable that the finances of the house of austria are reduced to such distress as to make impossible to pay ... and thus unable to pay me and other state creditors.

by all this it appears that i have been charged to pay much more that i would have received if even my salary had been regularly payed.

Mr Drummond the banker who received all my money may testify the above mentioned relation if that was found absolutely necessary and attest that since jan. 1798 i have not received a single remittance to the contrary that i have sent to vienna all that was in his hand. and that he overpayed my account of £143 - 11

i make no difficulty to declare this to a committy of respectable gentlemen, who certainly will not, by divulging it, diminish that degree of respectability which i

have allways enjoyed in this country from the time i was in greater prosperity,
that is to say, before i lost all my income in France, in Holland and in Spain by
the consequences of the french revolution.
i thought, with mr drummond, and respectable men whom i consulted, that
according to art. VIII of the act of parliament i could be charged as having
but a temporary and precarious residence under continued orders of departing,
if my health permits it, and having not exercised any function what ever, by
which i could have gained any thing.
can i be charged with a heavy tax for reason of not engaging a better state of
health, or for having been obliged against my inclination to spend my income in
this country without gaining a single half penny or being at charge to any body.

The drama is clear. Here is a man at word who stands with his back against the
wall and who sees the world crumble around him. IngenHousz was stranded
on an island far away from his family, wife, house and boss, with no money and
in bad health.

It may sound curious that not more correspondence between him and his
wife has been found. However, a few letters from this period have survived. It
is hard to judge the nature of the relationship between Jan and Agatha. There
is only indirect evidence that their marriage might have been unsatisfactory.
In the biography of Lord Shelburne it is stated that "the social charms of the
learned physician were not shared by his wife who seems to have been a second
Xantippe".[118] This must have been at best an indirect judgement, based on hearsay or an interpretation of IngenHousz own remarks, as his lordship could
never have seen Agatha, who never accompanied Jan on his travels.

On 17 July 1798 he sent a letter to Agatha "Ma chere femme" containing the
key to a cupboard in their house in Vienna and the instructions to open it in
the presence of his nephew Joseph, and nobody else, take out the obligations,
the golden snuffbox and possibly some bottles, close it carefully and hide the
key until he would be back. Joseph got a similar letter with the rest of the
instructions. Only a week later another letter went out explaining their financial
situation. All their money in Holland, Spain and France was lost through the
revolution. His obligations in the Austrian bank had lost most of their value,
only an annuity of some 800 florins was left, which would go to her after his
death. The coming years would be hard, but he hoped he would be able, with the
help of her brother and his children, to help her. He wanted her to sell all their
silverware. One could survive with just one fork and one spoon, he was doing
the same. She might have to do what he did in London, abandon his house and
live with friends. She could ask to live with her brother's family. Together with
her brother and with their banker Mr Stametz, she should explore all viable
possibilities to survive these hard times. He ended his letter with:

118 Fitzmaurice 1912

Quoique je ne perde pas mon courage, j'avoue cependant que ce qui arrive depuis quelques annees dans le monde m'affecte tellement, que je ne peux sonder? ni travailler comme je voudrois. Je sens même que ma santé en souffre et je ne vois pas comment retourner à Vienne avant la fin de la guerre. Mon bagage et ma voiture sont à Breda ou il n'est pas possible de venir ou d'en sortir si j'y etois. Si je meure on vous en avertira bientôt. Mon testament est déposé chez mon banquier ici et on vous en enverra une copie authentique de Breda après ma mort.

…

Faites bien mes compliments à toute la famille et dites leur que ma plus grande espérance et mon plus grand plaisir est de les rejoindre et d'être témoin de leur bonheur.[119]

He was not loosing his courage, but he had to admit that what had been happening during the last years in the world was affecting him so much that he couldn't work the way he desired. He even felt how his health suffered and he didn't see how he would be able to return to Vienna before the end of the war. His luggage and coach were in Breda, and if he would be able to get there, he would not know how to get away again. If he died, she would be informed right away. His testament was deposited with his banker who would send her an authenticated copy after his death. … Only remained for him to ask her to greet the whole family and to tell them that his greatest wish was to be with them and share their happiness.

Agatha's answer was written in a very clumsy and poor Dutch. She excused herself for her poor writing skills, but she had forgotten what she had learnt when she was young and never did write many letters. She was 63 and ill herself, throwing up blood, while the letting of blood had weakened her. She couldn't climb stairs anymore and was trembling so hard it took her three days to write this letter.[120]

Teer Beminde Man
Ue briev hebbe ik met groote aandoening ontfangen en verlang niets meer als dat Ue hier komt om in ue ongeluk deel te nemen en dat ik u blyken kan geeven van myne opregte vriendschap joseph is gelyk by Mijnheer Harmetz geweest en die heeft hem gesegt dat hij geen obligaties van Ue heeft en dat hij gelooft dat de tyd tot langer uytstel zal gegeeven worden maar die obligaties moeten in de bank gegeven worden de naam helpt niets als de sleutel van Ue kast komt zal joseph in myn bysyn de goude doos en de obligaties neemen en de kast weder toesluyten en de sleutel meede neemen en by Ue te rugkomst selve te rug geeven

---

119 Breda, Collection IngenHousz B4
120 Breda Collection IngenHousz B5

*U kan verseekert syn dat geen schriften van U sal geleesen worden jospeh sal
het silver in de munt selve verkoopen wyl hy segt dat men daar het meeste
bekomt en niet bedroogen word*

...

*verontschuldigd joseph dat die zo weinig schrijft maar die is ook zo druk,
deze somer is hij met zijn familie en haar brier in Slissing en heeft de tweede
verdieping die stank gehad heeft hij moet in de stad komen om zijn callegie en
weegens examineren waar hij veel tijd verliest van bouwen in Botani Tuyn*

...

Weenen den 24 Augustus 1798
Vrouw de Mr IngenHousz

There was one more letter from Vienna to London in November to inform Jan
IngenHousz that Agatha's brother and his wife have been putting their best
words forward with the people in the administration and the court to secure his
income. But the verdict was clear: anybody in Imperial service abroad had to get
a three-year permission and after that time he was to reapply for an extension
otherwise one would lose his stipend. So the Hofrat urged IngenHousz to
come back in order not to lose all his income.

The next letter dated from the summer of 1799. Agatha was full of sorrow
to hear that her husband was ill again. She was just skin and bone herself, as
she had been ill too all winter. For half a year she had had fevers, an ongoing
cough, no appetite and little sleep. The troubles and misery undermined her
health. But the key of the cupboard had been underway for a year and had
finally arrived by diligence. On top of all that, it had cost 1.4 gilders to obtain it.
She added "please write (to) me again, because it is my only solace to know all
goes well with you". She signed with "Your affectionate wife of Mr Ingenhousz".

To Dr J Ingenhousz
Physician to the Emperor
London
Teer Beminde Man
*ik heb ue briev met groot leedweesen ontfangen en gesien dat ue weeder siek syt
ik ben ook gansche winter siek geweest ik hebbe een halv jaar alle dag de koors
gehad en gehoest en geen slaap ik hebbe niets gebuykt en tog ... maar hebbe
niets meer als vel en been de roog en kommer doet my als ue myn gesondheyd
verliesen.*
*De sleutel van ue commode is een jaar naar dat ue die gesonden hebt met de
deligence gekoomen en hebbe moeten betaalen een gulden 40 centen*

...

Weenen den 13 augustus 1799
U Geneege Vrouw de M Ingenhousz

It was probably the last thing he heard from her. The final remaining letter from IngenHousz' pen left for Vienna on 1 June and was addressed to his nephew Joseph Jacquin.[121] He sounded quite desperate: "j'ai perdu tout mon embonpoint, toute ma bonne couleur de visage, tout mon gout pour l'etude, tout courage et force de corps, j'etois sur le point de tomber d'apoplexie", and described how he had lost weight, the colour of his face, any drive to study, all courage and power, and had become apoplectic. He knew how much his aunt would love to see him back in Vienna, but he was so ill and weak, that even when it would be practically possible to make the trip, he would not be able to do it. Especially when he heard that the pensions were going to be scrapped and that pensioners would have to leave town, he lost all courage and abandoned his plans. What would be the sense in making the long and dangerous journey (via Hamburg and Gottingen, that he already had planned) to arrive in your home town where you were not welcome?

IngenHousz had spent Christmas 1798 in his rooms in Marylebone Street 37. It was a very harsh winter and he was only able to return to his friend in Bowood in March. By that time IngenHousz was in very bad shape, diagnosed by Dr Falconner in Bath as being terminally ill. The Marquess did not want his friend to return to London alone, so IngenHousz stayed in Bowood, writing his last letters to Vienna. He died on 7 September 1799. A letter[122] goes out to Breda from Lord Lansdown in Bowood.

> Monsieur,
> mes anciennes rélations d'amitié avec votre digne et respectable parent, le docteur IngenHousz, me rendent bien sensible au triste evenement que j'ai à vous communiquer.

It is a long and almost stately letter, describing IngenHousz' last months. He stayed at Bowood over the summer, but his health steadily deteriorated. Both Dr Faulkner from Bath and Dr Drew who passed some days at Bowood did not hold much hope. His forces gradually left him but his intellectual faculties remained.

> Jusqu'au dernier moment, il a montré la même vivacité et le même plaisir à répéter quelques expériences philosophiques en présence d'une compagnie nombreuse. Il n'a gardé le lit que deux jours et même dans cet état il ne souffroit point, on l'a toujours vu tranquille et serein, ferme sans ostentation avec le calme de la vertue er de la vrai philosophie. Après un asoupissement de quelques heures, il a rendu le dernier soupir sans se reveiller.

---

121 Breda Collection IngenHousz B6
122 Breda Collection IngenHousz A15

Only recently he and IngenHousz had had a long and intimate conversation in which he had talked with affection about his family in Holland, as well as about Mdme IngenHousz of whom he spoke with "sentiments d'attendrissement et de la plus haute considération". Lansdown had learned from his servant Dominique that everything had been done to leave no debts and to take care that all belongings were distributed honestly and wisely to all involved.

> *Je suis persuadé que c'est conforme aux intentions de votre digne parent que de mettre dans la dernière cérémonie la simplicité qu'il avoit mise dans toutes ses actions. Il serra enterré à Calne comme il l'avoit desiré, avec bienséance et sans faste. Mais je me reserve à moi-même d'élever sur sa tombe un monument simple avec une inscription qui rend hommage à son génie et à son caractère.*

IngenHousz was to be honoured in a last ceremony with the same simplicity as he has always lived. He was to be buried in Calne according to his wishes, with decency and without pomp. He, however, will allow himself to erect a simple monument on his grave with an inscription to pay tribute to his genius and his character. A few days later, another letter[123] left London, from the bankers who took care of the estate.

> to Mr A J IngenHousz Breda
> London 12 Sept 1799
> Drummond
> *It is with great concern we have to communicate to you the death of your respectable relation, Dr John Ingenhousz, which happened at the Marquis of Lansdown's seat in Wiltshire on the 9th. His Lordship informs us that he has written to you on this melancholic event but desires us to inform you that he was interred in the church of Calne in a vault of his lordships attended by his son, Lord Henry Petty, in his place, as he was not well enough to go himself;*
> ...

It was generally believed that he was buried in the graveyard of St Mary's Church in Calne, as the burial register recorded his name on 9 September. The above letter and recent evidence, collected by Norman and Elaine Beale,[124] indicates that he probably was buried within the church in a burial vault, although the exact location remains unknown.[125]

In a document extracted from the registry of the prerogative court of Canterbury, IngenHousz' will (a handwritten copy signed June 10 1796)[126] states: "Will of John Ingen Housz body physician to the emperor I declare the following to be

---

123 Breda Collection IngenHousz A15
124 Beale & Beale 2000
125 Schierbeek 1954
126 Breda Collection IngenHousz 18b

my will in case I should die in Great Britain and to be applicable only for as far as regard the possessions which I may have at my death in Great Britain". The Mrs Drummond bankers were the executors of his will. All his possessions were to be kept until one of his nephews from Breda had come over. All his possessions in Great Britain were left to his nephews. Further he stipulates "to direct my funerals which I desire to be simple and of little expense and I desire to be buried in the church of the parish where I will die". His servant Dominic Tede, native of Parma, should receive three months salary and enough money to travel back via Breda (to pick up everything that was left there) to Vienna, where he would find out which arrangements IngenHousz had made for him. His widow Agatha, alone in Vienna, was to die a year later.

Memorial plaque in Calne, Wiltshire, UK, inaugurated in 2000

rewind

A life in letters, the story continues 1780-1799

IngenHousz' last years reflect the tragedy of a life in turmoil on a continent in chaos. Nevertheless this did not prevent him from persuing his scientific ideals. We see him disappear in the mists of time, almost without leaving a trace. Even his burial place remains unknown.
War, uncertainty and lack of means can be prohibiting for anybody. Even a solid and stable mind like IngenHousz is not invulnerable to the dramatic events in society.

Still, IngenHousz demonstrates in his comments on quacks and charlatans and in his discussion with Jenner that he knows very well what evidence based knowledge is all about. He expresses his doubts or unease with bold claims by people such as Beddoes and is merciless with proven nonsense such as Mesmer's. He shows living proof of some epistemological virtues that have been shown to be the fundaments of trustworthy knowledge.

In his description of the agricultural system some forebodings can be seen of later theories of ecology. Not having words for photosynthesis or ecosystem did not hinder IngenHousz to describe in a preliminary but nevertheless astonishingly accurate way how life on earth is intimately intertwined with its environment.

EXPERIMENTS

UPON

VEGETABLES,

DISCOVERING

Their great Power of purifying the
Common Air in the Sun-shine,

AND OF

Injuring it in the Shade and at Night.

TO WHICH IS JOINED,

A new Method of examining the accurate
Degree of Salubrity of the Atmosphere.

By JOHN INGEN-HOUSZ,

Counsellor of the Court and Body Physician
to their IMPERIAL and ROYAL MAJESTIES,
F. R. S. &c. &c.

———————

LONDON:

Printed for P. ELMSLY, in the Strand;
and H. PAYNE, in Pall Mall. 1779.

Content, contest and context

Jan IngenHousz was a productive investigator and prolific writer, although he only got started late in life. He published his first article in the *Royal Societies' Transactions* in 1774, when he was already 44. In the 25 years that followed, he wrote one book, which was translated in three languages, in editions that were improved and expanded to many volumes. He published tens of articles in the *Journal de Physique*, later on and in different combinations, bundled in French and German editions in various volumes. He wrote and read several letters to the Royal Society, among which two Bakerian Lectures, all published in the *Philosophical Transactions*. His last substantial scientific work was the essay by order of the Board of Agriculture, later translated in Dutch and German and published separately. His work covered a bewildering range of topics: from all things medical, via electricity and plant physiology, to chemistry, technology, physics and agriculture. Reconstructing his life on the basis of his published works and notes he appears as the near prototypical man for the Enlightenment at the brink of the Romantic era. He demonstrates, in Holmes words, a powerful "sense of individual wonder, the power of hope, and the vivid but questing belief in the future of the globe."[127]

The secret life of plants

It was his debut and it took him right to the frontline of scientific innovation. *EXPERIMENTS UPON VEGETABLES discovering their great Power of purifying the Common Air in the Sun-shine and of Injuring it in the Shade and at Night* was to be the key work in his whole oeuvre, very typically exploring the laws of nature on the borderlines between chemistry, physiology and botany. In an undated note on an unattached slip of paper from the archives in Breda, IngenHousz himself described what this book represented, reflecting on it later in his life.

> As no body before me ever asserted that vegetables correct vitiated air and improve good air by the assistance of the sun's light, or that they vitiate or degrade good air when the light of the sun is withdrawn from them, it could scarce fail that such a new doctrine should excite the attention of the public.

---

127 Holmes 2009

The uncommon sudden sail of the book was an unquestionable proof of the impression it made, the whole edition sold out in a few months, and even before any of the reviews analysed it; it was exposed to sail in october 1779 and my bookseller desired me as early as in the month of february 1780 to prepare for a next edition and to send such notes or additions I should wish to have inserted in it. being at that time occupy'd at Paris with the translation of it in french i had no time for preparing a new english edition, and i was even not very fond of undertaking it as i could not my self take care of the impression. a second edition never having appeared in english the copies have been sold till now at an extravagant price and were seldom to be met with.

a french edition much completer than the original english edition came out in july 1780 at Paris; in the mean time i found a german translation of it published at Leipzich as also a dutch one translated by Dr Van Breda an eminent physician at Delft and member of the senate of that city. Thus in the time of little more than half a year four different editions of it have seen the light and have all been received very favourably by the public. The french and german translations have been reprinted several yearce since.

The book was based on a very long and concentrated series of experiments which IngenHousz performed in his country house near London in the summer of 1779. Even whilst preparing the publication of the book, he continued to perform experiments. His log book testifies that he kept on measuring the quality of the air and the effect of plants and plant parts during his trip back to Vienna and long after he had arrived home again. The book turns out to have only been the end of the beginning.

This book deserves a place on the shelf of the classics in the history of science. Not just because it describes a discovery of the first order, but also because it demonstrates how science works, how hypotheses grow in one's mind, how they are tested, how measuring apparatus and circumstances influence methods and results, in short: how one finds a trustworthy answer to an intriguing question, after having formulated the question properly.

And there is more to make this book a rather unique phenomenon. Not only do we have the published version. We can also study his lab-book as well as a copy of the first printed edition with notes in the margin containing his thoughts and comments in view of new editions, probably a second revised English edition, that was to be compiled somewhere during the 1790s, as he refers to articles and books published at that time. There is also an extensive folder of folio pages with notes, additions and amendments in preparation of an upcoming French translation (on which he is working right then) and in view of further editions in English or other languages. On top of that there are the translations in Dutch, French and German. The French version went into a second edition, which was published in two volumes: the last volume appeared in 1789. All these materials allow us to follow the development of his thoughts and how they are checked against reality in experiments and defended against criticisms and accusations, by performing more experiments. Moreover, it also allows us to see what experiments he actually performed, what results he got and what he finally put on paper. In the philosophy of science there has been considerable interest in how science communicates. Science can be manipulative in the autobiographical stories it tells when it creates a mythical image of itself by selective publishing of advantagous results, leaving out the failures. As if scientists know the direct route to the truth. It sells the cliché of the scientist as a hard working nerd who fanatically follows a rigid logical method to arrive at the final eureka-moment, for which he (hardly ever a she) is awarded a Nobelprize. It sounds heroic but does not resemble reality at all.[128] In this respect too, the case of IngenHousz offers the right materials to investigate these and other claims about the workings of science.

What he did & what he wrote

The first entry in IngenHousz logbook of experiments is dated 27 July 1779. These notes will continue almost without interruption, ending with the last experiment, numbered nr 546 and entered in the books in "Paris le 9 juin 1780". The book based on these experiments was published in September 1779 and was only to contain a fraction of the data that IngenHousz ultimately would have collected.

The book[129] itself is constructed almost like a present-day scientific paper. The title presents exactly what one will find inside: the findings and the method to obtain these data. In the preface, one finds the 'abstract'. IngenHousz describes what he is setting out to investigate. To explain this, he refers comprehensively to the history of the problem and what others did before. The Reverend Doctor Priestley is the "excellent philosopher and inventive genius" who discovered dephlogisticated air, among other kinds of air, who developed a method to measure the salubrity of the air, and who discovered that plants correct air fouled by a burning candle. "But in what manner this faculty of the plants is excited remained still unknown", and that's what IngenHousz wants to find out. Many philosophers doubt if it has anything to do with the plants at all (Scheele is one of them) and even Priestley himself is doubtful. IngenHousz indicates why: Priestley was getting contradictory results. Sometimes the plants produce the good air, sometimes they produce no air at all. And in some puzzling cases a green kind of matter had appeared on the inside of the phials in which the experimental plants were growing. In those vessels air was produced even after

---

128 Theunissen 2004
129 IngenHousz 1779

the plants were removed. Therefore the plants could not have been involved in the production of this air. "This far this matter was carried on when I took it up in June last." He suspected that plants were capable of actively changing bad air into good air (as opposed to Priestley's hypothesis that plants absorb the phlogistic matter and leave the remainder as pure air). Then, on page xxxiii, comes the 'abstract':

> I observed that plants not only have a faculty to correct bad air in six or ten days, as the experiments of Dr. Priestley indicate, but that they perform this important office in a compleat manner in a few hours; that this wonderful operation is by no means owing to the vegetation of the plant, but to the influence of the light of the sun upon the plant. I found that plants have, moreover, a most surprizing faculty of elaborating the air which they contain, and undoubtedly absorb continually from the common atmosphere, not real and fine dephlogisticated air; that they pour down continually, if I may so express myself, a shower of this depurated air, which diffusing itself through the common mass of the atmosphere, contributes to render it more fit for animal life; that this operation is far from being carried on constantly, but begins only after the sun has for some time made his appearance above the horizon, and has, by his influence, prepared the plants to begin anew their beneficial operation upon the air, and thus upon the animal creation which was stopt during the darkness of the night; ... that this office is not performed by the whole plant, but only the leaves and the green stalks that support them; that acrid, ill-scented and even the most poisonous plants perform this office in common with the mildest and the most salutary; ... that young leaves, not yet come to their full perfection, yield dephlogisticated air less in quality, than what is produced by fullgrown and old leaves; ... that all plants contaminate the surrounding air by night, and even in de day-time in shaded places; ... that all flowers render the surrounding air highly noxious, equally by night and by day; that the roots removed from the ground do the same ... that fruits have in general the same deleterious quality at all times;... that the sun itself has no power to mend air without the concurrence of plants, but on the contrary is apt to contaminate it.

He ends with: "These are some of the secret operations of plants which I discovered in my retirement, about which I will endeavour to give some account in the following pages." He further warns people that so many factors influence the measurements of that quality of the air, that it is not easy to obtain stable results. That is why he would spend quite a bit of time on explaining the right procedures.

When the book was entirely printed and only the last pages of the preface had to be finished, he added a handful of extra pages (which he mentions and which illustrates the production process of a book in those days). In these pages he announced a finding just communicated to him by Fontana, who constructed a device by which to produce dephlogisticated air (easily and cheaply from nitre) and make it available for breathing by a patient through a tube, while capturing the exhaled fixed air in lime-water. To IngenHousz the medical doctor, this sounded terribly promising, opening up clinical applications to this new science of airs. Years later these ideas would be worked out ambitiously and enthusiastically by Dr Beddoes, with whom he would have intense contact. Here, in his own copy, he wrote in the margin that, however promising it would sound, he first wanted to see the experimental evidence for the claim that inhaling dephlogisticated air was to proof beneficial to certain patients.

Next follows an explication of some technical terms. This concerns mainly the various airs that are mentioned in this book, summarised for those who have not yet read Priestley's book or who are unfamiliar with this new doctrine of airs. He also briefly explains another new concept, the eudiometer, which was invented by Priestley, but perfected by Fontana and named by Ladriani.

> Eudiometer is a new word. It signifies an instrument by which we may judge of the degree of salubrity of the common air. The invention of such an instrument belongs to Dr. Priestley. It consists of a glass tube, divided in equal parts; for instance, in two large divisions; each of which is divided, and each of these ten sub-divided again into ten parts: a glass measure, containing exactly one of the great divisions of the tube. One measure of common air and one of nitrous air, put together in a separate glass vessel, and left by themselves till the diminution of the bulk of the two airs is completed, and afterwards let up the glass tube, indicates at once the exact diminution of the two joint measures.
>
> The degree of goodness of the common air is found to be in proportion to the diminution of the bulk of the two airs.

[in the margin he later wrote]
> if dephlogisticated air is to be put to the taste, he joins two measures in stead of one of nitrous air to one measure of the dephlogisticated air.

The rest of the book falls apart in two parts. The first describes "The nature of plants" and gives a general overview of IngenHousz' theory about plants. The second is "containing the series of experiments made with leaves, flowers, stalks, and roots of different plants, on purpose to examine the nature of the air they yield of themselves, and to trace their effects upon common air in different circumstances". A modern science paper probably would do it the other way round and first of all give an account of the experiments performed with their outcome and only then expand on these results and interpret them to reach a conclusion. Considering the manner in which we often read research papers nowadays (pragmatically going from the abstract to the conclusion, then the discussion and finally, if it still interests us to the data), IngenHousz' sequence might be just as logical.

He opens this second section with a description of his 'methods'. He describes in detail how he worked with the eudiometer, the instrument perfected by his friend Fontana.

In the margin on page 152 he later writes, "for those who are not interested in doing these kind of experiments themselves", a short guide to interpreting the results that will follow:

I *the purer the air under examination is, the greater will be the diminution of the bulk of both airs.*

II *if the quantity of both airs destroyed is found to be between 170 and 210 subdivisions, the air examined is to be considered as respirable air, of the same goodness thereabout as common air.*

III *if the quantity of both airs destroyed is much less than 170 it signifies that the air under examination is s of an inferior quality than common air and so much the less fit for respiration*

IV *if the quantity of airs destroyed is much above 210 subdivisions, the air examined is the be considered in goodness to common air and so much the more approaching to real dephlogisticated air*

In the book he describes the eudiometer in detail, with all the necessary parts, referring to a plate at the back of the book, he also gives a step-by-step description of the handling of each measurement, producing a sort of protocol to follow for anybody who would like to reproduce these experiments. On top of that he gives an extensive list of instructions (which he learned from Fontana) to reduce the various sources of errors. This is a significant part of the book. He goes into a lot of detail in almost thirty pages, ten percent of a book of three hundred. IngenHousz must have thought this rather important. The first error described is caused by touching the glass tube with the bare hand, which communicates heat to it, so that the air inside is likely to expand. The result of this error may amount to two subdivisions. Another error is the result of not roughing up the inside of the glass tube with emery as water drops form on the inside of the glass, which diminish the volume inside the tube. This may amount to at least three subdivisions. Another error is the result of using a phial without a slider to close it. Philosophers who do not use this slider may get a difference amounting to ten or even more subdivisions. The first seven errors he lists add up to 25 subdivisions, which may not sound much, but IngenHousz warns the reader and would-be experimenter, that when one makes the same mistake five times in a row, each time a next measure of air is let up into the large tube, "these errors may, if they were all committed, amount to five times this number, or to 125 sub-divisions". And these errors have got everything to do with the handling. The list goes on. Much depends on the time the two airs are left to react. Body heat or environmental temperature may influence the results. Reading the subdivisions may be done carelessly. Not holding the tube perpendicular can influence the reading. This kind of accidental errors may occur, but a fundamental problem is the quality of the nitrous air, which is used in the measurement. But even more important is the way in which the two airs are mixed and how long one lets them stand before the diminution of the resulting air is measured. In total he lists twenty sources of error. In each case he clearly indicates how he and in the first place his teacher in these matters, Abbé Fontana, have examined how the results are influenced by any of these variables. In the later chapter on the eudiometer, more time will be spent on this instrument and its importance in the development of this new scientific discipline.

After the 'methods', comes the 'data part'. He is not just recounting his experiments in the sequence he performed them, but in the way he could logically introduce his findings. The first experiment is a handful of grass, put in an inverted jar in the sun from eleven o'clock until two o'clock. By the time he was writing the book he knew that he did not have to put plants in jars for 24 hours or longer. In the course of doing his experiments, he learned that plants produce very much vital air in a short time and that's probably what he decided to tell or demonstrate first in his manuscript. By the time he started writing, he knew much more about what he was investigating and decided on the most didactic or convincing way of sharing his story. The first experiment in his book is in reality the 258th experiment in his long run of experiments, executed on 30 August. This is what his notebook says:

> 257 2 *handfulls of leaves of vine were placed in a jarr of a gallon in the sun shine between elleven and one. a good deal of deph. air was obtained of a good quality. one measure of it with 2 meas. of n. a. occupyed 0.45. with 3 meas. 1.83. with four measures 2.81. By Fontana's method 192. 179. 168 ½ 167 ½ 285 res. 415*
>
> 258 *two handfuls of <u>gras</u>, the roots being cut of, in a jar of a gallon full of water in the sun between elleven and two. a great deal of dephl. air was obtained of a good quality. one measure of it with two of n. a. occupyed 0.88 with three 1.73 with four 2.70 by Fontana 190 . 177 . 165 . 167 = 265 res. 435*
>
> 259 *two handfull of willow leaves were again put in a gallon of water in the sun shine from 10 to one. 2 n a 0.85 3 1.75 4 2.72 fontana 196 183,5 171 164 = 255 res 445*

His book,[130] published less than two months later, says the following:

> Exp. I. TWO *handfulls of grass, the roots being cut off, were put in an inverted jar holding a gallon, filled with pump-water, in the sun between eleven and two o'clock; a great quantity of dephlogisticated air was settled at the inverted bottom of the jar, in which the flame of a wax-paper became very brilliant. By the test of nitrous air, according to the present method of Abbé Fontana, the result was as follow: two measures of it being let up in the glass tube, and one measure of nitrous air joined to it, the mark stood after shaking and reposing at 1.92; a second measure being added, I stood at 1.79; after a third measure a 1.68; after a fourth at 1.87½; after a fifth measure at 2.85.*
> [415]

[The margin forms a separate column in which is indicated: Quantity of the two airs destroyed. At this particular place, next to 2.85, stands: 415. In the French edition he gives some explanatory comment on this. See further.]

> By trying it in the other way familiar to Dr. Priestley, the result was as follows: one measure of it, with two measures of nitrous air, occupied 0.88.
> By pushing this manner of trial farther, as I do, the result was this: by adding to the two former measures of nitrous air a third one, the mark was 1.83; by adding a fourth, it marked 2.81. Thus it appears that by this last method the quantity of both airs destroyed amounts to 96 subdivisions less than by Abbé Fontana's method.
> [319]
> 2. Two handfuls of leaves of a willow tree were put in the same way in the sunshine, between eleven and two o'clock. The dephlogisticated air obtained gave, by the nitrous test of Abbé Fontana, the following results:
> 1.96; 1.83½; 1.71; 1.64; 2.55
> [455]
> By the other test it gave the following result: one measure of this air with two nitrous air occupied 0.85; with three 1.75; with four 2.72
> [328]

Remarkable but true: IngenHousz makes a mistake in the first experimental data he publishes about his crucial finding. What he wrote down in his book in which he brought his data out into the public domain, was wrong! Compare the data from his logbook with the ones that were finally published.

---

130 IngenHousz 1779b p 185

The page from IngenHousz' logbook with the infamous experiment no 257

Experiment 257 was performed with vine leaves and was set up in completely the same way as experiment 258, performed with grass. IngenHousz used the measurements from the former in the report of the latter. When copying his data into his manuscript he must have swapped them. All too human a mistake. One starts on the first line of exp. 256 and one continues on line four of 257. Without personel or assistants to check and double check on these things, a mistake like this is easily made, never to be debugged until 230 years later. This error was reproduced in all subsequent editions and versions of this book.

This may be remarkable but not deadly. As anybody can verify, the next experiment IngenHousz recites, is reproduced exactly as he had noted it down. And as far as some sampling in both log book and printed book may indicate, IngenHousz was as meticulous as he appears to be. Moreover, the swapped results are in the same order of magnitude and support his claim just as well.

Still, he knows he is proposing something new, based on new instruments and calculations and he wants to make sure that his readers understand what he is doing. He wants to get his message across as clear as possible. That is why he gives some additional explanation on his calculations in the French edition, which came out a year later, in his own translation. The citation continues in the English text above, after "after a fifth measure at 2.85."

> ... après la cinquième, à 2.85. On voit, par cette experience, que des sept mesures d'air employées, savoir, deux d'air déphlogistiqué & cinq d'air nitreux, il ne restoit que deux mesures & 85/100 de mesure, ou deux cents quatre-vingt-cinq subdivisions, lesquelles étant déduites des sept cents employées, il se trouve quatre cents quinze subdivisions de détruites des deux airs.[131]

If one adds up the amount of air that is left in the glass tube after one has added all five measures of nitrous air, 2.85 measures are left. If one deducts that from the 700 (7.00) with which one stated, the result is 415. That is the result he gives in the margin with every measurement.

The one and main thing that changes in the second French edition, published in 1787, is that he omits the comparison with Priestley's method. He concluded after all this effort to perform every test twice, in order to be able to compare the two methods, that Priestley's was just as good but required more time and effort. Time and experience made it redundant.

Good air and bad air, in all shapes and sizes

He measured the goodness of the air produced by grass and willow, brooklime, thistle, beans, tobacco, juniper, cedar, laurel, pear, apple, artemisia, walnut, oak, elm, ash, strawberry, mulberry, black berry. It shows how many plants he

---

131 IngenHousz 1780 p 201

must have had in his own garden or in those of the neighbours. He traced the differences in the air produced by the leaves of the same plant at different times, during different periods. So he observed that midday, when the sun is hottest, is not the best time for plants to produce dephlogisticated air. He investigated whether the effects of the amount of leaves made a difference (a hundred Indian cress leaves vs three cabbage leaves). Next came the crucial experiments "tending to investigate the quality of air yielded by plants in the night, and by day in dark or shaded places" and in a next section investigated to what degree plants could affect common air in the dark. IngenHousz worked with leaves and whole plants, puts them under water or in plain common air (under inverted vessels in a dish full of water or mercury to cut the communication with the atmosphere).

He got consistent results. Take e.g. the experiment in which he puts an equal quantity of branches of lime, willow, lime, oak, vine and walnut each in a separate jar of the same size, inverted over a plate with water and leaves them there over-night. He measures the air between eight and nine in the morning. He mixes one measure of the air from the jar with one measure of nitrous air and reas:[132]

lime-tree	1.24
walnut-tree	1.25
vine	1.30
oak	1.26
willow	1.23

After this, he placed the jars in a fine sunshine and the air inside was examined again between ten and eleven that morning

lime-tree	1.08
walnut-tree	1.07 ½
vine	1.05 1/2
oak	1.12 1/2
willow	1.07

And he put them in the sun again, for another round of measurements at three in the afternoon:

lime-tree	1.06
walnut-tree	1.05
vine	1.05 1/2
oak	1.12 1/2
willow	1.07

132 IngenHousz 1787 p 214

The same air which the plants had fouled in the night, so IngenHousz concludes, was again restored to its former purity, and even rendered better by some of these plants than common air. The common air he measured at that moment was 1.07.

He also tried all kinds of fruit, from all those fruit trees already mentioned, and showed how they spoil the air. Roots do something similar. Only green leaves and stalks improve the air. He used different types of water, to see if there was any difference between water from a pond, boiled or distilled water, water from a pump or from a source. He put the plants under inflammable air, common air or air already fouled by flame or animal expiration.

An important experiment, although he did and could not see its importance at the time, was when he used water impregnated with fixed air. He produced fixed air with apparatus from Dr Noot or Mr Parker - making sparkling water was a household trick since Priestley had made it popular. He found the resulting air often inferior in quality, though sometimes also better than common air. His provisional conclusion was that the fixed air disturbs the natural operation of the leaves. He was here very close to another crucial aspect of photosynthesis, but did not recognise it. Because he did not know he could have been looking for it. Fixed air (carbin dioxide) did not look like a piece of the puzzle he was trying to solve.

When testing plants and their effect on inflammable air, he was forced to conclude that the nitrous air test failed with this kind of air. It explodes with a loud bang, makes chickens sick, so is not wholesome and still gives sometimes rather good results with nitrous air. He wondered if plants have the power to change inflammable air into some kind of air, which was not known until then. Here he stumbled on another complication which was unsolvable at the time, as it was not even recognised as a complication: the difference between breathable air and healthy air. The nitrous test only made visible that there was oxygen in the tested measure or air. All other gases with their various effects on animal and vegetable life could not be detected by this method. Just to make sure he treated all possible parameters, he also recounted the experiments in which he shows that it is not the sun on her own who purifies the air.

A man of letters, a life in books

When IngenHousz started travelling back to Vienna in the autumn of 1779, he was not only busy doing his salubrity tests on the way there. In his mind he was preparing or organising the French, Dutch and German translations of his work. He stayed in Paris before heading back home and delivered there his home-made translation in French to the premises of printer and bookseller Didot le

Jeune at the Quai des Augustins.[133] At that moment the English edition was already out of print and the German edition,[134] translated by Dr. Scherer, was about to be published in Leipzig. The Dutch translation was being prepared by Van Breda, for printing in Delft. Thanks to a very intense correspondence with his friend in Delft, he reckoned that the Dutch edition[135] would be the most comprehensive and exact.

These translations were not, as we would suppose now, accurate and faithful to the original.[136] All these editions were really parallel productions that told the same story and presented the same experiments and results in roughly the same structure. But each had additions, such as an introduction by the translator, who sometimes added an essay on IngenHousz or on the eudiometer, specifically aimed at the local audience. Or other additions, such as some words of thanks to Fontana for his support and expertise, as well as Mr Eton, the botanist from Kew Gardens, who helped him to obtain the many (exotic) plants for his experiments. He also added that the eudiometer he was using was made by Mr Martin in London and that the same instrument could also be found recently at Mr Sikes, who owned a shop of British made mathematical and physical instruments at the Place Royale in Paris. Another difference was the addition of the official stamp of approval by the registration and censorship authorities in Paris that allow the book to be printed, signed and dated "15 juin 1780".

In the Dutch version the translator Van Breda delivered a preface to introduce the book and its writer to the Dutch public. He sketched the preceding history of the research of airs and quotes Priestley who supposedly said that maybe they had reached the brink of great discoveries of wonderful phenomena in nature. He translated: "These investigations are not just about the air in general but about sublime principles. By working with a trough of water or a saucer of mercury, we may discover the fundaments that are maybe on a par with the discovery that has brought eternal fame to Newton's name." And, could we not, so he continued, say that this philosophical prediction has been fulfilled? As Mr. IngenHousz has succeeded to equal the splendid discovery of Newton, by doing some clever and precise experiments, with nothing more than a trough of water and some bright thoughts.

He further reproduced a letter from IngenHousz written in Brussels on 20 May 1780 in which he authorized Van Breda to take care of the translation, having enjoyed reading Van Breda's translation of his letter to the Royal Society on the electrophorus. He promised to publish for long a second volume in English and already gave a preview of the contents, asking the readers to take this on face value: preliminary and sketchy information, not to be seen as the perfect truth. In a nut shell:

133 IngenHousz 1780b
134 IngenHousz 1780c
135 IngenHousz 1780
136 Bensaude-Vincent 2000

In the second volume to his work he will talk about how plants in ill health lose their capacity to purify the air. How the exhalation of leaves by night lessens during winter, but this is only a change in quantity not in quality. Evergreen trees keep on producing dephlogisticated air, also in winter, although the quantity diminishes. Some of them even keep up this production in the shade. When spring is in the air, the capacity of plants to produce good air during the day and bad air during the night starts up again. Just as some plants are better than others at producing dephlogisticated air, others are more proficient in worsening the air at night. These bad emanations have nothing to do with the scents some flowers produce abundantly. All sorts of moss are capable of purifying the air, all mushrooms however spread bad airs around them. The nightly air produced by plants is a mixture of two kinds: fixed air and phlogisticated air. Both are noxious for animal life. Both are also of the same kind as the exhalations of animals. The mephitic exhalations of plants are extremely important for the state of the atmosphere. This fixed air is a kind of acid vapour, which is an essential element of the common air. (IngenHousz is here very near what Lavoisier would demonstrate some years later.) You don't discern it however, as it is invisible just like water vapour and only becomes visible under specific circumstances. The atmosphere filled with common air is of such a constitution, "designed by Nature", that it is on average good for both plant and animal life.

The other major addition to the Dutch version of the book is the conclusion to part one. It is a philosophical essay in itself in which IngenHousz developed his never halting train of thoughts, based on the many discussions he had with his fellow natural philosophers and on the many experiments he has performed since the English book went to press. He opened with some praise for the new experimental method by which we were able to learn so much more about nature than before. He even blamed the chemists of the past that they did not do more experiments but were satisfied by building systems not based on verifiable facts and evidence but on sophisticated argumentations only. His discovery, thanks to this experimental method, of the salutary effect of plants on the atmosphere, which makes the life of animals possible, was a source of great wonder. The indifference by which the vegetable kingdom had been treated was wholly undeserved. He was very impressed by the astonishing harmony between all parts of nature, destined to the conservation of the Whole, and magnificently created by the Almighty. He was also amazed to realise that man is the only organism on earth that is clever enough to understand this intricate coherence.

But after all those deep thoughts he returned to the hard reality. He disagreed with Fontana who claimed that the common air is at its best in high summer and in deep winter when either plants are producing the most dephlogisticated air or when the cold is limiting all processes of rotting and fermentation that produce foul air. This idea would explain why most illnesses are recorded in spring or autumn, before the leaves are fully developed and before the winter calls a halt to all processes of fermentation, and this especially in low laying regions with marshes. IngenHousz tended to be cautious and wonders if the eudiometer would help to determine this phenomenon. The only thing he knew for sure was that when he measured the air in Passy near Paris in January and February 1780 when it was very cold, he found the air to be of very good quality. When the temperatures started rising, the quality diminished. (And he added in a footnote that it was not the temperature of the water in his eudiometer that made the difference, as he compared warm and cold water. He never forgot the rigid elimination of irrelevant parameters.) He tested the air during the whole of his trip, hoping to find some evidence for his medical intuition that the quality of the air we breathe is detrimental for our health. Even the best foods cannot prevent people from becoming ill in countries with unhealthy air, while we can survive on sober food in very pure air. He held a plea for ventilators, on ships or in buildings such as hospitals and jails to refresh the air inside. In the countryside, marshes or a lack of trees and crops were a certain recipe for bad health. He compared the planes around Rome with the hills of Tuscany and how easily one can become ill in the Roman capital and how the reclaiming of the marshes in Jamaica or Pennsylvania improved the health of its inhabitants.

Next problem however was the production of this neat and healthy air. One can't rely on plants, so we should look for a chemical way of producing it in useful quantities. Metal calcs might offer an interesting approach because they contain plenty of this air. Lavoisier deserved all honour to have determined that in the forming of these metal calcs, they increase in weight. This is one of the Achilles heels of the phlogiston theory and IngenHousz was clearly on Lavoisier's side, although he finely pointed out in a footnote that Lavoisier demonstrated it beyond all doubt but that it was Jean Rey who described it for the first time in 1680. How to extract this favourable air from oxidised metals (because that's what these calcs are), that was the pragmatic question that bothered IngenHousz. And maybe saltpetre was the best alternative.

Wij verteeren, by elke gewoonlyke inädeming, omtrend 30 cubic duimen luchts, het welk 450 cubic duimen in ééne minuut uitmaakt; onderstellende, dat wy in ééne minuut 15 maal inädemen: wy hebben dus in één uur omtrend 27000 cubic duimen luchts nodig; en dus volgd uit deeze bereekening, dat één pond salpeter, ten naaste by voor een half uur, gedéphlogisticeerde lucht genoeg zoude opleveeren...[137]

---

137 IngenHousz 1780b p 150

He calculated, on the basis of the estimate that if one needs 450 cubic thumb of air per minute to respire, one pound of saltpetre should be enough to deliver enough oxygen for half an hour, taking 27,000 cubic thumb as the amount of air needed in one hour. But as the air would be so pure, it should be possible to breathe it several times, as this air could take up many times the phlogiston that is loaded from our blood onto the air in the lungs. The blood unloaded not only phlogiston but also fixed air, and that air could easily be captured if we let the breath go through a vessel filled with chalk-water that would absorb it. The amount of dephlogisticated air for half an hour could then be used for maybe seven or eight hours, definitely according to Abbé Fontana, who communicated this to him. Fontana already had sketched what this breathing-machine could look like, using partly the technology used in the vacuum pumps and other pneumatic instruments. IngenHousz suggested in a footnote a possible improvement by way of using tubes from 'gom elastique', as he has seen these caoutchouc tubes in Paris where Dr Bernard, surgical instrument maker in Rue de Noyens, uses them as hollow probes for various applications.

Still, there was a long way to go before this technology could be put into practice, as first one had to understand in which vessels to brew this air and, even more importantly, for which diseases this treatment could be useful. Although

> alle proeve schynen aan te toonen, dat men er de voordeeligste verwagtingen van hebbe moge. - Laat ons de beslissing aan de ondervinding, by het ziekbedde, overlaten.[138]

Although all trials until that time seemed to indicate that it could be beneficial, but one should leave the final decision to the clinical experience, at the bedside of the patient. He had met already more than one physician who doubted the universal advantages of this purified air. Animals inhaling this air seemed fortified and full of vigour, but their hearts seemed to beat faster and they are dying more quickly, as if their body was wearing out faster. And he warned:

> Laat ons echter niet trachten de natuurlyke perken van ons leeven te verschuiven; maar zulke vruchtelooze poogingen overlaten aan lieden, die, vol van dwaze inbeeldingen, onöphoudelyk het Pharmacum Immortalitatis, het algemeen Geneesmiddel, en den Philosophischen Steen zoeken. - De perken van ons leeven zyn door de zelve Oorzaak gesteld, die ons het bestaan gav; en de wetten der natuur zyn onveränderlyk.[139]

Let us not try to move the natural limits to life and leave such impious attempts to those people who, full of silly fantasies, can't stop searching

---

138 IngenHousz 1780b p 158
139 IngenHousz 1780b p 161

for the Philosopher's Stone, the pharmacon for eternal life. The limits to our lives are determined by the same Cause, as gave us our existence, and the Laws of Nature are unchanging.

IngenHousz' trust in the power of medicine didn't stop him from being critical about bold claims by wild charlatans, nor losing sight of the essential ethical consequences of new technologies. Although those could have been at least partially inspired by his belief in the Almighty.

intermezzo

The birth of the electroshock

IngenHousz scientific explorations led in many different directions. One of his interests can be seen as the origin of electro-convulsive therapy.[140] His friend Franklin was one of the theoretical and practical pioneers in electricity research. IngenHousz had been doing electrical experiments already at the time he was practicing physician in Breda. It was a common field of interest that would be a topic of fascination in their correspondence and conversations. IngenHousz was one of the most prolific and enthousiastic promotors of Franklin's theory on electricity.[141] They found electricity as a curiosity and when they left it, it was a science. But playing around with electricity proved not to be without danger. Many of these early electricians received powerful shocks from their charged Leyden jars. These were glass bottles, covered with foil and filled with water of lead shot. They could be charged and used as a primitive battery unleashing the charge when connecting the wires coming from the inside and the outside of the jar. This created a jolt of electricity and Franklin had more than once been knocked down by this discharge, suffering from retrograde amnesia about the incident itself. In addition, he witnessed a patient's similar accident and later 'shocked' six volunteers experimentally to show how humans could endure shocks to the head without serious ill effects, other than forgetfulness. He did not go any further though than registering these facts, not seeing any practical aplication for this knowledge.

Almost thirty years later, in 1783, IngenHousz wrote a letter to his friend in Paris to inform him that he had suffered a strong electrical shock to the head, which had severly affected his memory and his cognitive faculties (unit 209, volume 40 in the *Papers of Benjamin Franklin*). He described how he felt unusually elated the next day. And he wanted to know more about "the circonstances and consequences of the two electrical explosions, by which you was hit by accident and struck to the ground."

---

140 Finger 2006
141 IngenHousz 1778

He recalls:

The yarr by which I was Struck, contained about 32 pints. It was nearly fully charged when I received the explosion from the conductor supported by that jarr. The flash enter'd the corner of my hat. Then, it entered my forehead and passed thro the left hand, in which I held the chaine communicating with the outward Coating of the yarr. I neither saw, heared nor [sensed?] the explosion by which I was Struck down. I lost all my senses, memory, understanding and even sound judgment.

My first sensation was a peine on the forehead. The first object I saw was the post of a door. I combined the two ideas together and thought I had hurt my head against the horizontal piece of timber supported by the posts, which was impossible as the door was wide and high. After having answered unadequately to some questions which were asked me by people in the room, I determined to go home..; yet I was more than two minutes considering whether, to go home I must go to the right or the left hand.

having found my lodgings, and considering that my memory was become very weak, I thought it prudent to put down in writing the history of the case. I placed the paper before me, dipt the pen in the ink, but when I applyed it to the paper, I found I had entirely forgotten the art of writing and reading and did not know more to doe with the pen, than a savage, who never knew there was such an art found out.

He went to bed and woke up the next morning with a headache and a red mark on his forehead. He felt strangely elated and believed his mental capacities were now significantly better than they had been before the accident.

My mental faculties were at that time not only returned, but I felt the most lively joyce in finding, as I thought at the time, my judgment infinitely more accute. It did seem to me I saw much clearer the difficulties of every thing, and what did formerly seem to me difficult to comprehend, was now become of an easy solution. I found moreover a liveliness in my whole frame, which I never observed before.

He felt so good that he wrote to several "London mad-Doctors" to convince them to try administering electric shocks to the heads of some of their patients. Cranial electricity might be "a remedie to restore the mental faculties when lost". He suggested it might be good therapy for melancholia (what might well have been what we now call depression).

Franklin wrote back to IngenHousz to tell him about his experiences, referring to his book on electricity, and that he too thought it worthwhile to conduct some trials on mentaly ill people. He told IngenHousz that he had recommended the treatment to a French medical operator. It was after

IngenHousz' and Franklin's reports and proposals that the use of cranial electric shocks for treating melancholia started to be used on a wide scale.[142] "Because the idea was generated and presented by two of the greatest minds of the Enlightenment, this proposal was probably further circulated by word of mouth."[143] It was the beginning of a long history of clinical research that would finally lead to the electroconvulsive therapy we know today. The equipment has been greatly improved, but the retrograde amnesia is still recognised as one of the side-effects.

Some "farther considerations" on plants and animals

While IngenHousz was in Vienna the fifth volume of Priestley appeared in London and comments in a London magazine criticized IngenHousz claims. To refute these and further reinforce his theory, he wrote a letter to the Royal Society, read 13 June 1782. "The influence of the vegetable on the animal kingdom" opens with:

> Upon being informed, a few months ago, as well by private letters as from the Critical Review, that my doctrine was over-turned by the fifth volume of Dr. PRIESTLEY, and by an experiment quoted in the book of Mr. CAVALLO on Air; I invited some of my friends here to assist at some decisive experiments, of which I will give here an exact account. I told them the whole result which was to be expected from them, if my system was founded on nature, explaining to them before-hand the theory of the results, and promising, at the same time, that, if the result should fail, I should myself be the first to discredit my own system. The experiments are the following, all made in the hot-house of the Botanical Garden in the winter of 1782.[144]

It may sound as a casual introduction to the report on some experiments, but IngenHousz here displays the workings of science in its simplest of forms: I have a proposal for a theoretical explanation of some aspect of nature; if I am right, I can make such and such prediction; this is my hypothesis (if a and b then we will see c happen); I set up an experiment in which a and b, so in a moment we (he invited witnesses) will see c; if not c, then there is something wrong with my theory and I have to think again. Crucial is that he did this beforehand. In contrast to Priestley who often seemed to be doing experiments at random and tried to understand the results afterwards, instead of formulating the hypothesis to be tested upfront - or at least was pretending to do so.

---

142 Beaudreau 2006
143 Lanska 2007
144 IngenHousz 1782

The experiment was straightforward. He filled six vessels with water boiled for two hours, poured it in while hot. In two vessels he put *Conserva rivularis*, a waterplant. In the next two vessels he put pieces of wool and silk cloth, in all colours, tied to pieces of corks. In the two remaining vessels he put nothing. He took a seventh vessel in which he put fresh pump-water and some *C. rivularis*. He inverted these vessels and immersed their orifices in quicksilver, so effectively isolating them from the atmosphere. He got very good dephlogisticated air from the plants after some days, no air from the cloth, and no air from the plain boiled water, not even after two months. In the seventh control vessel he saw air bubbling up on the same day already, much faster than in the vessels with boiled water. His explanation was that the boiled water took up the first dephlogisticated air produced and after it was saturated, one could see the air gather in the vessel. In the pump-water already a lot of dephlogisticated air was naturally dissolved, so the air started to gather above the water much quicker. The plants all began to wilt after a few days, according to IngenHousz because they didn't grow in an atmosphere with an excess of dephlogisticated air, while the water had lost all its stock of common air which the plants needed to survive. Replacing the water with fresh water prevented this and allowed the plants to continue producing dephlogisticated air (as he had shown in other experiments). In fully exhausted water, plants and the green matter stop producing dephlogisticated air "because this water has lost its own natural air, and together with this air the nourishing and phlogistic particles which are necessary to keep up the full vigour of the plants." He continued by neutralising the criticisms on his theory as if the air he found was dislodged from the water, as Cavallo, Priestley &co maintained. He did experiments in which he isolated a plant in a glass tube with common air, sealed from the water in which its roots stood with wax. The plant produced dephlogisticated air as well. And for those who did not yet believe him:

> *I should advise them to be present, at least once, at the most beautiful scene which they will behold, when a leaf of an agave Americana, cut in two or three pieces, is immersed in a glass bell or jar full of pump-water, inverted and exposed to the sun in a very fair day in the middle of the summer, when this plant is in full vigour; and when they shall have seen those beautiful and continual streams of air, which rush from several parts of this vegetable, principally from the white internalm substance of it, I will be answerable for laying aside all further doubt about the truth of my doctrine.*

It was a glorious vision as he saw in these streams of bubbles the "means to preserve from destruction the living beings which inhabit our earth".

The economy of plants

In June 1784 IngenHousz published an article in the *Journal de Physique* entitled "Réflexions sur l'économie des plantes", later bundled in his *Nouvelles experiences et observations*.[145] This reflects nicely how he was transposing what he had learned in Edinburgh about the 'animal economy' (the name for physiology) to the vegetable kingdom. By some straightforward reasoning he came to the conclusion that the air was the most important factor in the nourishment of plants. They expose much more leaf surface to the air than roots to the soil. They cannot be totally dependend on the soil, as many plants (can) survive without soil but not without air. Water couldn't be the prime nourishment for plants, only the carrier of nourishing substances. Water in itself did contain only very little nutritious particles. Even more, distilled water did not contain anything and was sufficient to let plants grow and flowers bloom. Which did not imply that they didn't need water or soil for other reasons. But his conclusion was firm: plants eat air.

Together with the atmospheric air, plogiston entered the plants, and maybe other nourishing principles that were contained in the air. The leftovers of this food was damaging to the constitution of the plants and was expelled, with the help of the light of the sun.

> *Le phlogistique, toujours inhérent à l'air atmosphérique, & peut-être encore d'autres principes qui y sont contenus, y demeurent comme nourriture de la plante. Le reste étant privé de ces principes nutritifs, devient inutile ou superflu, ou nuisible à la constitution du végétal: mais la plante ne sauroit se défaire de ce superflu, au moins dans l'état d'air déplogistiqué, qu'à l'aide de la lumière du soleil.*

Phlogiston, always part of the atmospheric air, as well as maybe other principles of the common air, are retained in the plant as nourishment. The rest of the air becomes superfluous and even injurious, so the plant gets rid of it, as dephlogisticated air, with the help of the light of the sun. Again, IngenHousz proved to understand the mechanism of photosynthesis perfectly well, without ever using the word. Plants, he continued, do not have reservoirs to stock these by-products as animals do. They do not have alimentary tracts, stomachs or intestines, so they have adapted the whole of their surface to imbibe and expel all they need from the air, in a continuous process. He assumed that two streams of fluid never stop flowing in plants: the flow of air entering and leaving the leaves, and the flow of water, pumped up through the roots, and partly taken in through the leaves from the aerial humidity (especially in plants living in arid climates). One of the airs that were expelled by the plants was the fixed air they produced in the darkness. This fixed air was heavier than atmospheric air and

---

145 IngenHousz 1785b

descended towards the earth. He conjectured that this fixed air is taken up by the soil where it can serve as food for the plants and would be very favourable for all animal life, getting rid of this noxious gas.

> ... selon mon opinion, cet air fixe, dégagé de l'air commun, imprègne la terre d'un principe de végétation qui met les plantes en état de continuer à nous rendre le service important que nous en derivons, c'est-à-dire, à entretenir l'état de l'atmosphère dans un dégré de salubrité nécessaire pour notre conservation.

He also mused a little on the controversy with Senebier, but mainly added a few more hypotheses on the physiology of plants on the basis of further experiments and reflexions. He thought plants produced more air, both during the day and at night, than he had been able to measure until that time. Plants covered with water or mercury were not in their normal habitat and encouter probably too much pressure on their leaves to push out the air they produce.

He also mused about the fact that there would come a day when people would be bewildered that people could deny that plants produce air, just because that air was invisible. He had difficulties understanding how people in his time kept denying the simple facts. He repeated again his standard experiment. Take a tube, fill it with mercury, inverse it over a bath of mercury (and one has a barometer), in which one enters a plant and puts the whole set-up in the sun. One would soon see how air starts to press down the mercury. This air was produced by the plant. On measuring the air, one found fixed air. (The plant did not have any atmospheric air to photosynthesise into oxygen.) When one repeated the experiment with some common air in which the plant could aspire, one saw the air diminish to start augment again after some time. The air that was then found is dephlogisticated air (in the sun) or fixed air (in the dark). He paid a lot of extra attention to this mephitic air that was produced during the night as this idea got a lot of resistance among physicians. He had to admit that he had been quite astonished when he first saw the nightly effect of plants, but after so many experiments, he was convinced that that was the way nature worked.

But he was pleased with all these critical reactions.

> De la collision des opinions résulte souvent la lumière; & en tous cas, les connoissances ne sauroient qu'y gagner. Je ne puis donc voir qu'avec satisfaction mon système contrarié par les hommes respectables & les plus éclairés, tels que MM. Senebier, Priestley & Cavallo. Ces deux derniers nient absolument toute émanation d'air des végétaux, même au soleil.

On the other hand he was pleased that so many refused to accept the fact that plants produce mephitic air at night, because it proved that they never saw, and were still unable to see, that the influence of the sunlight was crucial in this process. Light was the secret behind the 'vegetable economy'.

> J'obstinai à chercher, de toutes les manières imaginables, l'objet qui m'avoit occupé depuis tant d'années; & en cherchant, même dans les ténèbres, ce que j'espérois d'y pouvoir trouver, la lumière à la fin d'ouvrit à mes yeux.

For years he researched this object of his curiosity in every possible fashion. He even went looking in the darkness, which is where he found the light to open his eyes.

Ten years after the first edition

IngenHousz' first book would see many more print runs. The German version came out in a new revised edition in Vienna in three volumes between 1786 and 1790, again translated by Scherer. The French saw a new edition, of which the first volume was published in 1787 and the second in 1789. This intense translation activity was typical for the eighteenth-century chemistry.[146] IngenHousz took care of this translation himself and it therefore gives us a good idea of how his ideas about his discovery developed over the years.

Volume one is more or less an updated re-issue of the first edition of 1780. The second however shows great advances in his thinking. It should be noted that it was published in the same year as Lavoisier's landmark publication in which the new chemical system was publicly presented in full. That is one of the reasons why it is well worth taking a closer look at IngenHousz' book. One of the reasons why he wrote this second volume is that he had, after nine years of often vehement opposition against his "doctrine of nocturnal emanation of mephitic air by plants", no reason to change his opinion. On the contrary, he had amassed in the mean time only more experimental facts that confirm his first hypotheses. As he said before: one single observation is not enough to build a theory. One needs repeated comparative experiments in order not to be deceived by a single pure chance event.

> Je ne suis jamais contenté d'une seule expérience pour en tirer une conclusion, & j'ai eu soin d'y joindre toujours des expériences comparatives, sans lesquelles on s'expose à prendre un pur accident pour une règle générale.[147]

This book is an assembly of "mémoires" and some overlap in subjects is inevitable, so he warns in the preface. Two connecting themes run through the

146 Bensaude-Vincent 2000
147 IngenHousz 1789 p iv

book: the dispute with Senebier and the new chemistry from Paris. The anger about the former and the excitement about the latter turned the book into an endless torrent of ideas that overlap and digress needlessly. Some editing would have done wonders. But it shows the passion that was driving IngenHousz' pen. The row with Senebier had become almost an obsession, especially because he thought Senebier was not using the appropriate scientific arguments and in doing so was damaging IngenHousz' reputation and harming the progress of knowledge. IngenHousz excused himself to his readers for all this unfruitful bickering and uninteresting palaver. More about this dispute in a later chapter. More important for the development of IngenHousz' scientific theory is how he handled the new insights that came from France under the flag of the 'New Chemistry'.

In his Avant-Propos he describes how he came to his investigations, put on the right track by Priestley. He tried to find out why the Reverend Doctor failed to find uncontradictory results. By systematically investigating all possible parameters involved he came to the conclusion that plants improve the air in the sunlight and make it bad in the absence of this light. It showed him how plants live by 'eating air'.

> *Il paroit donc que toute leur substance dérivé de cet océan immense qui enveloppe la terre; en un mot il paroit probable que la plante est un air transformé, metamorphosé.*[148]

He was formulating the hypothesis that plants are a form of air, transmuted in an as yet unknown way. He even dared suggest that humans and all animals, as they are eating plants, were equally built from the same substance. This conjecture seems to accord with the chemical analysis that can be made of these organisms. The new chemistry offered the right framework to understand this continuous transformation of elements.

> *Le composition & décomposition de l'eau appuyées par les MM. Cavendish & Lavoisier sur les expériences qui seroient difficiles à invalider, paroit ajouter de la probabilité à ce système. Quoi qu'il en soit, il n'est pas moins vrai que l'air peut être consolidé & entrer comme une espèce de ciment dans la composition des corps les plus durs.*[149]

Whatever the true nature of water and air, he continued, they play a main role in the life of plants, although this role has not yet been clarified. And while Lavoisier's system got more admirers every day, his principles already permitted the deduction of some theoretical reasoning. According to this system common

---

148 IngenHousz 1789 p xxxiv
149 IngenHousz 1789 p xliv

air exists of 28/100 of vital air, of 72/100 of air which is totally incapable of sustaining life, even for the shortest of moments, and of a very small portion of fixed air. The basic element of the vital air was oxygen, which allied itself with either hydrogen, to form water, or with carbon to form fixed air.

> Les végétaux dont la substance a beaucoup d'analogie avec celle des animaux, soutirent de l'air commun, de même que les animaux, ce principe du feu, ce calorique, comme M. Lavoisier l'appelle: il ne reste donc que la base ou l'oxygène, qui, en se combinant avec les principes que je viens de nommer, forme l'air fixe. La plante donc ne trouvant plus dans cet air ce qu'il lui faut pour vivre, doit y périr à la fin, si elle y est enfermée dans un endroit ombragé ou obscur; mais elle s'y soutiendra très-bien au soleil; parce que la lumière de cet astre, qui paroit être la matière de feu, le calorique dans toute sa pureté, & mise en mouvement, entre dans la substance de la plante...
>
> ...
>
> Ce pouvoir singulier de s'assimiler, pour ainsi dire, la matière de la lumière, ne réside dans sa perfection que dans les seules feuilles: les racines & les fleurs sont à cet egard comme des animaux; ils ne participent en aucune manière de cet vertu.[150]

"Plants that are able to capture the matter of light, a capacity which is built into their leaves...", would there be a better way of describing photosynthesis? Only, IngenHousz had no inkling of an idea about the existence of photons or the actions of chlorophyll inside the leaves he was studying. The leaves were a black box: the only thing known was that gases went in and out through pores in their surface.

Here, whatever controversy may exist about other things, he referred to work by Senebier who had "thrown some great and new light on the wilting of plants". Science is science and anybody contributing was welcome. And one could be right on one point and mistaken on another. A theory should follow the facts, not the other way round:

> Les théories doivent êter appuyées sur des faits bien observés; & si une théorie déduite des expériences particulières, ne s'accorde pas également biens avec d' autres faits observés avec grand soin, il faut changer cette théorie, & non pas nier les faits.[151]

He reformulated his findings. He was able to assert without doubt that his experiments had shown how plants purify the air in the sun and mephitise it in the shade or darkness. But one could just as well say that plants caused the

---

150 IngenHousz 1789 p xlv
151 IngenHousz 1789 p xivj

deterioration of their proper element, the air, (as all organisms do), except in the sunshine. They grow and, in doing so, use oxygen, like all other organisms. What makes them special is that they can restore this oxygen with the help of the sun. One can feel how IngenHousz got more and more into the essence of his findings and felt at home in this world full of wonderful coherence between animal and vegetable life. Thanks to the new chemistry he was able to understand this better. On the next page he does it again, giving a little "précis", a neat little overview, of the new system with regards to the vegetation, for those who reject the phlogiston doctrine:

> Ceux-ci qui disent que l'eau est décomposée par les forces de la nature dans les végétaux, surtout par l'influence de la lumière solaire. La base de l'air inflammable ou l'hydrogène contenu dans l'eau, se combine avec la substance charboneuse pour former de l'huile, tandis que la base de l'air vital ou déphlogistiqué ou l'oxygène, autre principe de l'eau, s'unit également avec la substance charboneuse, & forme de l'air fixe ou acide carbonique, qui entre dans la composition des acides végétaux. Une partie de cet oxygène s'unifiant au calorique, est chassée au dehors & fort par les feuilles dans l'etat d'air vital, sur-tout lorsqu'elles sont exposées au soleil. Ainsi, dans ce système, l'eau & la substance charboneuse sont presque les seules principes de la végétation.[152]

In other words: those who reject the phlogiston theory thought that water was decomposed by the forces of nature in the plants under the influence of the light of the sun. Water falls apart in inflammable air of hydrogen, which recombines with carbon to form organic oil. And the vital air, dephlogisticated air or oxygen recombines with carbon to form fixed air or carbon dioxide, which enters in the composition of vegetable acids. Part of the oxygen links up with caloric (an as yet vague substance, postulated by Lavoisier to explain heat effects in all these chemical reactions, and to some just as mysterious as the old phlogiston) and forced out of the leaves when exposed to the sun. In this system water and air were practically "the sole principles for vegetation".

Not everybody agreed. But there was no way of deciding where the oxygen that plants produced came from: from the water they absorbed through their roots or from the carbon dioxide they inhale through their leaves? IngenHousz had no final verdict to deliver. He gave the example of De La Métherie who doubted the carbon could come from the atmosphere, as there is so little carbon dioxide in the common air. IngenHousz could only finish by saying that time and more research would tell.

The rest of the book did not bring much new. No more pages full of data, as in the first volume. But everything in a more extensive framework, as IngenHousz can now combine a lot more experimental data, also on the growth of plants

---

152 IngenHousz 1789 p xlviij

or the germination of seeds. The book is one long and meandering exposition of what he proposed in the foreword, built from chapters that are only loosely linked. One feels that IngenHousz was still up against a lot of people who were not convinced yet that plants actively absorb air, transform it and expel it. Again and again he referred to experiments that show how undeniable these facts were. Still, he found the interactions between airs and plants intricate and sometimes difficult to understand. While a lot of his results are uniformly indicating some general rule,

> je rencontre quelquefois, parmi mes notes (toutes dressées dans le temps que j'ai fait les expériences), d'autres faits, qui me paroissent jetter du doute sur ce que je croyois pouvoir conclure des autres. [153]

When he looks into his notes, he found results that are inconsistent, and he did not know why. He feared that the phenomenon he was studying is so complex, with so many determining factors that he was not able to oversee them all. That's undisputable, as we know now that it took a century or more after IngenHousz before plant physiologists started understanding the complex system of photosynthesis. Moreover, his measuring methods were rather crude, to say the least. The fact that he realised as much is a big plus. Slowly but surely he was able sometimes to separate the wheat from the chaff. He realised e.g. that there is a difference in the consumption of airs between germinating seeds and flowering plants. Without oxygen, seeds don't sprout and plants don't grow. Without light and carbon dioxide plants wilt. Plants are organisms that clearly need different things at different stages of their life, at different moments of their twentyfour-hour cycle. The temperature influences the proper function of plants, too. Too much heat in the middle of a hot summer's day incapacitates the production of oxygen. Locking up plants, even in big glasshouses, in too much heat and without refreshing the air, is deadly for them. He took his time for digressions, inspired by leafing through his notes of all those experiments from the past years. Sunlight reflected via a mirror is often more beneficial for the plants and their oxygen production than direct light, which gives too much heat. He remembered the summer of 1783 when from 18 June on a light mist (the Laki dust cloud) shielded the sun for days. This lowered the temperature from the otherwise fierce sun, yielding better results in the production of dephlogisticated air by the plants in his vessels. His logbooks turn out to be an invaluable database in which he was able to discern patterns, which became clear looking backwards and knowing more in 1789 than he could have ever imagined in 1779. A series of experiments from July 1781 suddenly started to make sense. The temperature influenced the activity in the leaves, just as well as the clarity of the incoming light. This made him wonder if it would be possible

---

153 IngenHousz 1789 p 13

to make artificial light of the right intensity by which one could make plants produce oxygen by day and by night.

And all the time, IngenHousz was playing around in his mind with the possibilities offered by the new chemical system to understand the relationships between soil, air, water and living organisms. A fact he was struggling with the role of phlogisticated air, azote or mofette in Lavoisier's terms, or nitrogen (N_2), making up 72 percent of the atmospheric gases. It does not contain oxygen and he couldn't imagine how it reacted with the other elements. It had to remain a mystery for a long time coming. He wondered about the role of fixed air in the grand scheme of things.

> *Est-ce que l'air fixe, en se répandant dans l'état de nature parmi l'air commun, se précipite réellement vers la terre, & s'incorpore, comme un sel, avec l'humidité du sol? ou se change-t-il en air commun?*

The cycles of carbon dioxide through the atmosphere, the oceans and the soils would take a lot more time to unravel, but IngenHousz was hinting in the right direction. He referred to Senebier again, for whom fixed air plays a major role in the vegetation of plants. Senebier claimed that rain falling from thunderclouds contained a lot of fixed air, that it was also to be found in the humidity of the common air and in the substance of plants where it was transformed by the sun into dephlogisticated air. IngenHousz admired the eloquence by which Senebier has been proposing his theory, while he had never found any fixed air, in rain, nor in dew, nor in common humidity, nor in plants. He had to admit that he did not understand how plants do what they do, the only thing he knew for sure is that they produce fixed air at night and were capable to repair that air, back into its common state with the help of the sun. He preferred to admit that he did not exactly understand the role of fixed air in this scheme of things, instead of following Senebier who asserted that the fixed air came from the clouds, while there was no evidence to substantiate that claim. He referred to his experiments described in his book from 1779, where he got inconsistent results when growing plants under water in which fixed air was dissolved. He knew that many physicians asserted that plants did better in water saturated with fixed air, but he maintained they needed more experiments on more and on different plants from the same species and at the same time, in waters with different degrees of fixed air dissolved in them. Still, he thought there could be some truth in Senebier's idea. He thought however that the water would only contain a little fixed air, just like the natural atmospheric air in which plants grew, instead of water saturated with fixed air, as proposed by Senebier.

Senebier was not all bad. It is surprising how IngenHousz gave honour where honour was due to Senebier, after the angry treatment he gave him in the preface.

L'ingénieuse théorie de M. Senebier, par laquelle il fait jouer à l'air fixe un rôle très-important dans la végétation, pourroit peut-être servir à expliquer l'enigme, quoique lui-même soit un des plus grands antagonistes de ma doctrine.[154]

It shows how scientists can disagree about certain things, while appreciating one another about other topics. As long as there are facts to fall back on, and not just thoughts or intuitions. The facts, that's what it is all about. Senebier maintained that plants in water which was saturated with fixed air, or to which acids were added, always delivered bad air, while IngenHousz had found the opposite. He performed these experiments again as soon as he arrived in Vienna in 1780 and set up a new series of experiments in the summer of 1783. On top of that, when analysing what Senebier claimed, citing from Senebier's books, the "Savant from Geneva" appeared to be contradicting himself, sometimes obtaining bad air, sometimes (a little) good air. IngenHousz asked himself where he, or Senebier, or both of them had made a mistake. IngenHousz had tried to reproduce the experiments in the way Senebier described them and obtained opposite results. He got "de l'air déphlogistiqué d'une grande bonté", while Senebier obtained "un air mauvais". IngenHousz wondered why. Would it be possible that "un Physicien et un Ecrivain aussi éclairé" was not able to distinguish "un air absolument mauvais d'un air infiniment meuilleur que l'air atmosphérique"? Was he using different kinds of water? Was he making mistakes in using the eudiometer? Or did it depend very much on the plant species one uses? He had himself experienced that certain types of plant, such as the grasses, produce far less dephlogisticated air under the influence of increased levels of fixed air. This is one of the few places in this book where he reproduced the results of tens of experiments. He did not want to speculate about the explanation for their differing results but was content reporting the facts. He tended to conclude that everything depended on the amount of fixed air dissolved in the water in which the plants thrive, keeping in mind that not all plants reacted in the same way to the same concentrations of fixed air.

The big picture

Although not all relationships or connections between plants and gases are clear to IngenHousz, a picture starts to emerge from the multitude of facts and figures. It all fits nicely with the broader picture of life on earth as IngenHousz conceives it. It is logical, when one thinks of it, that plants get their nourishment out of the air through their leaves. Plants can't move and therefore can't go looking for food, as animals do. Neither do they have reservoirs to stock their food. So, plants have to eat on the spot where they grow and they have

---

154 IngenHousz 1789 p 107

to eat continuously, as they have no stomachs to store their food. They are in contact with the earth and with the atmosphere, so these must be the sources of their nourishment. Diurnal and seasonal variations in the composition of the atmosphere, powered by the movement of the earth around the sun, will influence directly how plants (can) 'eat'. Seeing mountains composed of fossilised remnants of crustaceae, he believes living organisms can turn into inert and inanimate stone, while inert chemical substances such as the elements of the air can be build into living beings. These organisms are suited to the place where they grow. To IngenHousz it is logical that plants from the tropics are better at producing oxygen at higher temperatures than plants from temperate climates. IngenHousz wondered about the incredible variation in living forms and how they all seem to be designed to suit certain circumstances. Dried plants can't ameliorate the air, but mosses, first having dried out and afterwards rehumidified, continue doing so. After such a grand vision, he remained humble and recalled where it all began, sixteen years earlier.

> C'est à M. Priestley que nous devons la grande découverte, que les végétaux possèdent le pouvoir de corriger l'air mauvais, & d'améliorer l'air commun; c'est lui qui nous en a ouvert la porte.
>
> ...
>
> On peut voir, dans la Préface de mon ouvrage sur les végétaux, combien mon esprit étoit, depuis 1773, occupé des idées que je formois sur la manière dont ce merveilleux ouvrage s'opéroit. Je brûlai d' envie, depuis ce temps, de parcourir le vaste champ dont j'entrevoyois les beautés, & dont je voyois la route ouverte.
>
> ...
>
> Je m'obstinai à chercher, de toutes les manières imaginables, l'objet qui m'avoit occupé depuis des années, & en cherchant, même dans les ténèbres, ce que j'espérois d'y pouvoir trouver, la lumière à la fin brilla à mes yeux.[155]

He honoured Priestley who opened the door to this new and vast unexplored territory, that he, since 1773, wanted to enter. He described almost lyrically how he dreamed about the many experiments to perform and the many unknowns to discover, as he was prepared to go into the most obscure depths in order to see the light in his eyes - literally the light of the sun proving to be the motor of all life on earth.

And in the mean time, some IngenHousz Motion

IngenHousz' research in plant physiology had not been the only thing he had been doing since 1779. In 1785 IngenHousz published in Paris a bundle of

---

155 IngenHousz 1789 p 338

various articles on a wide range of topics *Nouvelles experiences et observations sur divers objects de physique*. The book was dedicated to "Benjamin Franklin Ministre Plénipotentaire des Etats-Unis de l'Amérique auprès de la Cour de France", a dedication which was written "1re novembre 1783, a Vienne en Autriche". It was to be followed by a second volume in 1789. Similar bundled volumes appeared in German as *Vermischte Schriften* in 1782 and 1784 translated by M. Molitor in Vienna. Part of the French first volume, on the system of electricity, had already appeared in German in 1781. At the beginning of the French preface IngenHousz expressed his frustrations about the fact that the German translations all were sold out long before the French original finally had become available and that the second German volume was at the printers at the moment the French first volume appeared at the booksellers. He hoped the reader would understand how displeased the author had been about this unnecessary delay. Putting pressure on the Parisian printer had taken a lot of his time and energy over all those years, the only advantages being that the present edition was fully corrected and enhanced.

The main theme is electricity. The first essay is an extended version of his lecture for the Royal Society on Franklin's system. The following essays treat the electrophore (Volta's machine) and topics such as the optimal form of conductors (pointed or spherical). Electricity and chemistry were close partners in those days: IngenHousz described the making and handling of electrical pistols fired by igniting inflammable gas, ways of obtaining that gas from swamps, ways of making lamps burning in inflammable gas, as well as themes that we have seen popping up in his books on plants, such as the measuring of the salubrity of the air in various places, and the production of dephlogisticated air and how to administer it to patients. He also included an extended essay on the production and use of cannon powder, magnetism and the properties of different metals.

The second volume appeared in 1789 and is dedicated to Baron Dimsdale, his friend and colleague who introduced him to the techniques of the variolation twenty years earlier, which was the decisive fact that paved the way to Vienna and a life of independent research. In a first Mémoire he discussed the 'green matter', in a second part he wrote about platinum and its place amongst the metals. But, more importantly, he opened this book with a short article on the use of the microscope. This is particularly interesting because here he not only gave the first description ever of the Brownian motion, but also provided a description of the use of a cover slip (also a première).

The Brownian motion gets its name from its supposed discoverer Robert Brown. His observation of the vivid motion of small particles in a liquid was published in an edition of the Ray Society in 1827[156] and lay dormant for some thirty years before physicists started to show interest. They did not realise

---

156 Brown 1827

that the idea had been dormant already for much longer. Van der Pas[157] has unravelled the history of this forgotten discovery. He describes how Brown had been looking into the history of the microscope and searched the books of all those famous men before him who had been looking through microscopes: Leeuwenhoek, Gray, Needham, Buffon, Spallanzani and many others. Some of them had observed some irregular motions under their lenses, but none had performed the decisive experiment to put some very small, dead or inorganic particles and checked if they exhibited the motion. Brown clearly overlooked IngenHousz paper hidden at the beginning of his *Nouvelles Experiences / Vermischte Schriften*. The fact that IngenHousz published his remarks on the use of the microscope together with his essay on the mysterious green mater suggests that he wrote them while working with the microscope to study the green stuff. He had developed the technique to cover the microscopical liquids with a thin film of glass to avoid premature evaporation. The idea to do this was not entirely new, mica or talc had been in use for many years, but mostly, objects were viewed without them. In his paper IngenHousz described how he wanted to observe little insects in droplets of water or alcohol but sees how the liquid evaporated quickly. The curved surface of the droplets was also causing colored borders in the image. To avoid these inconveniences he recommended covering the droplets and their inclusions with thin glass films that were produced in glass blowing workshops as leftovers from the glass blowing process. According to Van der Pas, this was the first time anybody used a thin glass film for this purpose. Remember IngenHousz mentioned this in a letter to Van Breda in 1783. Here is what he writes in 1784, as printed five years later in the French edition:[158]

> ... durant tous le temps que la goutelette subsiste, son évaporation continuelle met nécessairement toute la liqueur, & par conséquent aussi les corps qui y sont contenus, en un mouvement continuel; & ce mouvement peut, dans quelque cas, en imposer, & en faire envisager certains corpuscules comme des êtres vivans, qui n'ont cependant aucun principe de vie. Pour voire clairement qu'on pourroit se tromper dans son jugement à cet égard, faute d'attention, on n'a qu'à mettre au foyer d'un microscope une goutte d'esprit-de-vin, & y mettre un peu de charbon pilé, on verra ces corpuscules dans un mouvement confus, continuel & violent, comme si c'étoient des animalcules qui se meuvent rapidement entre eux.

As long as the droplet lasts, its continuous evaporation keeps the droplet and everything inside it in continuous motion. This movement can easily give the impression of living bodies in action, although they have no principle of life in them. To demonstrate how easily one can be deceived, one puts a drop of alcohol in the focal point of the microscope and throws some finely ground

157 Van der Pas 1968
158 IngenHousz 1789b p 2

charcoal in it. Then one sees these little bodies move around in a confused, continuous and violent manner.

These were just the opening sentences of a paper that intended to cover real problems with microscopy such as the refraction and the evaporation.

> Mais ce qui surpasse de beaucoup ces lames de talc, ce sont les lamelles de verre des plus minces, qu'on foule aux pieds dans toutes les verreriers; j'en couvre la goutte que je veux examiner. Cette lamelle étend la goutte, l'amincit, & la rend par-tout d'une égale épaisseur. L'évaporation s'en fait si lentement, qu'une goutte qui s'évaporoit en quelques minutes ne s'évapore alors à peine qu'en autant d'heures, si on la couvre d'une telle lame de verre; de façon qu'on peut, par ce moyen, contempler à son aise les objets les plus minutieux, assez longtemps pour suivre les changements ou métamorphoses, auxquels quelques-uns sont sujets.[159]

These little slices of glass are superior to those of talc. They are very thin and are thrown away as waste in the glass factories. Put one on the drop you want to examine, and it flattens the drop and gives it an equal thickness. The evaporation is slowed in such a way that while a drop would not last longer than a minute, now you can study a drop and what it contains on your ease during an hour or more, long enough to be able to follow the changes or metamorphoses that are of interest.

Clearly, IngenHousz did not realise the importance of his observations and techniques. It shows the pragmatism of the man. Problems deserve solutions and if he had stumbled upon something that could help, he would publish it in order for the rest of humankind to benefit from it. His description of the Brownian motion shows how he was interested in the first place in eliminating a possible confusion between living and non-living matter. He demonstrated that non-living materials could exhibit a very life-like behaviour and published it to warn his colleagues-microscopists that one could easily be mistaken. To him this delusion was of extreme importance as he was busy investigating the true nature of the green matter. Seeing inorganic corpuscles move as if they were living creatures must have made him ponder again and again about the vegetable or animal character of the green matter.

It should not come as a surprise that this microscopic phenomenon is not known as the 'IngenHousz Motion'. IngenHousz has been forgotten for many reasons. In this particular case, it did not help that he did not see it himself as an important discovery. Looking at the way he described it, he was also not interested in a physical explanation for what happened under his lenses. On the other hand, Brown, whose name was indeed tagged onto the microscopic

---

159 IngenHousz 1789b p 4

movements, did not realise its importance either. Whereas his paper was completely dedicated to describe the phenomenon, IngenHousz gave it only a casual description in passing to more important topics. On top of that, IngenHousz opened his paper with the message:

> N.B. J'ai jugé à propos de faire précéder le Mémoire suivant, de ces réflexions, quoique d'aucune importance pour ceux qui se sont exercés beaucoup dans l'usage du microscope; elles pouront guider au moins les autres.[160]

He warned that his remarks were of no importance to those who were familiar with the use of the microscope. Too modest to be true, but also pragmatic to the point of diminishing his own importance, he did want to prevent that experienced microscopists would lose time by reading things they already knew. It reminds one of how he warned his readers of his English book that a French edition was forthcoming so that they would not spend money on the English if they would prefer the French.

The IngenHousz Motion disappeared from the historical accounts. The phenomenon he described really became an important physical fact, when Einstein brought it into the limelight. Einstein seems to have known nothing about his curious predecessor when he wrote in 1905 his paper on the molecular-kinetic description of Brownian motion (as translated by Hänggi):[161] "In this work we show, by use of the kinetic theory of heat, that microscopic particles which are suspended in fluids undergo movements of such size that these can be easily detected with a microscope. It is possible that these movements to be investigated here are identical to the so-called Brownian molecular motion, the information available to me on the latter, however, is so imprecise that I cannot make a judgement." Since then most accounts of the history of the Brownian motion start in 1827.[162]

In the tale of IngenHousz' life it was only a little excursion in the margins. His thoughts raced from the mysterious green matter under his microscope to the green fields and pastures on the other side of his window. In a world with an expanding population and a growing need for more food to keep all these people alive, he was turning his pragmatical mind towards practical applications of his theory of the vegetable production of airs.

---

160 IngenHousz 1789b p 1
161 Hänngi 2005
162 Van der Pas 1971

planche III: fig I appareil pour respirer l'air déphlogitistiqué, fig II appareil
pour tirer l'air déphlogistiqué du nitre ou du mercure précipité rouge
Plate to illustrate one of the many topics in *Nouvelles experiences et
observations sur divers objects de physique* from 1785.
This shows one of the pioneering applications of inhalation therapy.

About plants and soils and the future of the world

In 1796 he published his last major scientific treatise. It was an essay written
on the demand of the Board of Agriculture, based on series of experiments
IngenHousz had been performing over the previous few years. IngenHousz
had been declared honorary member of that board in the beginning of that
year, after a regular correspondence with its president Sir John Sinclair. From
Sinclair's address to the board[163] on 24 May 1796, one can understand the
pressing needs of the government at that time. He remembered all the attendees
how they decided to promote the cultivation of potatoes the year before and
that this had to have the happiest of consequences:

---

163 Breda Collection IngenHousz A11

In fact, in times of scarcity and distress, there is no article comparable to potatoes. - They will grow in the poorest soils; they can be taken up in detail as they are wanted; they require no manufacture or drying, milling &c. previous to their being used; and they can be prepared in various ways for consumption.

The Board promoted all kinds of measures to increase the harvests and to improve the general production of farmlands. One of the things the Board successfully did in the previous year was to encourage the people to abundandly sow winter wheat (instead of sitting on their stock of wheat in the hope of selling it later for higher prizes). One of their aims at the time is to pass a general Bill of Enclosure by which the cultivation and improvement of the waste, unenclosed and unproductive lands of the kingdom could be enforced. Whatever they do, so stressed Sinclair, the circulating of agricultural information should be one of their main goals, as that is the only true foundation of improvement. He referred to the printed paper on manures and to reports on the size of farms, on the manufacturing of barley flour, on meals and breads made from pearl barley &c. He also mentioned the work of the "able chymist Dr Fordyce" who is working on a farm in Hertfordshire on ascertaining the principles of vegetation and the effects of manures. In any case, so he states, they should adopt the wise principle "not to print books of reference but books for use, not massy volumes on a variety of different subjects, beyond the income of the generality of the people to purchase, or their time to peruse, but if possible, distinct publications, each of them on one article, exclusively of very other, avoiding the intermixture of various topics, and districts in the same work." He continues:

Agriculture, though often treated of, has hitherto never been discussed; and it can never be much improved, until information respecting it has been collected from all quarters, has been afterwards thoroughly canvassed, and has ultimately been condensed and systematized.

He was not afraid of making political statements and expressed his hope that soon the hostilities would come to an end, knowing that a single additional acre cultivated at home would be really more valuable, than the most extensive possessions acquired abroad at an enormous expense of treasure and blood, and retained in difficulty and anger. This repugnance of the war in Europe and overseas was something Sinclair must have shared with IngenHousz. He too was against violence and bloodshed, knowing there were so many things that could be done to improve the fate and health of the people. Sinclair's question to write up his thoughts on methods to improve the yields of agriculture based on sound scientific reasoning must have sounded very attractive to IngenHousz.

IngenHousz' interest in agricultural problems went back some time. In one of his notebooks[164] dating from 1784, one can find the beginning of another long series of experiments, concerned mainly with the germination of seeds and the circumstances in which they grow best.

> *experiences sur le developpement des graines et l'accroisement des plantes*
> *1784*
> *1 dec je mis mes semences de cressons sur 4 compres de lieges enveloppés de toille,*
> *et enflottent dan l'eau*
> *2 dec les semences etant gonflées*
> *sur ces lieges dans vesseau en 12 puces de l'air*

He put the watercress seeds on floating islands of cork inside vessels under various kinds of air: one in inflammable air, another in dephlogisticated air of 300 degrees, a third in respired air of 44 degrees, a fourth in common air. He put them near the window and registered which ones fared better. Those in the dephlogisticated air grew best: they were the greenest and biggest, better than those in fouled air. These experiments were in line with the experiments on airs and plants, only extended to the field of germination. Later he would perform experiments in which he let electricity run through the floating seeds to conclude that this didn't make any difference. The experiments in this diary ran until 6 August 1785, eight o'clock in the morning.

The real agricultural experiments probably started after talks or letters with Sinclair, a copy of one of these letters dates from July 1795. IngenHousz started a new series of notes on a new series of experiments

> *1796*
> *20 april*
> *Experiments made with seeds not steped, the ground being sprinkled, after the seeds were put in the ground, with the following liquids*
> *N VI six grains of wheat, six of barley, six of oats, six of rye. After sowing the ground was sprinkled with half an ounce of bitter salt.*
> *VII do the ground, after sowing, was sprinkled with half an ounce of glauber salt.*
> *VIII do with half an ounce of salt of tartar*
> *IX dothe ground was after sowing, sprinkled with half a pint of oxygenated muriatic water, such as is employed for bleaching.*
> *X do the ground was, after sowing, sprinkled with about three drachms of oil of vitriol diluted with water*
> *N. XI, XII & XIII were sowen with hordeum nudum, without any preparation*

---

164 Breda Collection IngenHousz A11

22 april experiments to judge the proper depth in sowing

N XXI seeds sown one inch deep in 5 rows, five in each viz. 5 of barley / 5do/ 5 of wheat / 5 do / 5 do

XXII seeds sown 2 inches deep

XXIII three inches deep

XXIIII six inches deep in four rows

XXV one inch deep, sown in two rows about one foot broad viz 120 grains of barley, and 120 grains of wheat. this last experiment serves only to see the size of the plants when crowded together, compared with the size of others when sown at the distance of 9 or 10 inches.

26 april sown in Dr Morton's garden British Museum wheat and rey steeped 18 hours in

oxygenated marine acid water of each 10 rows in sandy hard ground, neither manured nor dug upthe last 10 years.

Next to these towards the south west side of the seeped seeds a similar quantity of these two seeds were sown without having been seeped in any liquid.

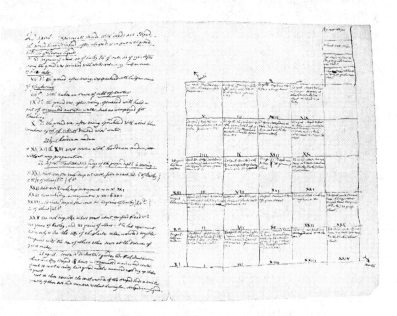

Notes by IngenHousz on agricultural experiments with the map of a field divided in seperate plots that got different treatments and seeding regimes, dated 20 April 1796.

Now the focus had shifted completely to the practical situation in the field. He started as systematically as ever, dividing the plot in squares of four by four, all the soil treated in a similar way. Firstly he checked two things: did it make a difference if one treated the seeds before planting them? And did it make a difference to treat the soil with various chemicals? And it did not stop there: he also varied other parameters such as the depth of sowing &c. He was also careful to perform his seeding experiments in a soil that had not been treated or ploughed for many years, in order to have a neutral control. He found it in the garden of the British Museum. The series of notes continued until September when he was in Bowood and

sept 1796 at Bowood
new soaking exp triticum maceratum in 11 different liquids

Attached to the paper is a little package, containing some grass leaves wrapped in the *Medical Journal* dated 1770 with as headline "Some medical directions concerning children's food". IngenHousz scribbled on the outside: "specimen of plants of wheat sown at 1 to 6 inch deep by which it is evident that deep sowing is very bad and make the plants perish before they reach the surface".

Over the summer of 1796 he performed experiments at different places, in the gardens and on the fields of friends he used to stay with on his many travels through the countryside. In his note on 3 June he mentioned he was continuing the experiments he did in 1795 in Hertford, on soaking seeds in various liquids and where he already had concluded that too much manure could be damaging, although the different kinds of manure all seem to show a little benefit. As one can see in his notes, he was systematically investigating all kinds of variables that could influence the production of biomass. His findings were neatly summarized in his essay "On the food of plants and the renovation of soils", which appeared first as a twenty-page appendix, among several others, to "the Outlines of the Fifteenth Chapter of the Proposed General report from the Board of Agriculture, On the Subject of Manures" in 1796. This original English version seems to have been hard to find,[165] but a Dutch translation by Van Breda was published in the same year, a flawed German translation was published in 1798.

This essay is a clenched and purposeful piece of thinking, which is interesting in many ways. It is a sort of roundup of IngenHousz' active scientific life. It shows how the new chemistry had thoroughly been absorbed by IngenHousz and was being used to develop further his theories on plant growth. It shows how IngenHousz had changed his mind on the role of fixed air. And it illustrates how old resentments as well as familiar inspirational sources and new liaisons continued to play a role. His starting point was that air is the true nourishment

---

165 Gest 1997

for plants. As they did not survive *in vacuo* and so many kinds of plants survive in dry and arid soils, water or soil clearly were not essential food for plants. He recalled how he discovered in the summer of 1779 that "all vegetables were incessantly occupied in decomposing the air in contact with them, changing a great portion of it into fixed air, now called carbonic acid, which, being specifically heavier than atmospheric air, tends naturally to fall downwards, and being miscible with moists, salts, and different sorts of earthy substances, is apt to combine with them". This decomposition was what the roots are constantly doing, just as the fruits and flowers, while the green leaves cease to do so when in the sunshine, as they then produce the finest vital air. He did not doubt that

> this continual decomposition of the air must have a general utility for the subsistence off the vegetables themselves, and that they derived principally their true food from this operation, by changing the decomposed air into various juices, salts, mucilage, oils, &c. much the same as the graminivorous animals the simple grass changes, in the various organs, into the numerous and very heterogeneous fluids and solids.[166]

He cannot explain how exactly these transmutations happen, but can we explain how grass can become a cow or how rice can be made into skin and bone and fat and even grow an embryo inside? How to explain the transmutation from simple food to the complexity of the body? But the new chemistry from the French chemists can help to throw some light on these phenomena. Oxygen with carbon becomes fixed air, with sulphur if becomes sulphuric acid, with azote it becomes nitrous acid. So:

> If plants imbibe fixed air, or carbonic acid, it is not more difficult to believe that this substance may be transformed, elaborated, or modified into various other substances and salts in the organs of plants, than to believe that the above-mentioned changes take place in the human body.

Just as it had been difficult to believe before demonstration proved it, that limestone consists for 40 percent of carbon dioxide. In passing he downplays emphatically the idea of Mr Hassenfratz who thinks that the carbon is taken up by the roots. Hard to conceive says IngenHousz, that a huge tree could find enough carbon on the one spot where he stands for hundreds of years. Remarkable in this quote is also that IngenHousz was by now finally convinced about the important role of fixed air or carbon dioxide for plants. While in 1789 he had still ridiculed Senebier for his suggestion that fixed air contributed to the nutrition of plants, he now reversed himself and recognised this uptake

---

166 IngenHousz 1796 p 3

as the only source of carbon contained in the organic matter of plants.[167] He demonstrated how a rational mind sometimes has to change his opinion and surrender to the facts.

The boldness of his assertions shows how confident IngenHousz had become over the years in the validity of his theory. IngenHousz thought it very probable that plants take in oxygen in the dark and carbon in the light, then keeping the carbon for itself as nourishment while expelling the oxygen. It made him think that airs, which did not contain oxygen were useless to plants, such as inflammable air or azote, contrary to what Dr Priestley or Mr Scheele might think. Just like he thought carbonic acid was only useful to plants in the right concentrations, also "in contradiction to the two celebrated philosophers just mentioned". This brings him back to the days when it all began.

> When I discovered in 1779, that all vegetables decompose the common air by night ... the new doctrine of chemistry was not yet published, and being ignorant of all the beauties of this system, I was unable to reduce these facts to a proper theory. But since we have been instructed in the analysis of water and air, it is become much easier to explain the phenomena of vegetation.

Still, not everything was clear-cut. How to explain that plants grow so abundantly almost everywhere, while the atmosphere contains only a minimal amount of carbonic acid? How do all these chemicals decompose and recompose into the wonderful profusion of solid and liquid substances? Would this argue for a defect in the new system? "Let a better judge than I am decide this", proposes IngenHousz. To him the new system was better than the old one and he had plenty of reasons to use it for his purposes. One counterargument against the critics is certainly that calcareous stones deprived of all their carbonic acid by fire, regain it just by exposure to the air. The same source, the atmosphere, could therefore rightly be seen as the general store of all the substances which enter into the composition of organised bodies of the animal and the vegetable kingdom.

Next he discusses the amount of food an organism needs to flourish. He recalls the customs in the Low Countries, where they used to give limited amounts of well-prepared food to the cows and horses (barley bread or boiled turnips), which seemed to benefit them greatly. Not enough attention is paid in England to the quality, quantity and preparation of the food necessary for healthy or productive animals. Lots of food was thus wasted, while the condition of the animals could be better. He saw an analogical process possible in the nourishment of crop plants. To achieve this, air and light in the right amounts at the right moments were essential. Seeds need darkness and oxygen,

---

167 Rabinovitch 1969

plants need light during the day to develop. So, a continuous transmutation takes place, an old doctrine, well know in the time of the Ancients, and he cites Lucretius:

> Aëra nunc igitur dicam, qui corpore toto
> Innumerabiliter privas mutatur in horas.
> Semper enim quodcumque fluit de rebus, id omne
> Aëris in magnum fertur mare, qui nisi contra
> Corpora retribuat rebus recreeque fluenteis,
> Omnia jam resoluta forent et in aëra versa.
> Haud igitur cessat gigni de rebus, et in res
> Recidere assidue, quoniam fluere omnia constat.
> De Rerum Nat, Lib V 274

> I shall now talk about the air, the composition of which
> is changing every hour, in many different ways,
> because all that comes streaming out of the things,
> ends always up in the wide sea of air; if she would not
> return particles to the things, if she would not replace
> what had run out, then everything would now be dissolved
> and changed into air. So air originates from the things and
> returns to the things, because it is clear that everything always flows.

He refers to a letter he wrote to John Sinclair 2 December 1794 in which he quoted his proof that carbonic acid is the principal food of plants. He suggested then also that azote entered the plant and had also some share in feeding it. He thought so because plants constantly absorb the whole of the respirable atmosphere and expel whatever fluid they cannot digest. At night or in the dark it's the azote and the carbonic acid, during the day it's oxygen. He admitted that this theory had not all the clearness he would wish for it.

> The facts, however, quoted to support it, though contradicted during twelve years, are now admitted publicly, even by those who have been the principal champions and the most violent (even to declare my doctrine to be a downright calumny against nature, to be revenged by nature itself) against it.

It is mentioned almost as an aside, but he clearly couldn't refrain from venting his feelings one more time over the dispute with Senebier, without mentioning his name. Another element that illustrates how this article summarizes his professional life. But it took him straight to the element that pervaded his whole theory: carbonic acid. Why is it, so he speculated, that this essential ingredient, abundantly created, synthesised and recombined by both animal, plant and mineral kingdom was only scarcely found in the atmosphere? He suggested "it sinks to the ground, becomes easily miscible with moist, different salts, &c and thus it disappears almost as soon as it is produced, and becomes, perhaps the first step towards the transformation of common air into solid bodies". He felt supported by several men of high reputation, such as Erasmus Darwin who described "the economy of vegetation" in part one of *The Botanic Garden*, his amazing poem in 2000 couplets, backed by scientific notes running to some 100,000 words[168] and by Lavoisier who developed an experimental method to decompose water, the natural version of which IngenHousz had already demonstrated to exist in plants:

> As water itself is a composition of two airs, vital and inflammable, or oxygen and hydrogen, in which two substances, Mr. Lavoisier found means to analyse water, and which analysis, as far as regards oxygen, I affirmed in my first volume on vegetables, to be performed by Vegetables, with the assistance of the sun, even before Mr. Lavoisier, as I think, published his Analysis.[169]

Which brings him to the core of the matter. How can we manage the economy of vegetables, knowing that some plants, such as flax, are well known to exhaust the soil quicker than others? Which is why the value of one crop of the plant in question, in places like Waas in Flanders or Valenciennes, is equivalent to the value of the land itself. This impoverishment of the soils has induced the farmers to leave the ground at rest for a whole year. These fallow lands however would be found full of weeds, originating mainly due to the dung of cows and horses containing a vast number of undigested seeds of various plants. There is a lesson to be learnt here from the Chinese and the Japanese who exclusively use human dung to fertilise their fields. Van Breda added a note in his Dutch translation[170] that he knew some communities in Holland and Flanders where families collect their dung to fertilise their fields. IngenHousz expressed his doubts about the utility of the old traditional ways of fallowing the land and ploughing in the weeds. He thought the soil on its own was capable of consuming oxygen and so producing carbonic acid in great amounts and referred to some of his experiments from 1779 in which the earth in a flowerpot communicated carbonic acid to the air and absorbed oxygen. He could not have known what we now know, i.e. that populations of microorganisms take care of these metabolic processes in the soil. But what he did when putting a pot with compost under his glass jar, was a distant forerunner of the 1993 Biosphere 2 project where eight people locked themselves up in huge terrarium hoping to create an independent ecosystem. It turned into a fiasco because they had overlooked the microbial flora in the soil

---

168 Darwin 1789 p 434
169 IngenHousz 1796 p 14
170 IngenHousz 1796b

that consumed oxygen and produced carbon dioxide. While many philosophers before IngenHousz had tried to analyse the soil before and after cultivation in order to find what had gone in or out, he had followed an alternative route. He had investigated what had changed in the air above the ground.

> Now as all acids derive their acidity from the oxygen, of which the common store exists in the immense ocean which surrounds our globe, the atmosphere, is there not some probability that it would be possible to restore, in a moment, to the soil, what it can acquire in no less than a whole year from the incumbent air when left to itself?[171]

All these considerations lead him to the idea that it might be possible to restore the oxygeneous principle to an exhausted soil by pouring acid over it, in diluted form or mixed with earth. He performed the first experiments to find out in the garden of Baron Dimsdale's father in Hertford, as we have seen in his notes above. He also performed it in flowerpots. He often used huge quantities of acids and saw no deleterious effects on the germinating seeds. The plants growing in the treated soils seemed to grow more vigorously than controls. Instead of loosing a whole year of crops while waiting for the soil to restore itself in the long and slow natural process, here proposes itself a method in which "all that the air can grant in a whole year may be communicated all at once to the ground". He found it almost unbelievable that Britain, one of the richest, enlightened and best governed countries in Europe, had to import grain worth one million sterling in one year while one-third of all arable lands lies uncultivated.

Still, the summer of 1795 was too short to perform all the experiments he would have liked, but he suggested plenty of them, in the hope that many people would take them up. In a postscript he added that also the additional use of alkaline salts could be promoting the growth of plants. There was still so much work to be done.

One small sentence depicts his research programme. "In the sequel of this paper, the manner by which roots of trees beget carbonic acid will be further traced." There would, sadly, be no second paper on the food of plants. Three years after the publication of this paper, the diseases that undermined IngenHousz' health would finally bring a long and fruitful life to an end. His mind must have still been full of plans, papers to be written, experiments to be performed, letters to be written with questions for clarification or information from his friends and colleagues all over Europe.

Testimony are the notes he made on the inside of the cover of his own copy of his 1779 book on *Experiments on vegetables*. There he wrote down his first thoughts on further experiments to be done:

---

171 IngenHousz 1796 p 17

Experiments to be made.
1. to keep a jar well corcked full of pumpwater on purpose to see whether the
green matter will be formed in it.
2. to keep a similar jar full of water in the sun till the green matter be formed,
and that to shut it by a ground stopper on purpose to see whether the dephlog.
air continues to be disengaged.
...

He summed up thirteen possible experiments, e.g. with the bark of trees and
with pump water to which salt is added.

Inside the back cover he made notes and copied some fragments from books
and articles he had been reading. He was taking in the new theory about
chemistry as if taking in deep breaths of a new and fresh air. He also carefully
copied quotes from relevant books he must have been reading recently.

From *Nature des differens airs selon la nouvelle doctrine, tiré du Journal de*
Physique de chimie et d'Histoire naturelle, par Jean-Claude Lamétrie, janvier 1794,
discours préliminiare pag 32 he notes:

air nitreux dephlogistiqué est composé d' azote 0,63 oxygene 0,37
air atmospherique d'air pure oxygene 0,27 azote 0,73
air nitreux rdinaire oxygene 0,66 azote 0,33
acide nitreux rouge fumant oxygene 75 azote 35
acide nitreux ordinaire blanc oxygene 80 azote 20

From *Nature de gaz nitreux, tiré de Elements d'histoire naturelle et de chimie par*
Mr. Dr. Fourcroy, edition quatrieme a Paris 1791 he copied a lengthy quote of
particular interest to him as it concerned the reliability of the eudiometer:

Le gaz nitreux avoir été entrevu par Hales, mais c'est Mr Priestley qui l'a bien
fait connaitre ce fluide elastique se dégage pendant l'action d'un grande nombre
des corps combustibles sur l'acide nitrique et surtout des métaux,... il éteint des
bougies, il tue les animaux. il n'est ni acide ni alcalin. il n'est point altéré par
l'eau pure. Avec l'air vital il reforme de l'acide nitrique, parce qu'il n'est lui-
même que l'acide nitrique privé d'une partie d'oxygène, mais contenant plus de
la première et moins du seconde que l'acide nitrique. Ce gaz contient souvent
une portion tres variable du gaz azote qui depend de la décomposition plus
ou moins abondante de l'acide nitrique par les matieres combustibles que l'on
prend pour les dégager. De la vient l'incertitude sur les effets eudiometrique de
gaz nitreux. on concoit d'apres cela pourquoi dans plusieurs cas, et specialement
lorsqu'on emploi pour obtenir un gaz nitreux un corps tres aride d'oxygène et qui
en absorbe beaucoup pour la saturation, on obtient un gaz nitreux contenant
du gaz azote à nud et quelquefois meme on ne retire que de gaz azote; ce

gaz nitreux composé d'azote et de l'oxygène contient encore plus de ce dernier qu'il n'y en a dans l'air atmospherique; on demontre le fait en le decomposant par un sulfure alcalin liquide (hepar sulphuris). Une dissolution de ce sulfure mise dans une cloche pleine de gaz nitreux, en absorbe promptement une partie; bientôt ce gaz ne rougit plus par le contact avec de l'air, il entretient la combustion des bougies, mieux que l'air atmospherique. c'est un effet de l'air un peu plus pure que l'air commun ; la proportion de l'air vital au gaz azote y est plus considerable que dans l'air atmospherique. mais si l'on continue à renouveller et à laisser agir le sulfure sur le gaz, tout l'air vital en est bientôt absorbé, il ne reste plus que du gaz azote. remarquons que le gaz nitreux donne à la flamme une couler verte avant de l'éteindre, et que dans un grand nombre de cas, cette couleur est produite par les composés, dont l'azote fait partie. Ces propriétés principales du gaz nitreux, et en particulier sa combinaisons rapide avec l'air vital indique son analogie avec les corps combustibles, et Nacquer avait remarqué que la formation artificielle de l'acide nitreux qui a lieu dans le melange de ces deux gaz, est une espece de combustion; mais comme celles, n'est point accompagnée de flamme, je n'ai pas cru devoir ranger le gaz nitreux dans la classe des gaz inflammables. Il differe de l'air atmospherique et par la proportion de ses deux principes et parleur etat de compression. Dans le gaz nitreux l'oxygène et l'azote sont privé de toute la quantité de calorique et de [?] qu'ils contiennent dans l'atmosphere. L'oxygène retient cependant assez de l'un et de l'autre de ces principes, pour que plusieurs corps combustibles y brulent avec flamme, comme ce fait le pyrophore.

And out of *Extrait d'une mémoire sur la nature de l'oxide gazeux d'azote nommé par Priestley gaz nitreux dephlogistiqué J R Deiman, Paets van Troostwyck, P Neuwland N Bonn & A Lawerenburg (Journal de Physique cahier de Nov 1793 p 321)* comes:

Depuis que les decouvertes de la nouvelle chimie ont fait connaître l'oxygene comme une substance élémentaire qui en se combinant avec la plupart des autres substances, sait a former un tres grand nombre de differents combinaisons, produits de la nature ou de l'art, on n'a pas tard a remarquer qu'une seule et meme substance elementaire peut former avec l'oxygene plusieurs composés, dont les proprietes sont tres differentes, a raison du rapport different de l'oxygène à la base [corrected in red: *avec l'azote*]. *Un premier degree d'oxygenation produit des oxides, un degree plus fort des acides, chacun de ces deux etats est encore plus susceptibles de differentes modifications dues a la meme cause, à la variation de rapport entre l'oxygène et la base simple à la quel il se trouve uni. Il n'y a pas de substance qui nous presente une succession de ces combinaisons plus complexes et plus variées, que celle qu'on connaît sous le nom d'azote. Mêlé a l'oxygène dans un état gazeux sans qu'elle lui soit entierement unie, elle compose un grande partie de l'atmoshere dans laquelle nous vivons et de laquelle nous tirons l'oxygène dont nous avons besoins, tout pour entretenir la vie animale elle-même, que pour mille autres usages qu'elle exige.*

It shows the extent of the intellectual background at the time of writing his essay on the food of plants. He used his own book - literally - as the template for new ideas. He could not stop learning and was constantly absorbing new hypotheses, methods and insights. Just like the plants he studied which constantly absorbed carbonic acid, to grow and flourish, driven by the everlasting power of the sun. He shed the light of his mind on some of the most complex processes in this living world.

Postscript of letter to Van Breda on 25 December 1781. As an after-thought IngenHousz formulates a very bold conjecture: would it be possible that part of the absorbed water is changed into air? That would explain how the green moss produces so much dephlogisticated air, because during this process the air quality in the water does not change. Nitre changes into dephlogisticated air through fire, so why would not the light be able to change water into air inside the plant? In the article on Mutabilitas aeris in his book (*Experiments upon plants*) he already mentioned this.

The first page of a 32-page notebook with "Corrections & additions for the second edition of the book on vegetables". He notes that "the words <u>a new method of examining the accurate degree of salubrity of the atmosphere</u> have to be put in better english".

rewind

Content, contest and context

IngenHousz was a true polymath in the eighteenth century tradition of the Enlightenment. Fundamental research and applied science in many different fields flow naturally over into each other.

The mainstay of his investigations was plant physiology. He became inspired by the first observations of Priestley on how plants purify the air. This led him to discover the essential role of the sun, and to unravel the different vegetable activities during the day and at night, which would finally lead to applications in agriculture and a grand vision on the connections between all forms of life on earth.

On the way he suggested improvements to scientific tools such as the eudiometer and the microscope and described the phenomenon of the Brownian motion. And he explored the worlds of electricity (the conductivity of metals as well as the generation of electricity in animals) and chemistry. Being a doctor, health was always on his mind. The first applications of electroshocks and of inhalation therapy for treating patients were just two other 'inventions' that logically flowed from his natural curiosity and experimental passion.

Communicating about all these findings and about his theories was an essential part of his activities. Trustworthy knowledge had to be the result of debate. Communication and discussion were essential. And debate was what he got, in fierce disputes with Senebier, Priestley and others. No matter how hot the dispute, part of the game for IngenHousz was the ability to change one's mind when the evidence refutes one's ideas or assumptions. IngenHousz often explicitly realized how methodological and epistemological issues were relevant or necessary in amassing reliable knowledge.

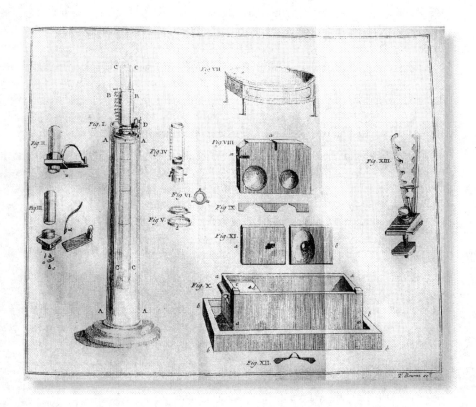

Plate depicting eudiometer in *Experiments upon vegatables*, 1779.

A crucial instrument:
the rise and fall of the eudiometer

Measuring is understanding. It is one of the motto's of empirical science. Scientific knowledge stands or falls with evidence: measurements or observations. They deliver the data, the building blocks of a theory. Measurements are the way to test hypotheses. Measurements are the staple diet of the scientist at work. No matter how nice a theory might look, without observations or measurements it is an empty box. Still, the instruments to make these measurements will hardly ever see the limelight. The attention in the history and philosophy of chemistry and other scientific disciplines has been overwhelmingly focused on the history of the theory, with practice getting little consideration and with the apparatus that rendered that practice possible all but ignored.[172] Most attention goes to the theory, or to the experimental results. Rarely does one hear the story of the actual performing of the experiments, of the hardship, the trial and errors, and the messy work in the laboratory. Still, in all scientific publications, even nowadays, one finds the methods and the set-up described, in full detail, in order for all the readers to be able to understand what has been done, and if necessary, to replicate the experiment independently. At least, that's the ideal. Those descriptions are often detailed enough, but they don't tell about the many decisions that have been taken while setting up the experiment. How does one decide on the accuracy that is necessary to measure one or another? How often does one measure? How does one know that what one does is right? Does one measure what one thinks one is measuring? In many cases a scientist or student can rely on the experience or knowledge of peers or elders. But in a new exploratory field, when one enters uncharted territory, everything is new, including the measurement methods. That's what IngenHousz and his contemporaries had to do when they began measuring the quality of airs in the middle of the eighteenth century. What exactly did they do? What did they try to measure? Did they measure what they thought they measured? How reliable were their methods and instruments? The measurements that play a major role in the beginning photosynthesis research have been done with the eudiometer. IngenHousz has been promoting and defending this instrument as a crucial tool, not only for his own research but also for setting the first steps towards what we would now call environmental health.

---

172 Levere & Holmes 2000

The story of an instrument

It all started in 1772 when Joseph Priestley read his "Observations on different kinds of air"[173] to the Royal Society in London. In this lecture he presented a method to measure the salubrity of the air. His method was based on the fact that nitrous air (NO) reacts with dephlogisticated air (O_2), forms a red vapour and is absorbed by water. So, the amount of air above the water diminishes. This allowed him to deduct the amount of dephlogisticated air in a particular measure of air by looking at diminution of that air in his test bottle. A small decrease of the air tested implied that that air contained little dephlogisticated air and one could conclude the insalubrity of this air. So universal was the belief that the larger the amount of dephlogisticated air in the air the better the air, that the instrument designed to measure this variable almost automatically served as a health tool. As dephlogisticated air supported respiration (mice) and combustion (candles) much better than ordinary air, it was natural to attribute the healthy state of the latter to the amount of dephlogisticated air contained in it.

Marsilo Landriani, professor at the Scuole Palatine in Milan, was one of the first to see the technical possibilities and scientific promises of this method. Milan was notorious for its bad environmental conditions. The Milanese air was considered insalubrious because of the exhalations of the rice fields around the city. As documented by Beretta[174] this unhealthy atmosphere was the subject of serious debate in periodicals such as Il Caffè, the most enlightened periodical in Italy in the eighteenth century, engaged in political and social reforms. Since the 1750s, long before Priestley published his article, the Milanese poets, naturalists and intellectuals were seriously debating the effects of the bad air of their town on public health. Landriani combined these social concerns with his scientific interests. The analysis of gases and the salubrity of the air came together in his 1775 booklet Ricerche fisiche intorno alla salubrita dell'aria. In this essay he coined the word "eudiometro" (from the Greek for 'goodness of air' and 'measuring') and kicked off an ambitious research program to measure the salubrity of the air and use that knowledge to leverage campaigns to prevent epidemics and contagion. And he knew he was not alone, because he wrote: "Pneumatic chemistry occupies all physicists in Italy."

Bad air was not only a problem in Milan. In Tuscany, in 1761 an epidemic of fever struck a third of the inhabitants of a town west of Firenze and caused the demise of 566 people.[175] People believed that the fever was caused by the pestiferous inhalations from stagnant waters of a brook, which had become more active and poisonous due to the excessive heat and dry season, combined with the equally pestiferous emissions of the marshes and swamps. The

173 Priestley 1772
174 Beretta 2000
175 Knoefel 1979

sickly constitution of the inhabitants and bad food and drinking water were not considered prime causes as these often occurred without resulting in a pestilence outbreak. The conclusions of the local physicians were that to be healthy and live long it was necessary to live in good and salubrious air and avoid the bad. Stagnant water was seen as the culprit, so drainage was seen as the best precautionary measure. Avoiding bad air (mal aria) was the message.

Felice Fontana was one of the influential figures in the ongoing debate. He was born in 1730, the same year as IngenHousz, in a small village in Alte Adige in Northern Italy. By 1760 he had become one of Italy's most eminent scientists. Following his studies in Parma, Padua and Bologna he was appointed in 1766 as physicist to the Florentine court of Peter Leopold, where he would spend the rest of his life doing research in anatomy, histology, embryology, physiology, pharmacology, pathology, physics and chemistry. He wrote a book on rust of wheat, one of the causes of famines. It was dedicated to Van Swieten, emphasizing the connection with the Viennese court. Fontana was one of the many chemical philosophers in Italy and elsewhere who picked up Priestley's method. He wrote an article about nitrous air[176] ending it by stressing the importance of investigating how to improve the air we breathe: "this research interests all men equally, even sovereigns who share its use [...] most useful would be an instrument that might show us the goodness of air." Later that year he wrote a book[177] in which he elaborated the idea of making instruments for measuring systematically the salubrity of the air. He opens with what one could call his research program, his "Analisi dell' Opera":

I.   *The importance of publishing these new instruments.*

II.   *Of air, which we should know more than other elements, we know less.*

III. *Priestley discovered that the more air is healthful, the more it is consumed by nitrous air.*

IV.   *The utility of finding an instrument that measures with the greatest exactness the absolute amounts of nitrous and respirable airs consumed in their union.*

V.   *The advantages that may be drawn from such an instrument depend on its perfection.*

VI.   *The instrument can give only the simplest information on the salubrity of the air.*

VII. *It is not necessary to know the absolute amount of air consumed to know its degree of healthfulness.*

VIII. *We lack the first and the last items to mark on the instruments the salubrity of the air.*

---

176 Fontana 1775
177 Fontana 1775b

IX. *Wishing to have the minimal value of salubrity we may have it immediately and mark it on the instrument.*

X. *The maximum value of salubrity seems impossible to fix.*

XI. *For good results it is enough to begin with free air from fumes; this air will show on the instrument the maximum of salubrity.*

XII. *With time we will learn to evaluate the risk run in breathing air less good.*

XIII. *Meanwhile we can also immediately learn that airs of certain places are dangerous to breathe, and less healthful.*

XIV. *The real and immediate utility for medicine, and the public.*

XV. *Perhaps we may even predict epidemics.*

XVI. *The utility of these instruments in physics.*

In his book Fontana presented eight instruments. Eight different designs, four of which depended for quantisation on the weighing of mercury, four were volumetric, based on a vertical glass tube. Fontana provided no results analysis in the book and knew himself that they were premature designs. This can be understood from his letter to Bonnet of 21 June 1775[178] accompanying a parcel with his book which:

> *...contains the description and use of various instruments for measuring the salubrity of the air [...] I have been obliged to publish it thus devoid of experiments because already more than one person thought to take away from me these little instruments as they did also with other things, in such a way that for ten years and more I saw others publish and appropriate my labours.*

Plagiarism clearly is a thing of all times and Fontana was not the only one bothered by it. He published his premature designs in order to cut short possible copy-cats. A precise description of Fontana's eudiometer and its use would only be published for the first time by IngenHousz in his book on *Experiments on vegetables* in 1779, as we have seen with all due respect and credit to its inventor. IngenHousz and Fontana had met in 1769 when IngenHousz visited Firenze to inoculate the members of the royal family. They had become close friends and collaborators. IngenHousz took Fontana's eudiometer to a higher level of practical use and theoretical insight.

> *The method was the same which the celebrated Abbé Fontana makes use of now and which he himself has not yet given an account to the public. As I had no right either to claim the invention of this method, or to anticipate the publication of it without his leave, I have asked his consent an this head.*

---

178 Knoefel 1979

He agreed to my request very readily, gave me his notes to consult, and even permitted me to get his instrument engraved, for which purpose he allowed me to make use of his drawings. As he had already shown me his method of examining the different kinds of air in regard to their degree of salubrity, or fitness for respiration, when I was with him in Paris in the beginning of the summer of 1777, and as I have since he rejoined me in London in 1778, seen a very great number of like experiments, on purpose to imitate his method of examining air, which I found so accurate, then in ten experiments made one after the other with the same kind of air the result differed seldom above 1/500; that is to say, that the remaining bulk of the three measures of nitrous air, which he joins one after another to the two measures of nitrous air, that the difference will seldom amount to more than 1/500 of the whole; which accuracy in exploring the degree of goodness of respirable air surpasses the judging the degree of heat and cold by the thermometer of Reaumur. The Abbé has, since I saw him at Paris, changed somewhat his instruments and the method of using them, or rather corrected them a little; but they remain still materially the same as they were before.[179]

The instrument IngenHousz used does not resemble any of the designs by Fontana from 1775. By 1777 Fontana too had returned to the design described by IngenHousz, which actually resembles very much the original one of Priestley. Fontana never published a new work with his final model, except as represented in IngenHousz' writings.[180]

In the mean time Landriani published the design of his eudiometer, of which only the name would be kept. In England Henry Cavendish examined the nitric oxide eudiometer finding that the instrument invented by Abbé Fontana is by much the most accurate of any hitherto published.[181] He noted that the results could vary according to the way in which the gases were mixed. And about the diminution in volume of atmospheric air he wrote: "The highest that I ever observed was 1.100, the lowest 1.068, the mean 1.082" which is the same value IngenHousz would confirm in his article dedicated to the use of the eudiometer.[182]

The various models of eudiometer, by Volta, Fontana, IngenHousz, Spallanzani and others were studied by Capuano and Cavalchi,[183] who also led a team of volunteers that built replicas of the instruments, which are exhibited at the Centro Studi Lazzaro Spallanzani in Scandiano.

---

179 IngenHousz 1779
180 Knoefel 1984
181 Cavendish 1783
182 IngenHousz 1785
183 Capuano 1998

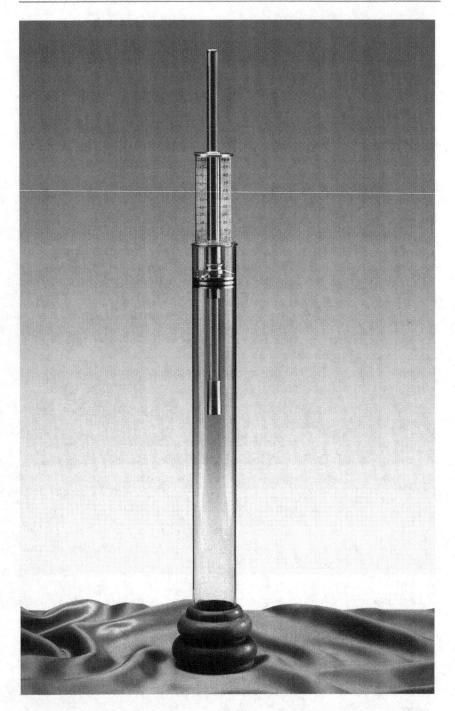

A modern replica of IngenHousz' eudiometer. The small eudiometric tube is
46cm long with an external diameter of 1cm and 10 divisions (1 cm each).
The large tube is 51.5cm long with an external diameter of 6.2cm.
(Courtesy of Centro Studi Lazzaro Spallanzani, Scandiano, photography by Davoli Franco)

Eudiometry for beginners

In an article published in the *Journal de Physique* in May 1785, IngenHousz used twenty pages to situate the technique of the eudiometry in its historical development, illustrated and described the equipment needed to do proper measurements and explains the results one can expect.[184] It is the reprint of a letter addressed to Dominique Beck, professor in mathematics and experimental physics and counsellor to the Prince Archbishop of Salzburg. Beck had been in Vienna in 1782 and 1783 and assisted IngenHousz in performing these eudiometric experiments. In his introduction IngenHousz stressed that this collection of reliable and accurate facts was the only way knowledge could advance. He admitted that many instruments were called eudiometers although they strictly speaking, didn't deserve the name. He defended Fontana's model as being the most reliable and understood that misunderstandings could exist when not everybody was using the same instruments or methods. This article was written to remedy this. Still, until then, he had not seen any fundamental proof against the usefulness of the eudiometer, apart from some vague insinuations, sometimes made by people who did not even possess a Fontanaian eudiometer. Many physicians had visited him in Vienna and attended his demonstrations and all were impressed by "l'exactitude & l'uniformité des épreuves faites avec cette instrument". And the same was being reported to him by Van Breda, the only one who possessed such an eudiometer in Holland. After honouring Priestley, the first to come up with the idea, and Fontana who developed the current and most workable model, he referred to other publications for the reasons why he preferred Fontana's model to Priestley's. Then he started describing the different parts of the eudiometer and how to use them, with the help of a plate (see page 204).
Summarised:

> Fig II. a glass tube, closed at one end, with an even diameter all along its length, between 14 and 18 pouces (inches) long and a diameter of 1.5 pouce (some 40 centimetres long and 1 cm diameter). This tube is divided in sections of each 3 pouces long, each divided in 100 equal parts. As it is difficult to mark all of these on the glass (what he used to do in the beginning with a diamond knife, often leading to breakages), so he uses a brass scale engraved on a cuff in which the tube can glide (cc). Only the big measures are indicated on the glass.
> This cuff fits the edge of a bigger tube with three pivots on a brass ring. This big tube (Fig. III) which can be made of glass, but IngenHousz prefers it in brass as it is less liable to breaking. (He travelled a lot, so he liked a sturdy version, in a custom-made case while the brass tube

184 IngenHousz 1785

served to protect the glass one.) Next, one needs a glass measure (Fig.V), containing as much volume as the volume of one segment of 3 pouces in the glass tube. This measure can be closed by a valve (Fig.VI) and has an aperture that fits neatly on the opening of the glass tube (Fig. II bb). The other thing one needs is a tub (Fig.VIII), filled with water in which is installed a platform under the water surface, in or on which one can fix or put tubes and vessels, and with funnels to transmit airs from one recipient to the other under water. This is the far descendant from Hales' pneumatic through. And it is more or less all one needs. He went into more detail on how exactly to make the tubes and how to fix the different parts together. He suggests e.g. to roughen up slightly the inside of the glass tube to prevent water from sticking to the sides. And he also advises to use distilled water (referring to Van Breda who compared all different kinds of water), as that ensures a standard quality of water for all physicians attempting to undertake these experiments. But with this basic ingredients one can start making eudiometric measurements, IngenHousz asserts : "Je n'ai pas encore vu un seul Physicien assez mal-adroit, qui après un quart d'heure d'exercice, ne sût pas manier l'eudiometre de Fontana assez bien pour faire des épreuves concordantes."

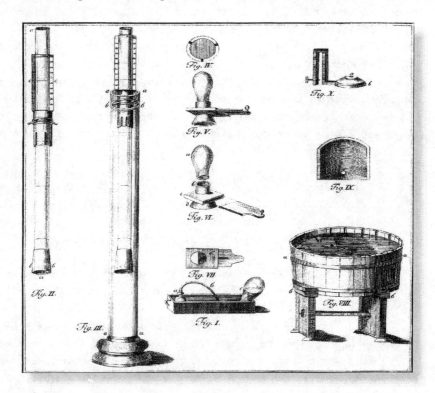

Illustration of eudiometer apparatus in *Journal de Physique*, 1785

The principle of the measurement is easy enough. The long exposé in his book of 1779 on the many errors one can make and repeated in all subsequent editions, however indicates that much can go wrong. This is how it works: One fills the small glass tube with water and puts it upside down in the tub. One puts one measure of the air to be tested in the glass measure, closes the valve, puts the measure on the opening of the glass tube, opens the valve and lets the air up the tube. In the beginning he did this twice, letting in two measures in the eudiometric tube. Later he would experiment with just one measure, especially if the air was of a good quality.) Then one put one measure of nitrous air (which he carried freshly made from copper and strong nitric acid in a flacon) in the glass measure, closed the valve, put it under water on the opening of the glass tube and let up the nitrous air. More measures of nitrous air were let up until no more reaction was seen (no more red fumes). One shook the tube, still keeping the opening under water, while taking care not to let any air escape. After shaking one put the glass tube in the gliding scale which rests on the big (brass) tube, which was on the bottom of the tub. (Remember these measurements were sometimes also done under mercury to avoid the interactions between the airs and the water. And such experiments were not done on ventilated workbenches and without protection for skin, eyes or lungs.) When all this was done, one could read the result from the scale. (IngenHousz advised to let the whole eudiometer in the water to avoid sudden temperature fluctuations that could influence the measurement.) One knew how many measures, equal in volume, of both test air and nitrous air had been put inside the tube. The more dephlogisticated air to be measured, the more nitrous air had to be added to destroy it, i.e. the more nitric oxide had to be added to saturate all the oxygen. Putting the zero on the measuring scale level with the upper meniscus of the water column, one could read out how much of the used measures had disappeared. In later versions of the instrument, a small magnifying glass would be built in to better read the exact volume.

Handwritten in the margin of his personal copy of his first book, on page 175,[185] one finds this stream of thought in which he tried once more to explain the use of the eudiometer again and to provide the reasoning for the implications of varying qualities of nitrous air.

> *after the saturation of the two measures of common air is completed, there will remain in the great measure, or tube, a column of air so much the longer as the nitrous air employed was the weaker. I will illustrate it with an example: let us suppose, that after the three measures of strong nitrous air are let up, and the saturation of the two measures of air under examination be completed, the remaining column of air be found equivalent to three measures, and eight sub-divisions, or 308 sub-divisions; this number, substracted from the 500 parts of*

---

185 Museum of Breda

sub-divisions of both airs employed, will give a result of 192, which is exactly the quantity of both airs employed. Let us now again suppose, that the nitrous air was so weak, that, instead of three measures, six were required before saturation was fully completed, and that thus the remaining column of air in the great tube occupies 608, instead of 308, sub-divisions; we shall find that the result will be just the same; that is to say, that, by substracting the 608 parts remaining from the 800 parts of both airs employed in the experiment, there will be found exactly 192 sub-divisions destroyed; and thus in both cases the accurate salubrity of the air is ascertained. If such bad nitrous air was only at hand as was just now supposed, it follows, that a longer tube ought to be employed.

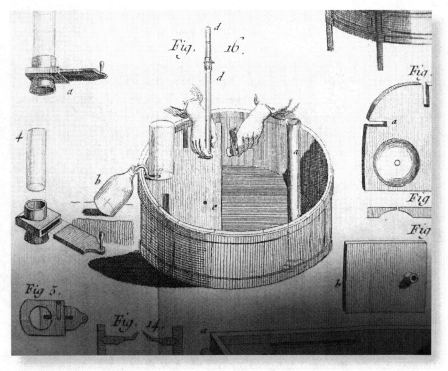

Illustration of manipulations of eudiometer in tub of water from *Expériences sur les végétaux*, 1787 (detail)

... for the better understanding of this important article, I will illustrate it with an example: Two measures of common air being let up on the long glass tube; one measure of nitrous air drawn from mercury was joint to it, and the tube was shook forcibly during ½ a minute in the water at the moment the two airs came into contact, or after two minutes repose the column of water stood at 179.

after the second measure of nitrous air, it stood at 210. after the third at 306, which number substracted from 500, being the 5 measures of both airs employed, the remainder was 194 which was the exact quantity of both airs destroyed.

after which the same nitrous air was weakened by mixing a quantity of common air with it. and after this mixture it was employed in the same way as before in examining the common air. the consequence was that the three measures of the weakened nitrous air are not sufficient to saturate the two measures of common air, but a fourth measure was required. the column of water stood at the following marks after each measure of nitrous air was let up: 205 . 217¾. 310¾ . 407.

thus the four measures of nitrous air being joined to the two measures of common air were so much diminished in bulk, that from the six measures or 600 subdivisions, there remained in the tube 407 subdivisions, which being substracted from the 600, the remainder was 193, so that in the two trials the difference of the result did not amount to more than 1/600, which is a minor [?] trifle.

It shows how he must have thought that the practical use of this instrument was very important for his theory, as only reliable results could support any trustworthy theory. It also shows that he saw the importance of the proper utilisation of the instrument by others and that that would only be possible if these users were properly instructed. His didactic efforts on the usage of the eudiometer would permeate his whole written oeuvre. In his first book and in all its subsequent translations, the eudiometer got a lot of attention. And time after time IngenHousz explored the limits of this eudiometric method:

..., en secouant, par exemple, dans le tube eudiométrique une égale mesure d'air nitreux & d'air commun, on trouvera la colonne d'air être réduite à environ la moitié (en supposant que le tube ait été rempli d'eau distillé), & dans un essai semblable d'un air déphlogistiqué très pure la colonne occupera environ 0.80, ou quatrevingt centièmes de mesure: il n'y auroit donc que la différence de 20 degrés entre la bonté d'un air déphlogistiqué de la meilleure espèce & de l'air commun: ainsi les airs déphlogitiqués d'une bonté moyenne s'approcheroient, en apparence, trop pour pouvoir les distinguer entr'eux avec exactitude. [186]

...when one mixes in the eudiometric tube an equal amount of nitrous air and common air, the volume of air (two volumes) is reduced to approximately half (one measure). If one does the same with dephlogisticated air of very pure quality, the volume of air is about 0.80, or 80 hundreds of a measure. Therefore, there is only 20 degrees difference between de goodness of dephlogisticated air of the best kind and common air. So, dephlogisticated airs of medium quality resemble, so it seems, are too similar to distinguish them with great exactness.

---

186 IngenHousz 1785

All the procedures IngenHousz prescribed are the result of repetitive series of experiments to identify the most accurate and reliable protocol. The way to hold the tubes and the measures, when and how long to shake the tube for, the period to let the tube rest before measuring the amount of air left, the quality of the nitrous air to be used and the kind of water over which to work, &c were all decided on after long runs of experiments.

IngenHousz stressed that even were one to decide to do it otherwise, it was necessary to always follow the same procedure in order to get comparable results. He himself had developed his method conscious of the reasons why he chose some procedures. He sometimes performed 50 experiments in a day and did not have the time to wait for hours or even days (as others did) before measuring the height of the water column in his eudiometer. IngenHousz ends his 1785 article with a reminder of the eudiometer's importance in providing insights in the unknown causes of environmental health or disease. All those practicalities were to serve a greater good and as long as no better eudiometer than Fontana's could be put forward, they would have to do with that model.

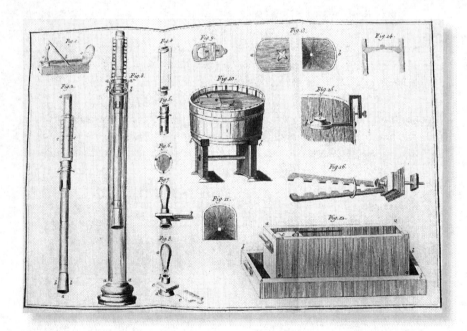

Illustration of eudiometer in the second edition of *Expériences sur les végétaux* from 1787, the copy dedicated by IngenHousz to Spallanzani, kept in the municipal library of Reggio Emilia. Fig 12 (bottom right) is a wooden through used when doing the eudiometric measurement over mercury.

Eudiometers as matter of discussion

The nitric oxide eudiometer would get competition from a new instrument by Alessandro Volta, developed in 1777. He used hydrogen as reagent and the gas mixture was detonated by an electric discharge. This proved to be a convenient method, though Volta himself expressed his doubts about the ideals of the "medicina aerea" or aerial medicine, as Landriani had baptised it. Volta pointed out the distinction that should be made between respirability and salubrity. There had to be other factors that determined if a particular air was healthy. Fixed air, nitrous air and dephlogisticated air were relevant for the insalubrity of the atmosphere in a certain place but could not be seen as the sole viable indicators. Some airs or gases did not allow the survival of animal life, while other airs were very breathable but unhealthy because they carried fevers and other disease. However how exactly they did this, remained a mystery. Those unhealthy airs could not be detected by the eudiometer. He was not convinced that the so-called bad air was that bad, as one could just as well get ill by sleeping out in the open air and stay perfectly healthy in a closed room corrupted by lamps and breathing.

Fontana turned out to be not the strongest defender of this doctrine either as he expressed his doubts in his letter to the Royal Society in 1779:

> I have not the least hesitation in asserting, that the experiments made to ascertain the salubrity of the atmospherical air in various places, in different countries and situations, mentioned by several authors, are not to be depended upon; because the method they used was far from being exact, the elements or ingredients for the experiment were unknown and uncertain, and the results very different from one another. When all the errors are corrected it will be found that the difference between the air of one country and that of another, at different times, is much less that what is commonly believed, and that the great differences found by various observers are owing to the fallacious effects of uncertain methods. This I advance from experience: for when I was in the same error, I found very great differences between the results of the experiments of this nature which ought to have been similar; which diversities I attributed to myself rather than to the method I then used
>
> [...]
>
> we clearly see, how little the experiments hitherto published about the differences of common air are to be depended upon. In general I find that the air changes from one time to another; so that the differences between them are far greater than those of the airs of different countries, or different heights
>
> [...]
>
> I do not mean to deny the existence of certain kinds of noxious air in some particular places; but only to say, that in general the air is good everywhere and

*that the small differences are not to be feared so much as some people would
make us believe.*

[...]

*I would not have any body suppose, that I think it of little importance to know the
goodness of the atmospherical air, and the changes it undergoes. On the contrary, I
believe it be a useful inquiry for mankind, because we do not yet know how far one
kind of air more than another may contribute to a perfect state of health; nor at
what time small differences may become very considerable, when one continues to
breathe that same kind of air for whole years, especially in some kind of diseases.* [187]

Anyway, natural philosophers and medical doctors all over Europe were
enthusiastic about this new tool. In Austria Johann Scherer, IngenHousz' friend,
would translate IngenHousz' works and promote this "Luftgüteprüfungslehre".
In Holland, Boudewijn Tieboel, pharmacist in Groningen, was enchanted by
the possibility to gain proper knowledge about airs of different kinds as the key
to solving the problem of epidemic diseases: "one may consider this invention
as one of the most splendid and useful discoveries in natural philosophy".[188]
Jan Rudolf Deiman, doctor and chemist in Amsterdam asserted that the
eudiometer was "one of the most useful instruments in human history". Other
Dutch researchers used the Fontana eudiometer with great fervour, Van Marum
and Paets Van Troostwyck among them. Van Breda was one of the eudiometer's
greatest proponents in Holland, inspiring many to start eudiometric research.
But not everybody was as convinced about its advantages and accuracy. Van
Swinden experimented with the eudiometer but found it "too difficult" for an
average meteorological observer. He doubted its reliability and even questioned
the results by IngenHousz, Deiman and Paets van Troostwyk. He stated in
a letter[189] to Tieboel that: "Ingenhousz had failed as much as he had made
progress". Across the Channel, Priestley's attacks on other eudiometrists, such
as Fontana and IngenHousz, concentrated on their illegitimate use of rival
techniques in assessing atmospheric restoration.[190] Talking about Fontana, he
wrote in 1780: "I am astonished and provoked by the little care with which some
persons make experiments, and the confidence with which they report them."

As we have seen in the reactions of Hassenfratz and Lavoisier, not everybody
was of the same opinion. Lavoisier was to become a fervent user of the
eudiometer too. The first trace of him doing so was on 10 May 1777 when he
presented a memoir on "Observations sur les altérations qui arrivent à l'air et sur
les moyens de ramener l'air vicié à l'état d'air respirable" to the Académie Royale
des Sciences, in the presence of Emperor Joseph II. This event was reported by
Bachaumont in the following words:

---

187 Fontana 1779b
188 Zuidervaart 2006
189 Zuidervaart 2006
190 Schaffer 1986

Monsieur Lavoisier eut l'honneur de démontrer, par des expériences multipliées comment on pouvoit décomposer l'air de l'atmosphère en demi-portions, l'une salubre, respirable, susceptible d'entretenir la vie des animaux, la combustion & l'inflammation; l'autre, au contraire, funeste pour les animaux qui le respirent. Il fit voir que la respiration des hommes & des animaux avoit la propriété de convertir en air fixe la portion salubre de l'air, de sorte que dans les salles de spectacles, par example, ou dans les dortoirs des hôpitaux, où l'air a été long temps respiré, il existe deux especes d'air nuisibles; savoir, la partie nuisible propre à l'air, & qui entre dans sa composition; & la portion d'air fixe qui s'est forméé par l'effet de la respiration.[191]

Mr Lavoisier did have the honour to demonstrate in many experiments how one can decompose atmospheric air into two portions: one good, respirable and capable to sustain animal life, combustion & inflammation; the other, on the contrary, fatal to animals that inhale it. He showed that the respiration of men & animals has the property of converting the salubrious portion of the air into fixed air, of the kind one can find in theatres or hospital dormitories, where the air has been inhaled for a long time. There one finds two kinds of noxious air: the part which is a natural injurious component and which enters into its composition, & the portion of fixed air as the result of the respiration.

While Volta published his letters on his new eudiometer and its application in physics and his doubts about its medical applications, Lavoisier was refuelling the idea to use the eudiometer to measure the salubrity of the air while adding the novel perspective on the procedures that the respiration of men and animals and the combustion of bodies transformed the respirable part of the air into fixed air.[192] The noxious atmosphere in rooms full of people and lighted by torches and candles could explain this phenomenon. Lavoisier had been using a eudiometer based on Priestley's principle. But it was a slow method and it had a limited precision. These limitations led Lavoisier to start thinking about ways to improve the eudiometric procedures. By the late 1780's he proposed a new method based on the combustion of phosphorus, inspired by his long standing knowledge of the effects related to the absorbtion of oxygen and the great affinity between phosphorus and oxygen.

Les eudiomètres établis sur ces principes, c'est-à-dire ceux où l'on profite de la grande affinité qu'ont pour l'oxygène le phosphore, les sulphures d'alcalis et le mélange du souffre et de la limaille de fer, sont bien préférables à ce luis de Priestley, Fontana et d'IngenHousz, dont la base est le gaz nitreux.[193]

---

191 Bachaumont, cited by Berreta 2000
192 Beretta 1998
193 Lavoisier 1790

Eudiometers working along these principles, i.e. those that use the high affinity of oxygen for phosphorus, alcalic sulphurs and a mix of sulphur and iron filings, are much to be prefered to those used by Priestley, Fontana and Ingenhousz, that work with nitrous gas.

This improved eudiometric technique allowed Lavoisier to follow up on Landriani's research and measure the salubrity of the air in different seasons. Moreover, it served him to do some crucial experiments on human respiration. His collaborator in these experiments, Armand Seguin, in 1791 wrote a "Mémoire sur l'eudiométrie" in which he admitted there were limits to the instrument and that eudiometry was still far from being a hard science. They were:

> ... encore fort éloignés d'avoir une science que l'on puisse appeler proprement eudiometrie. Les bornes étroites des connoissances que nous avons acquises jusqu'ici relativement à cet object, ne sont cependant pas une raison de les rejetter; nous devons chercher au contraire à les étendre, à les perfectionner, & tel a été le but des nouvelles recherches que je viens soumettre au jugement de l'académie.[194]

> ... still a long way off before they could speak of a proper eudiometric science. However, the milestones that are already erected in acquiring knowledge on this object give no reason to abandon the project. On the contrary, we have to continue investigating to improve our method, & that was the aim of my new research that I come to present to the academy.

Seguin and Lavoisier were also using Volta's eudiometer. Volta had been in Paris in 1782 and demonstrated his electrical eudiometer, showing how fluids could vaporise. The Parisian chemists started to use this instrument too and that allowed them to make water by electrically sparking two gases. This was a confirmation of his chemical theory by means of a very simple instrument. Lavoisier & co used different instruments, which had the same name, in different investigative projects. As Beretta clearly demonstrated: the measurement of the salubrity of air and the research on the composition of water were in fact both manifestations of Lavoisier's new theory of the chemical composition of gases and fluids. He concludes:

> The objective limitation of the instrument did not hinder these naturalists from elaborating and substantiating different theoretical standpoints. In the end, the projection of different theoretical expectations in the uses of the eudiometer

---

194 Seguin 1791

were fulfilled and the instrument successfully brought experimental practice from a utopian illusion to a revolutionary research program, the importance of which became apparent throughout the nineteenth century.[195]

These historical developments help to illustrate an interesting philosophical point. The developments in chemistry, proto-plant physiology and public health involved a constant feedback between theory, experiment and instrument design. Advances and improvement in these areas opened up new opportunities in each of the others.

This dependence of theory and experimental methods on the available apparatus (which depended in turn on the available budget to finance these expensive objects) leads nicely into Holmes' discussion of the instrumentation developed by Lavoisier. It is widely known that Lavoisier's laboratory was extremely well equipped and charges have been levelled that his experiments involved unduly complex, and expensive, apparatus. Holmes[196] refutes this assertion and states, "Lavoisier did not resort to complicated and unique apparatus fortuitously or to make chemical experimentation inaccessible to those less wealthy than he. He did so when confronted with problems that required new solutions." The prime aim was to solve a scientific problem. Costs and complexity were at best a limiting factor for someone like Lavoisier with very big (state) budgets. Still, complex instruments and complex procedures generally require more precautions and more interpretation of results while they increase the number of ways by which error can creep in. IngenHousz tried to keep his method as simple as possible to make it manageable even while travellling. His motto was to make it simple, but not too simple.

Instruments as the motor of knowledge

Instruments have played a minor role in the history and philosophy of science, and chemical instruments even more so. They are made of glass, which is fragile and easily damaged. IngenHousz was trying to send some glass tubes to some of his correspondents, to make sure that they were using the same instrument, but had to hear very often that they had arrived in splinters. These vessels and tubes would also break in the hands of the natural philosophers as they were exposed to sudden heat or cold, to violent explosions and damaging solvents. The quality of the glass was not really that of modern laboratory glassware, both in standard sizes (constant and even thickness of the vessel walls, smoothness of the inside, &c) or consistent composition. In the late eighteenth century there were no

195 Beretta 2000
196 Holmes 1985 p 148

standard fittings, stoppers and couplings. Rubber (caoutchouc) was introduced and used experimentally, with IngenHousz one of its first enthusiastic users (in contrast to earlier claims that rubber tubing was nonexistent in the eighteenth century[197]). Therefore, what can be found in museums or laboratory collections is neither representative nor comprehensive.[198] Old chemical glassware is not so glamorous as telescopes, microscopes or astrolabes. Lavoisier's laboratory contained around 6,000 pieces of glassware (known from his laboratory inventories) of which less than 1% survives. From Priestley's laboratory in Bowood, Wiltshire, where IngenHousz worked after Priestley left for Birmingham, nothing survives. Modern scholars have to be guided by manuscripts and engraved illustrations. Luckily these philosophers spend some time and effort in elaborating on their instruments and procedures. Moreover because the instruments were more than just measuring devices, they were also tools for public demonstration. As Golinski has convincingly argued the new scientific insights did not just become accepted because the natural philosophers really wanted it but because there was a receptive context in which patrons, allies and audiences could be readily recruited. The work of the experimental lecturers and the understanding of their courses by the audience helped to make their discoveries into established public knowledge.[199]

The didactic was just part of the natural philosophers' unwritten job description. Not only IngenHousz spent a lot of time explaining procedures and demonstrating his experiments, Lavoisier ensured that one-third of his *Traité élémentaire de chimie* in 1789 was devoted to the instruments and operations of chemistry. Apart from the theme of precision (we have seen IngenHousz' obsession with error), stability and change were described as the two opposing but essential tendencies in the development of instruments in the eighteenth century and beyond.[200] Innovation could be both technical as conceptual. Instruments and skills travelled swiftly once innovations were introduced. Dissemination of instruments and concomitant dissemination of practices and theories were important aspects of the chemical activity and an important product of the instrument maker's trade. Almost half of the addresses in IngenHousz' address book are from instrument makers. In the correspondence of IngenHousz many fragments allude to the shops or ateliers where to buy what instrument and on deals with friends and peers to buy instruments in London (the capital of instrument making at that time) or Paris and hand them on to travellers to destinations abroad.

197 Crosland 2000 p 93
198 Levere & Holmes 2000
199 Golinski 1992
200 Levere & Holmes 2000

The unwritten manual

A point of discussion among users of eudio- and other meters was that of reliability. Often the most elaborate eudiometers were the least reliable. The more manipulations that needed to be carried out, the more the chance for errors - something many the experimenters were well aware of. Tacit knowledge on how to manipulate these vessels, tubes and tubs was essential. That's why it was so often stressed that one had to perform these experiments many times in order to gain the necessary dexterity. There is a difference between reading about an experiment and seeing someone perform an experiment. Much better to do it oneself. It is like cooking. The difference between reading the recipe and doing it by the book, or learning by shadowing an experienced cook is immense. Moreover, many recipes in cookbooks are written by people who have done it so many times they can do it blindfolded. The result is that they tend to forget the little manipulations, the simple handlings and the manual dexterity to accomplish things that are very difficult to a novice. Some procedures are so evident to them that they simply forget to mention them. That must have been another reason why Lavoisier, Priestley, IngenHousz and so many others often invited people to their laboratories. Not just to have witnesses, not just to demonstrate their wonderful results but also to teach others how to do things properly. Golinski calls it "phenomeno-technics", the methods to reproduce the phenomena under investigation.[201]

Sibum calls it "gestural knowledge".[202] These particular capabilities were difficult to communicate. These embodied skills left few traces, even fewer than the material appartus. Still, IngenHousz and his peers were in the midst of a evolution in which instruments started to play an increasing part. By the end of the eighteenth century laboratory equipment was part and parcel of the scientific enquiry: thermometers, eudiometers, calorimeters, gasometers, and balances. The end of chemistry's ancien régime was marked by a shift from reliance on the senses and on informal estimates of quantities to a culture of increasingly precise measurement with a range of refined instrumentation.[203]

Questions about the role of instruments and their practical use have been asked before with regard to another influential experiment in the history of science. Heinz Sibum reconstructed the experiment which enabled James Prescott Joule to formulate the law of conservation of energy.[204] We have a fairly accurate description of this experiment, including drawings of the set-up, published in 1850. A paddlewheel driven by two weights that fall over a certain height churns water. The friction that is caused warms the water. The heat

---

201 Golinski 1998
202 Sibum 1995
203 Roberts 1995
204 Sibum 2000

the water gains should be the work of the weights while falling (or the work delivered by the man who pulled the weights to their starting height). In reality it turned out to be pretty difficult to repeat an experiment with similar results. Nature is full of variables and it is almost impossible to eliminate or circumvent them all, all of the time and in all places. Sibum described the methods and skills of Joule as a hybrid of craft consciousness and a gentleman specialism. One of the secrets of Joules was that he got his inspiration from brewers, whose business was a complex mixture of theoretical and practical tasks in which temperature control was of utmost importance. Coincidentally, brewers were also Priestley's source of inspiration when he discovered how to produce and collect fixed air, almost a century earlier.

IngenHousz and his contemporaries were also using an unwritten manual of tacit knowledge to perform their experiments. That's why reproducing these eudiometric experiments would be a intriguing task and an informative challenge. It was hard to perform even seemingly simple experiments, as testifies Thomas Beddoes, when lecturing at Oxford University in 1788, and writing to his mentor Joseph Black:

> What I find most difficult is to repeat some of those apparently simple exps which in your hand are so striking and so instructive. I have not yet learned how to show the gradual approach towards saturation by throwing slowly a powdered salt into water. What salt do you use? & how do you perform the expt? How do you contrive to make that capital expt which shews the burning of iron in dephd air? I mean to attempt it, but am told that the vessel has been frequently in other hands burst with great violence? do you put sand on the bottom? I know the form of the vessel &c. What salt do you use to shew the effects of agitation upon mixture? [205]

But reproducing IngenHousz' eudiometric experiments has to be a topic for further research. One hypothesis, though, may be proposed here: IngenHousz and his peers measured something with a degree of accuracy which was the subject of discussion, but which was reliable enough to be able to confirm some of their hypotheses and to reach a number of conclusions. The eudiometer was delivering results whose quantitative excellence was only partially important for the conformation of the assumptions about the photosynthetic activity of plants. It turned out to be crucially unreliable for confirming the claims in the field of public health, and would therefore disappear in the environmental discipline.

---

205 cited by Levere 2005

A question of precision

The eudiometer fits in a historical development of measuring subtlety that went from qualitative to quantitative to very precise. When the precision of measuring increased by orders of magnitude, new worlds emerged because the new accuracy revealed phenomena that went undetected before. With Priestley's proto-eudiometer one could start 'seeing' things in the air where only nothingness had been seen before. When Lavoisier started using highly precise scales, it became possible to weigh things that almost weighed nothing. Black's scales were able to weigh to one part in 200; Lavoisier's was able to go to an astonishing one part in 400,000.[206] It was an era when the invisible became tangible.

At the beginning of the aerial medicine, people were pleased that they could rely on something, which was more objective than just smell to determine if air was mephitic. The eudiometer in the form Priestley had presented it was providing exceedingly good results and many felt it had to be possible to produce a standard. Fontana wrote in 1779

> *The least variations of circumstances causes very great variations in the results of the experiments, ... so that the purest common air would appear to be noxious and phlogisticated air would appear less good, and even noxious... I have not the least hesitation in asserting, that the experiments made to ascertain the salubrity of the atmospherical air in various places, in different countries and situations, mentioned by several authors, are not to be depended upon; because the method they used was far from being exact, the elements or ingredients for the experiment were unknown and uncertain, and the results very different from one another.[207]*

Fontana and IngenHousz realised very well that the method needed to be standardised in order to yield any meaningful measurements. But none of their efforts, supported and enforced by men such as Cavallo (who published a description and illustration of the eudiometer[208] which repeated that of IngenHousz'), lead to a consensus. This was understandable as accuracy meant precision and the ability to reproduce. Small changes in the procedures could alter the outcomes. Leaving the gases mixed for a longer or shorter duration of time led to different results. Fontana, IngenHousz and Cavallo stressed this time and again. So, arriving at general standards about how to measure things was a point of debate. Standards in measuring or compatibility were then - and have been since then - a continuing and recurring problem, in science, technology and consumer goods.

---

206 Levere 2005
207 Fontana 1779
208 Cavallo 1781

As we have seen, IngenHousz devoted a complete essay[209] to the problem in which he carefully described the methods and procedures for good and reliable eudiometric testing. There he repeated in more detail what he had already described in his book of 1779. Five years later he confirmed that if one carried out the procedures always in the same way, "we may with as much precision judge of the degree of purity of the common air, as we are now able to judge of its degree of heat and cold by a good thermometer." The discussion between Landriani and the Fontanists would be ongoing. The latter mixed the airs in the testing tube, Landriani mixed nitrous and test air in a separate vessel, and only introduced it after mixing into the testing tube. Opponents of Fontana's method, such as Magellan, were having less consistent results but still kept defending their system because of their supplementary obsession with simplicity. Unnecessary elaboration of methods would keep eudiometrical procedures outside the range of competence of unskilled practitioners. The simplest was the best, thus repeating Priestley's epistemic rule based on politico-religious arguments, not on arguments to increase the trustworthiness of the knowledge. When the empirical basis gets lost in such a dispute, it turns into a battle between ideologies, a sociological game of rhetoric in order to convince as many as possible. Sociologists of science who reduce all scientific development to this battle tend to forget that in the end the instrument and the results it produces are what stays and not the people who were discussing them. It was the eudiometer that was still in use, even by people like De Saussure who took the science of plant physiology into another century, long after Priestley or Magellan had ceased to be important in the debate.

But it is a matter of fact that interpersonal relationship played a role too. Cavallo was one of Sir Joseph Banks clients, who had succeeded Pringle as president of the Royal Society. Magellan was having a heavy feud with Banks, as Magellan tended to communicate directly and independently with scientists on the continent, challenging the centralising monopoly of the Society. The scientific dispute over the value of the eudiometer was fuelled by a personal conflict between Magellan and Cavallo.

Others would cut straight through the ideological debate and improve the eudiometer on the basis of hard evidence. In 1783 Henry Cavendish published a letter in the *Philosophical Transactions* entitled "Account of a new eudiometer". He suggested new and rigorous standards to improve the accuracy of Fontana's method, which was definitely the most accurate to date, apart from the instrument that he had invented.[210] He proposed a gradual mixing of the airs. And he unmasked some sources of error: the size of the measure used (in Fontana's method too much water could stick to the sides of the measure and the sides of the tube - a possible source of error IngenHousz

209 IngenHousz 1785
210 Crosland 2000

had already mentioned from the beginning) and the variation in the ability of different kinds of water to absorb the nitric acid produced when the gases are mixed. Cavendish introduced in eudiometry what Lavoisier had been doing in chemistry in general. He started weighing the quantities of air under water instead of measuring volumes. All this would allow Cavendish to improve on Fontana's method. Fontana obtained results with an accuracy of 2%, Cavendish using even simpler apparatus and even greater discipline, got in 1785 results to approximately 1%.[211] Gradually refining the measurement method had some unexpected effect. The alleged differences in air quality between various locations were not any larger than the expected margins of experimental error. This was not new. Fontana had already indicated in 1779 that:

> When all errors are corrected it will be found, that the difference between the air of one country and that of another, at different times, is much less than what is commonly believed, and that the great differences found by various observers are owing to the fallacious effects of uncertain methods.

Senebier would find that when the mixed gases were made to sit for days and weeks over water, their volume would continue to diminish.[212] However, this was not considered to be a problem as the procedure was to shake the gases and measure them within half an hour. Senebier's fact was interesting but not relevant to the procedure, which continued to deliver consistent results.

We now know of course that they were making very crude measurements of the amount of oxygen in the air, which is roughly always the same, wherever you measure. Except for maybe in rooms where lots of people gathered, using up the oxygen and producing carbon dioxide. But the concentration of atmospheric oxygen in Ostend or in Brussels would not be any different, be there a storm, frost or sunshine. That's where the eudiometer was unable to provide some clarity in the aerial unknown.

Limits to the measurable

According to Golinski and others, the eudiometer was a failed instrument, because it did not deliver the right and reliable data.

> The eudiometrical apparatus and procedures diversified and became more complex. Individual practitioners staked their claims for the superiority of their own methods, and as a result no consensus emerged. In the absence of agreed-upon techniques, replicable results were not achieved. Instead disputes

211 Crosland 2000
212 Senebier 1783

*about the propriety of various methodological refinements multiplied. The nitrous air
eudiometer failed to live up to the original expectations vested in it as an instrument.*[213]

Priestley's contribution to this failed history was considerable. His ideal of
simplicity and economy in the design of instruments and procedures was going
backwards. Priestley was convinced that the potential of natural knowledge was
to stimulate moral and material progress. In order to realise these goals it was
essential that experiments were made as widely replicable as possible. That's
why he gave considerable attention to detailed descriptions of instruments
and procedures. That's also why he thought the more complicated and more
expensive eudiometers used by Fontana and IngenHousz were bad instruments.
Their price would make them inaccessible to the masses and therefore they were
to be despised. Not because they more or less delivered reliable results. Priestley's
personal situation is partly to be blamed for his opinion on the simplicity of
instruments. His financial situation did not give him enough leeway to install
any kind of expensive equipment in his laboratory. And over the years these
instruments became more and more expensive. Eighteenth century chemistry
started to become big science, just like physics. Historical reconstruction has
shown that some of the gasometers used by Lavoisier costed 5,774 livres (or
approximately 230,000 euro today).[214] These instruments were all custom- and
purpose-made, hand-build to the highest precision, often with precious materials.
To make a comparison: the yearly salary of Fourcroy as professor at the Jardin du
Roi was 1,200 livres. Lavoisier did not see any problem in these developments:

> *C'est un effet inévitable de l'état de perfection dont la chimie commence à
> s'approcher, que d'exiger des instruments et des appareils dispendieux et
> compliqués : il faut s'attacher sans doute à les simplifier, mais il ne faut pas que
> ce soit aux dépens de leur commodité et surtout de leur exactitude.* [215]

> It is an inevitable effect of the state of perfection to which chemistry is
> beginning to approach. This requires expensive and complicated apparatus:
> no doubt one should try to simplify them, but not at the expense of their
> convenience, and especially of their exactness.

This stands in stark contrast to Priestley who writes in his *Experiments and
Observations* in 1774:

> *...notwithstanding the simplicity of this apparatus, and the ease with which all
> the operations are conducted, I would not have any person, who is altogether*

213 Golinksi 1992 p 117
214 Crosland 2000
215 Lavoisier, 1789 p 267

without experience, to imagine that he shall be able to select any of the following experiments, and immediately perform it, without difficulty or blundering. it is known to all persons who are conversant in experimental philosophy, that there are many little attentions and precautions necessary to be observed in the conducting of experiments, which cannot be well described in words. ... Like all other arts in which the hands and fingers are made us of, it is only much practice that can enable a person to go through complex experiments, of this or any other kind, with ease and readiness.

However, this evolution was only the beginning of a trend in which smaller, simpler and cheaper instruments would enable delivery of accurate results, but some of these were only accurate enough to use in demonstration lectures, but lacked the precision that was rapidly becoming necessary for research. Sometimes nature is just too complicated for simple instruments. The construction of the 21st century Large Hadron Collider illustrates that the smaller the things you want to measure, the bigger your measuring machine has to be.

But back then in the 1780s the unwillingness of Priestley to budge for the arguments of others, such as Fontana an IngenHousz, would seriously hamper the development of more rigorous practices and refinements in instrumentation. Again ideological beliefs would stand in the way of trustworthier knowledge.

A failed instrument?

There has been discussion among historians and sociologists about the role of the eudiometer in science and society and the significance of the eudiometric research programme. Schaffer claimed that the programme to measure goodness of the air fitted in a wider movement in which the enlighted government of the Habsburgs was enthusiastically promoting social and political reforms but that such a program would meet its final defeat with the restoration.[216] Others have claimed the contrary. The eudiometer did not become redundant because its socio-political game of power was finished. In fact, the eudiometer would become even more important in the nineteenth century when capitalists would start really using it to measure the salubrity of the air in function of the productivity of the labour force.[217] Beretta has also shown how complex the sociological situation was in which the eudiometer was demonstrated and promoted. The lines between the natural philosophers and the government were sometimes short but not always decisive. Scientific discussions between men such as Landriani, Volta and Lavoisier about the real cause of the unhealthy effects of bad air, spilled over in concrete decisions to be made on the building plans for hospitals, theatres and prisons. Lavoisier expressed his

216 Schaffer 1990
217 Beretta 2000

concerns that it would be humiliating for the academy of sciences and for the nation if the building plans for new public buildings would not be based on the newest insights in the qualities of the air that were founded on sure and exact scientific principles. Golinski has argued that the connections between eudiometry and medical reform programs had been severed after the critiques of Cavendish and Volta. The project had failed to achieve anything and unravelled in disillusionment and disappointment.[218] Lavoisier however did not give in to the comments that the eudiometric measurements were illusory and continued to constitute commissions for the construction of a new hospital and to keep pointing at the importance of studying the effect of gases on public health.[219] On 15 February 1785 he presented a paper to the Société Royale de Médicine in Paris which title spoke volumes: "Mémoire sur les altérations qui arrivent à l'air dans plusieurs circonstances où se trouvent les hommes réunis en société". The composition of the air of common air was generally 25 parts vital air and 75 parts gaz azote, but the sensations and fatigue that appeared in closed rooms were due to the noxious action of the rising concentrations of fixed air, formed through animal respiration.

Interestingly enough, the eudiometer would still be in use long after the fall of the phlogiston theory. It was equally useful to the new chemists, and to natural philosophers investigating other parts of nature. People like IngenHousz and Senebier, who studied plants, would take the instrument with them into the era of the new chemistry. Sociologists of science have seen this kind of evolution as a proof for the fact that the utility of an instrument depends on subsequent applications or interpretation and on the contexts of practice in which it is taken up. The eudiometer story appears to suggest that the social context is not of ultimately determining importance, but that the inseparable and mutual relationship between a thing and its users leads its own life, being highjacked by social and other circumstances.

Can this explain that the eudiometer turned out to be a "failed" instrument? No user of the eudiometer in the eighteenth century seems to have used such a term, despite the fierce discussions about its application and importance. Golinski, looking backwards from a sociological viewpoint, did see only the disputes and the bewildering collection of various test results. His opinion is echoed by Lissa Roberts who wrote in her contribution about the eudiometer in the historical encyclopaedia of instruments of science[220] that the eudiometer could be classified as a failed instrument, as its structure and calibration were never standardized. It was indeed true that the eudiometer was tied by various uses to diverse fields of enquiry, from phlogiston chemistry to plant physiology and from aerial medicine to the new chemistry. Many researchers developed their own type of eudiometer, from Scheele and Magellan to Cavendish or De

218 Golinski 1992
219 Beretta 1998
220 Bud 1998

Saussure. Giobert used the phosphorus eudiometer as originally developed by Lavoisier and perfected it to the point that he could even measure the minute amount of carbon dioxide in the atmosphere.[221] He compared the air in his home-town Turin with the air in the spa of Vinadio, in the Piedmontese Alps west of Turin, at almost 1000 meters altitude. He built on two decades of eudiometric research and knew that, according to his great example De Saussure, the air on the top of the mountains was less pure (containing less oxygen) than down in the valley. In Turin he measured on average 26% of "air vital", 71% of "gas azote" and 2% of "gaz acid carbonique". The air in Vinadio contained slightly more oxygen and less carbon dioxide, and therefore was healthier. His eudiometer would be followed in the first decade of the nineteenth century by several new models, developed by, among others: Davy, Dalton, Gay-Lussac, Seguin, Guyton, De Marté, Berthollet, Allen and Pepys. They applied variations in the form of the vessels or tubes and used various chemicals such as iron sulphate of iron, sulphuret of potassium, mercury or phosphorus. At the same time variations on Volta's sparkling eudiometer were developed by Berzelius, Mitscherlich, Dobereiner and Ure.[222] In the course of the nineteenth century the eudiometer retreated from its public life as a tool in discussions about chemistry or public health into the chemical laboratories were it led a silent life, doing what it was first invented for: measuring the oxygen content of gaseous mixtures. Glass tube eudiometers were still sold in London in 1822. In a catalogue of optical, mathematical and philosophical instruments[223] one finds several eudiometers listed. Various models, from Fontana, Priestley, Davy &c are on offer at 4 pounds and 4 shillings. In 1838 Dalton was one of the last writers to make any considerable use of the nitric-oxide eudiometer. He compared the nitric-oxide method with Volta's s hydrogen eudiometer and the sulphide of lime absorption method, concluding:

> The nitrous gas eudiometer is of singular utility on many occasions. No other can exceed it in accuracy when mixtures contain very little, as one or two per cent of oxygen; or on the other hand when nearly the whole of gas is oxygen. But when the mixture of gases contains from twenty to eighty per cent of oxygen, as in the case of common air, it is not the best when great exactness is required. [224]

That explains why it is still being used, in the proper circumstances. Eudiometer tubes are still sold in shops today for scientific and laboratory equipment, for around 45 dollars.

In the eighteenth century however, despite its inherent limitations, the eudiometer had been used to support environmental insights in health

221 Giobert 1793
222 Brougham 1837
223 Jones 1822
224 Benedict 1912

and disease and succeeded in motivating the royal patrons of the natural philosophers to pursue public health projects. However, the variation in the atmospheric concentration of oxygen is (as we know now) too small for accurate measurement with the eudiometer and the instrument turned out to be useless as a device for tracing infectious disease. Meteorology and medicine developed further without the eudiometer. But its role in the discovery of photosynthesis proves also that a "failed instrument" can be very useful. The eudiometer was the motor behind three decades of inspired research, from the first experiment of Priestley to the final measurements by De Saussure. The eudiometer was first used in a phlogiston setting, but would survive the theoretical shift towards oxygen without any problem. Gas chemistry was central to the chemical revolution, and since this was a revolution based upon the consistent application of experimental quantification, measuring gases took pride of place.[225] The eudiometer was to be the stepping stone towards the gasometers. Instruments were the driving force behind all the conceptual changes. And the gasometers, just like the eudiometers, were not delivering the accuracy one would wish for. The radical changes in the chemistry of the 1780s are not as radical as has often been believed. The continuity, partly carried by the instrumentarium used, appears to be greater than many radical philosophers or sociologists have dared imagine. It is a fact however that an instrument such as the eudiometer can play a significant role in a wider social history, influencing developments in society and in turn being influenced by new insights in culture and policy.

Enthusiasm and criticism

Fontana and Landriani tried to use their instruments to enlist the support of their patrons, the Grand Duke of Toscana and the Austrian ruling house in Milan. They promoted the eudiometer as part of ambitious programs on environmental and social control; an approach that appealed to the enlightened leaders of the school of Joseph II. As well as with pneumatic medicine, Fontana & Co were very enthusiastic and saw in their method a new means to great ends, hoping it would enable them to fight famine and disease, improve agriculture and environment.

They tapped into a long and strong tradition. Public health and the possibility that bad air was contributing to the transmission of disease were an established topic since Sir John Pringle's seminal work *Observations on the diseases of the army* from 1752. Pringle was IngenHousz' friend and mentor, so it should be no surprise that IngenHousz, being a doctor too, had been very intrigued by the possibilities of ameliorating air in order to improve health. And he was not alone. There was a generally accepted belief that air and health were somehow

---

225 Levere 2005

connected. But how good was air? How good could it be? And how was one to tell good air from bad? With the beginning quantification of that aspect of reality, it is no surprise that IngenHousz, Lavoisier and many others seized this occasion with both hands. As a matter of fact, some went very far in their enthusiasm. Some pneumatic practitioners saw themselves as prophets of a forthcoming scientific enlightenment in medicine, when the causation of disease by material elements in the natural environment would be recognised and controlled. In this age of "philosophy and enlarged sentiment", as Adam Walker had said in his *Philosophical estimate of the causes, effects, and cure, of unwholesome air* in 1777, they would be able to locate in the air the causes of disease that a previous age of religious tyranny had ascribed to divine retribution.

The question is whether we have to go as far as Schaffer was so eager to go, to see the embodiment of the will to power and control in these ideas. Surveillance of the atmosphere was to lead to control of many aspects of human life; eudiometry would play a leading role in government-directed programs of improvement. Berreta has debunked these all too ambitious claims of the sociologists of science, and the history of the eudiometer as reconstructed here seems to support his idea. An instrument such as the eudiometer is having practical and theoretical implications that are far too complex to be reduced to a sociologically determined process.

The protagonists of this story themselves Fontana and IngenHousz, on the contrary were very well aware of the daunting task they had taken upon them. They stressed the immense complexity of the phenomenon to be measured and recognised that the multitude of errors that could be committed was only a reflection of the complexity of the reality they wanted to chart. Fontana, for one, realised there was more to the atmospheric air than he could dream of, and more than could be measured with the eudiometer, as he described in his unpublished essay *Science de l'air*, probably dated 1779:

> Il ya une vapeur aëriforme lourde, et pesante, qui ne s'eleve guere de la surface du sol, qui se trouve dans plusieurs souterrains, ou cavernes, et sur tout dans la Grotta del Cane près de Naples, dont les principaux caractères sont d'etouffer la flamme, de tuer les animaux qui le respirent, et de n'etre point diminué par l'air nitreux; mais il n'est point pourtant de l'air phlogistiqué.
>
> ...
>
> L'air qu'on croioit unique et seul s'est partagé en des nombreuses familles distinguées par des caracteres, qui ne sont pas moins saillans, que ceux qui distinguent le cheval de l'homme.[226]

There is a form of air, heavy and ponderous, that hardly elevates itself above the soil, and that can be found in several caves and subterranean spaces,

---

226 Abbri 1991

and especially in the Dog's Cave near Naples. Its main characteristics are that it extinguishes a flame, kills animals that inhale it, in no way reacts with nitrous air, but is definitely not plogisticated air.

...

Air, which was thought to be one unique substance, turns out to be distinguishable in several families each with their own properties, no less striking than those that distinguish a horse and a man.

IngenHousz described similar observations in his book. Anomalies like these did not stop their thoughts or works. These were just unexplicable phenomena waiting further investigation. This complexity with so much unknown variables, gave way to a multitude of approaches. And all criticisms and doubts fuelled the development of new eudiometers.[227] As we have seen, Volta came up in 1777 with his electric eudiometer in which a sample of air was mixed with inflammable air (hydrogen) and brought to ignition by an electric spark. This instrument was to be taken up by Cavendish and Lavoisier and many others. Especially in Lavoisier's terminology there was a nice rationale behind this procedure: the hydrogen and the oxygen reacted to form water, initiated by the explosion. It was to be just another in a whole series of 'improved' instruments that were to take the eudiometer right into the 1800s. The old nitrous air reaction would however stay as a trick very popular in chemical education.

Commentators such as Golinski see the eudiometer as a failed project as it "had failed to achieve anything like its originators' aims and had unravelled in disillusionment and disappointment." They should have read IngenHousz' and others' writings. They were enthusiastic defenders of eudiometry, even long after it had ceased to be the revolutionary machine that it indeed had promised to be. But it was their tool of choice because it gave them just enough leverage to unravel an unknown natural process and just the right kind of consistence to determine the effect of plants on the atmosphere.

Golinski continues his slightly tinged assault on eudiometry as a failed scientific phenomenon that should be understood as the result of a sociological construction. It failed partly because "the problems in building a network of disciplined practitioners to use the instrument". Priestley made no offers to install his eudiometric method as the one and only. His test was presented as an accidental discovery but he sought neither credit nor responsibility for it. Golinski: "Perhaps he was weary of the compromises with political or commercial power that an attempt to build a network of eudiometrical informants would require." He seems to oversee the fact that scientific procedures are always evaluated on the basis of the results they can produce. Priestley's ethics were for him and the people in his network a counter-productive factor. In 1800 he would write about Lavoisier's new chemistry that

---

227 Golinski 1992 p 126

"... till the French chemists can make their experiments in a manner less operose and expensive ... I shall continue to think my results more to be depended upon than theirs." Others, including Lavoisier, Fontana, Senebier and IngenHousz used the eudiometer to great effect. Results that many generations later could still inspire new generations of scientists. Measuring and defining the quality of the air is still one of the mainstays in environmental science. Be it that the methods used nowadays measure different substances in a different way. Just like the relationships between the climate scientistst and the politicians in 2010 have changed but have stayed fundamentally the same. The former struggle with complex questions, the latter want simple answers.

An instrument in history, the history of an instrument

A conventional view on the history of science identifies the end of the eighteenth and the beginning of the nineteenth as the moment when chemistry acquired the status of a discipline, the birth of a modern science. Lavoisier reorganised the subject and established a distinct scientific basis. This was primarily seen as an achievement of theory: the consequence of correct ideas grasped by a great mind. Golinski and others have rightly shifted this exclusive attention to theory from exactly this theory to the practice, and from individuals to communities. Science is not just a theoretical endeavour but just as well an activity grounded on an instrumental basis, which is a concoction of instruments and techniques wrapped in a social setting within a community of practitioners. The eudiometer has been shown to be one of the elements in the mortar that kept a whole community of researchers together, even while they were at odds about its significance and proper use.

Peter Galison is a philosopher of science who helped to open up this new additional and fruitful perspective on science. He chose to stay very close to the actual practice of science, in his case the physics of the twentieth century. From a fine and detailed reconstruction of the history of physics, he ended up with some fresh philosophical ideas about the way science works.[228] He describes a relative autonomy of the experimental versus the theoretical physics. Both disciplines are deeply intertwined but at the same time enjoy a great autonomy. Theories suggest experiments and experiments confirm or refute theoretical hypotheses. But essentially they both go through their own historical developments. And on top of that, Galison discerned a third layer of scientific activity: the instrumental dimension of science. This is where the instruments necessary for the experiments are designed, constructed, adjusted and refined. This field of activity is also relatively autonomous of the two other, still at the same time deeply interlinked, as theory and practice influence the development

---

228 Galison 1997

of instruments, and vice versa. This autonomy has grown in the course of the twentieth century. At the time of IngenHousz, the men who formulated the theories were often those who did the experiments and were just as well closely involved in designing the necessary instruments, if not in making these themselves. (One of the many things IngenHousz did was inventing the cover glass for the microscope.) Experiments and instruments are therefore no mere epiphenomena of abstract theories.

The evolution of science takes on different courses at different speeds in these three layers. As the case of IngenHousz demonstrates, the instruments (the eudiometer) stayed more or less the same, while the experimental data were refined, while the theoretical framework in which they were interpreted changed over the course of two decennia. The incommensurability question as phrased by Kuhn may look like a problem when looking at the level of theories, but seems to disappear when one looks also at all the interlocking activities which are displayed in the making of science, and of which the instrument is just one. (More on this and other topics of the philosophy of science in a later chapter.) No cement is needed in this 'dry wall' in which all components fit together and create stability because of their nice fit.

A forgotten instrument

It would be neither the first nor the last time that a scientific instrument only provided partial results. Every instrument, however precise you can make it, is prone to producing errors. The calorimeter, which was given that name by Lavoisier and Laplace in 1781, encountered similar problems. But the fact that calorimeters are still in frequent use in laboratories may have helped to give that instrument a more reliable fame than the eudiometer, which has been wiped from memories, just like one of its most important users.

rewind

A crucial instrument: the rise and fall of the eudiometer

The eudiometer is more than an instrument. It could be called a family of similar instruments designed to do the same but often based on fundamentally different designs, developed in the course of half a century, used by many natural philosophers. For Priestley the nitrous test with the eudiometer was the embodiment of the phlogiston theory (a direct measure of phlogiston in the sample of air) and as such a key empirical element in his world-view. To IngenHousz it was a crucial tool because it allowed him to quantify the degree to which plants ameliorated the air. The eudiometer was also a corner stone in the development of the new chemistry and was one of the main sources of argument in the new discipline of aerial medicine.

These developments in chemistry, proto-plant physiology and public health involved a constant feedback between theory, experiment and instrument design. Advances and improvements in these particular areas open up new opportunities in other areas. In the second half of the eighteenth century, the eudiometer was one of the instruments that supported the leverage of science from a qualitative towards a quantitative enterprise. The eudiometer, despite the fact that it wasn't all that reliable, was far from a failed instrument, as some have argued. It was essential in proving the existence of the process of photosynthesis. An achievement that hardly can be called a failure.

This history of the use and evolution of this instrument helps to show that science is not just a theoretical endeavour but equally an activity ground on an instrumental basis, which is a cocktail of instruments and techniques wrapped in a social setting within a community of practitioners. The eudiometer has been shown to be one of the elements in the mix that kept a whole community of researchers together, even while they were at odds about the significance and the proper use of the thing.

BOTANICAL GAZETTE

NOVEMBER, 1893.

On the food of green plants.[1]

CHARLES R. BARNES.

The constant tendency of biological science is to minimize the difference between the physiological processes of plants and animals, and to recognize, under the varying forms, a remarkable functional unity. External form and even function were studied long before it was known upon what, in essence, both depended. Dujardin, a zoologist, seeking in 1835 a name for the living matter of which some of the simplest animals were composed, selected the word "sarcode." Von Mohl, a botanist, seeing in 1846, in the cells of some plants, previously unnamed contents which he considered the simplest living material, called it "protoplasm." The acute Payen immediately suspected the identity of these two substances. Cohn in 1850 maintained this identity. But it was not till 1860 that Max Schultz definitely established it. This year 1860 marks an epoch, since modern biology takes thence its rise, thanks to Darwin and Schultz.

Having thus a common starting point in its physical basis, it was natural to expect that the manifestation of life in plants and animals should be essentially similar. Unfortunately, not only in the popular mind are the functions of plants and animals supposed to be radically unlike, but even in many scientific or pseudo-scientific text books they are either specifically or impliedly treated as belonging to totally different categories. And the popular notion is in reality derived chiefly from the text books; although the newspaper article is responsible for much "science falsely so called." This notion apparently depends in part upon the superficial, yet apparently radical, differences between the higher plants and the higher animals, and is deepened by the fact that we are con-

---

[1] Read before section G, A. A. A. S., Madison meeting, August, 1893.

30—Vol. XVIII.—No. 11.

Memory and priority:
who was first?
and why has he been forgotten?

The more one learns about IngenHousz and what he did, the more one wonders why his name has largely been forgotten. Why is he not better known - as the discoverer of photosynthesis, or at least as one of the people who contributed to the long process of trying to understand what plants and sunlight do to sustain life on earth.

Who discovered what?

As may be clear by now, photosynthesis was not discovered in one day. Apart from all the inter- and intra-personal circumstances that can partly explain IngenHousz' historical fate, a philosophical argument should be made. Although IngenHousz may have discovered photosynthesis, he didn't know he had. He rightfully concluded (which nobody else did) that the green parts of plants in the light of the sun produce oxygen, which is beneficial for other forms of life; and that they produce carbon dioxide when in the dark. He did not have the slightest idea on what really happened inside the plants. He did not -could not- know that a very intricate mechanism of green pigments inside the leaves, powered by photons from the sun, through a cascade of biochemical reactions, produces carbohydrates from CO_2 and water. All that was to come later.

In 1804 Nicolas-Théodore de Saussure recapitulated the whole botanical physiological knowledge in the new terminology, supported by state-of-the-art quantified experimental data.[229] He brought the studies on plant nutrition to temporary completion, while at the same time opening up new paths for research.

A new step forward was made when Julius Robert von Mayer describes in 1845 the metabolism of plants in more detail,[230] calling the light-driven process by which plants reduce carbon dioxide to organic matter "assimilation". In 1893 Charles Barnes was the first to coin the words "photosyntax" and "photosynthesis"

---

229 De Saussure 1797
230 Mayer 1845

for the biosynthetic process of green plants.[231] It is the latter that stuck and is still in use today. More than a century of further research would lead to an additional unravelling of the complex process of photosynthesis, a long story in which feature people such as Cornelius Van Niel, Robert Hill, Samuel Ruben, Martin Camen, Melvin Calvin, Andrew Benson and Rudolph Marcus.[232] Not that after all that the research on photosynthesis could be considered finished by the end of the 20th century. In 1993 Howard Gest succinctly gives this definition of photosynthesis:

> Photosynthesis is a series of processes in which electromagnetic energy is converted to chemical energy used for biosynthesis of organic cell materials; a photosynthetic organism is one in which a major fraction of the energy required for cellular synthesis is supplied by light.[233]

It had become clear that photosynthesis was not solely realised by green plants. By then, photosynthetic and semi-photosynthetic bacteria had been discovered, forcing a redefinition of the process. IngenHousz' description that the process happens in the green parts of plant becomes part of a wider ranging theory. Similarly to how Newton's laws nicely fit within Einstein's broader theory of physics.

And the quest for the essence of photosynthesis was not over yet. In 2008 scientists from the Carnegie Institute gave the story a new twist.[234] While plants, algae and bacteria produce food from sunlight, CO_2 and water, thereby releasing O_2, Arthur Grossman and colleagues found that certain marine microorganisms in nutrient-poor parts of the ocean have evolved a method without the uptake of carbon dioxide and the production of oxygen. The full significance of this finding remains to be seen, but it makes one thing clear: science hasn't finished yet with photosynthesis.

IngenHousz' discovery was just one small step in a long quest. The process he described would only get a proper name some hundred years later. And as is well known, things without a name don't get a proper and distinguished spot in people's minds. One could rightly defend the argument that the discovery of photosynthesis was a long sequence of events in which a multitude of researchers, in a transnational effort that lasted from the beginning of the 17th century till the early 21st century, slowly but surely unravelled this wonderful natural process.

Science is a step-by-step endeavour. Depending on how one defines photosynthesis and how one chooses the essential aspects of it, another person may be nominated as the discoverer. This nevertheless stresses the fact that

231 Barnes 1893
232 Govindjee & Krogmann 2004
233 Gest 2000
234 Bailey 2008

science is performed by many like-minded people, working as a team, spread over space and time. IngenHousz, Priestley, Senebier and many others may not have been the best of friends, they stimulated each other to reconsider their claims, improve their experimental set-up and refine their hypotheses. The competition may have been unpleasant at times but it spurred the drive to do better. The quarrels were all too human: the result transcended what any individual could have achieved in isolation.

Thus, the discovery of photosynthesis as outlined above took more than a century and the core of it involved one Englishman, one Dutchman, two Swiss, one American and one German, among others. Two were churchmen, two medical doctors, a professional chemist and a biologist. It is another characteristic example of the international and interdisciplinary nature of major scientific progress, driven by the colourful personalities and individual curiosity, sometimes hindered and then again propelled by social developments and tensions. This is not the story that is often told about science and scientific progress, with a fabulous 'eureka'-moment as the culminating moment of a rational discovery process.[235] Oxygen was not discovered on one day, photosynthesis took even much longer.

Paths to oblivion

It is however curious that IngenHousz, though he did play a significant role in this story, has been largely forgotten. His name does not feature in the text-books, not in secondary school, nor at university. A plant physiologist can easily get a PhD in the biochemistry of plants, without ever encountering IngenHousz' name. Of course, that's part of the scientific education. History of science is not often part of the sciences' teaching packet. But students readily learn to associate Newton with gravity, Darwin with biological evolution and Einstein with relativity. Others have been assigned a process, law, theorem or phenomenon to their name, such as Krebs, Coulomb, Fermat or Doppler, which helps to be remembered. Somehow IngenHousz has been forgotten, and it might teach us something about the enterprise of science to try to find out what might have caused this case of historical amnesia.

A sure way to be forgotten is to behave in such a way that nobody remembers you. IngenHousz was a humble person, not interested in fame, pomp or circumstance, enjoying his independent status, as testifies this passage from a letter to Van Breda on 20 February 1788, which was already alluded to earlier:

My werd aangeboden B. [Baron] Van Swieten op te volgen en dus het
opperbewint aller universityten en alle geleerde van een magtig en wyd

---

235 Schaffer 1986

uytgestrekt ryk te voeren. De Keyzerin bood niet alleen een grooter inkomen maar ook adelyke titels en publike decoratien. Ik sloeg dit alles af zeggende dat ik niets meer verlangde dan de continuaties van wat ik reeds ontfangen had ; en ik heb door een nu 20 jarigen ondervinding bevonden dat ik onyndig gelukkiger en geruster geleefd heb dan ik zou geleefd hebben indien ik door heerszugt en hoogmoed my had laten verlyden.

I have been offered to succeed Baron Van Sieten and thus manage the supreme command of all universities and all savants in a powerful and vast empire. The Empress not only offered me more income but noble titles and decorations as well. I declined saying that I desired nothing more than the continuation of what I had already received; and that I learned after 20 years of experience that I am infinitely more happy and serene than when I would have seduced by imperiousness and haughtiness.

He was low-key and introvert, enjoying friendships, shying away from stardom. This is in sharp contrast to some of his fellow researchers of that time. One of the strongest 'competitors' was Joseph Priestley. Colleague, sparring partner and source of inspiration, IngenHousz held him in high esteem. On many an occasion in his books and articles he sung Priestley's praise. He always started his story by referring to the work of the "Reverend Doctor" and how he first discovered the fact that plants ameliorate the air and that he developed the nitrous test to measure the salubrity of the air. But the appreciation was not reciprocal.

Lost in discussion

Joseph Priestley recognized the fact that IngenHousz rightly claims to be the first to have described the beneficial processes in plants. Priestley read IngenHousz' book shortly after publication and writes to Fabroni on 17 October 1779:

> *I have just read and am much pleased with dr Ingenhousz' work. The things of most value that he hit upon and I missed are that leaves without the rest of the the plants will produce pure air and that the difference between day and night is so considerable.* [236]

This would not prevent Priestley to repeatedly claim later that he was the first to observe and publish. Priestley promised in several letters to put things in order in the next volume of his publications (part IV). In the letter IngenHousz wrote to Van Breda on 24 May 1786 he mentioned that he has been informed that Priestley's new book came out in London. He was very curious to see if

---

236 Schofield 1966 p 160

Priestley mentioned IngenHousz at all, as he had promised two years earlier. He himself could not get hold of the new publication in Vienna quickly, so he hoped to hear from Van Breda, as he suspected it would be more easily available in Holland. On 6 July, Van Breda must have answered him and IngenHousz replies:

> *...geen reactie van Priestley omtrent het geen wy reeds hadden opgemerkt omtrent zyne ingewikkelde manier om sig het sonneligt op planten toe te eygenen. hy moet met de zaak verlegen geweest zyn en niet wel geweten hebben hoe sig te zuyveren. Dit denk ik te meer omdat hy volgens zyn brief aan my, zeer wel weet dat wy hem als een afgunstig physicus hebben doen voorkomen, zeggende zelfs, dat myne vrienden hem hielden als <u>den sultan die geen competiteur tot zynen troon oner zyn ogen dulden kan</u>.*

> ... no reaction from Priestley concerning what we already had remarked about his convoluted manner to appropriate the sunlight on plants. He must have been embarrassed about the situation and uncertain how to exculpate himself. I am even more convinced because he, according to his letter to me, very well knows that we exposed him as a envious physicist, even saying that my friends took him for <u>a sultan who would not tolerate a competitor to his trone in front of him</u>.

In a letter[237] written from Birmingham on 24 November 1787 Priestly wrote to IngenHousz:

> *Dear sir,*
> *... I thank you for the French edition of your work, which I received some time ago; but I am sorry to perceive, in the preface, something that looks as if you, or your friends, thought I wished to detract from your merit, which is very far from being my disposition. I do not indeed distinctly see what ground there is for any interference between us. That plants restore vitiated air, I discovered at a very early period. Afterwards I found that the air in which they are confined was sometimes better than common air, and that the 'green matter', which I at first, and several of my friends always, thought to be a vegetable produced pure air by means of light. Immediately after the publication of these facts, and before I had seen your book, I had found that other whole plants did the same. All the time I was making these experiments I wrote to my friends about them; particularly to Mr Magellan and desired him to communicate my observations to you as well as to others; but I believe you had not heard of them; so what you did with leaves was altogether independent of what I was doing with whole plants. The same summer and the same sun, operated for us both and you certainly published before me.*

---

237 Schofield 1966 p 248-49

This appears to be the true state of the case; and surely it leaves no room for suspicion of anything unfair, or unfriendly. But whatever your friends may say, I have no thoughts of troubling the public with any vindication. I value you and your friendship too much to wish to have any altercation on the subject. Indeed there is nothing to contend about; If, on any future occasion, you will do me the justice to give this state of the matter, I shall be happy. If not, I shall not complain.

Despite all those promises, Priestley never provided an accurate reference to IngenHousz' work or publications. Gest has reconstructed these sad facts meticulously.[238] In Priestley's book of 1790, IngenHousz is only mentioned once, in a footnote without reference to the date or any other publication details. IngenHousz is not listed in the index, although names of minor researchers are listed. IngenHousz, in contrast, is often giving reference to Priestley, honoured him in the foreword of his book and gives reference to experiments and publications of a man he described as one of his great sources of inspiration.

In July 1790, IngenHousz decided to try one more time and is so bold as to write to him.[239]

Dear sir
As the following intelligence may be interesting for you, i take this liberty to give you some account of it. Mr Seguin lately repeated at Paris the well known experiment of the composition and decomposition of water, by employing in the synthesis vital air disengaged from oxygenous salt, which air is certainly the finest that can be produced. twelf ounces of water were obtained without the least atom of any acid; the specific wight of this water was to that of distilled water as 18671 to 18670. the difference of the weight of the two airs employed and that of the water obtained was of 12 grains. i got this fact a few days ago from young mr de jacquin, my nephew, who is still at Paris. The french chymist think this experiment to be decisive. if it is not so, it is however by it self an important fact.

Receive my harty thanks for the present of your valuable philosophical book in three volumes.

As i doe not know, whether i understood rightly the VII section of the third volume, i was at first reading of it some what puzled, and thought that the contents of it might induce the reader to believe, that you, and not i, were the first, how have publish'd, that it is the light of the sun, which is the cause, why real plants correct bad air and yield vital air, tho all those of our common friends, whom i consulted on this head are of opinion that such an assertion is by no means intended by you. if you have realy publish'd this doctrine before me, i owe you the justice to aknowledge it publicely in the first volume of

---

238 Gest 2000
239 Breda Collection IngenHousz A13

my books, that will be publish'd or reprinted, and i will chearfully retract by aknowledging that by reading your philosophical works i have inadvertence and not design overlook'd this doctrine; and i will very readily quote the volume of your works and the page in which you will indicate me this doctrine is clearly and explicitly to be found; But if this doctrine was first publish'd by me, as i have till now been persuaded is the case, i will leave things as they are. The confidence i have in your liberal manner of thinking makes me hope you will not leave me long in this state of uncertainty, and that you will favour me soon with an answer, which will be very gratefully received by
your obliged humble servant and friend j. ingen housz
London july 7th 1790 direction N. 6 fleet street London

IngenHousz used the occasion of communicating the important news from Paris about a decisive proof for Lavoisier's system to write Priestley. He not only challenged him, politely but quite openly, about this doctrine, which he and everybody else knew Priestley wouldn't budge. But that was only the introduction to his polite but pertinent teaser about the priority question. He was willing to admit that Priestley was the first to clearly see, demonstrate and publish the effect of light on plants, but then perhaps he would be kind enough to give him the right reference? Priestley answered promptly. His ripost came within a week, strangely enough by excusing himself for not writing sooner:

Dear Sir
Absence from home has prevented my answering your letter sooner; An attention, however, to the frank and friendly letter I wrote to you by dr Pearson in 1787 would have rendered yours quite unnecessary; The discovery about which you seem to be so solicitous, I have aknowledged, was first published by you, tho made by me about the same time. But after my discovery that plants restore vitiated air, that the <u>green matter</u> (which for a short time only I doubted being a plant) gives pure air by means of light only, thou appears to me to have been ... degree of philosophical merit in you, or myself, trying whether <u>unquestionable plants</u> would do the same thing. .. was what I could not possibly have acceded trying after what I had done, and accordingly I proceeded to do it immediatly. For your sake, however, i wish the merit of the discovery had been greater. For with me philosophical matters are no great object, and i am less solicitous about them every day.
I should like to be informed of the circumstances of the experiment you mention of the decomposition of the two kinds of air and that, at present, I am an advocate for the doctrine of phlogiston, and against the decomposition of water, I have no great zeal in the cause, and .. conviction should very readily change ideas.
I am, Dear Sir, yours sincerely.J Priestley
Birm July 16 1790

He referred to a letter from 1787 in which all had been already long cleared up. Although the discovery was first made by IngenHousz, it was made by him at the same time. He also admitted that he had doubted that the green matter was a plant. No big deal, as IngenHousz too, along with so many others, had not been able to decide what it could be. He attributed all the merit to IngenHousz, tactically finding it a little sad that the merit had not been greater, rhetorically admitting that he was not so much interested in all these philosophical matters anymore anyway. He denied he was concerned with personal merit. But in the meantime he would continue his campaign for recognition of his discovery of the photosynthetic necessity of light for two decennia, as we will see shortly. And oh yes, he stuck to his phlogiston, as all these new experiments were not convincing. If they would be convincing, he would change ideas with pleasure. Leaving the question what was to provide convincing evidence wide open.

After IngenHousz' death and when he had already spent six years in America, Priestley cites this letter in the appendix to his 1800 pamphlet *The doctrine of phlogiston established.* [240]

> *Dear Sir,*
> *Receive my hearty thanks for the valuable present of your philosophical work in three volumes.*
> *As I do not know whether I understood rightly the seventh section of the third volume, I was at the first reading of it somewhat puzzled, and thought that the contents of it might induce the reader to believe that you, and not I, were the first who published that it is the light of the sun that is the cause why real plants correct bad air, and yield vital air, tho' all those of our common friends whom I consulted on this head are of the opinion that such assertion is by no means intended by you. if you have really published this doctrine before me, I owe you the justice to acknowledge it publicly in the first volume of my books that will be published, or reprinted: and I will chearfully retract, by aclnowledging that in reading your philosophical works I have tro' inadvertence, and not design, overlooked on this doctrine, and I will very readily quote from the volume of your works, and the page you will inform me this doctrine is clearly and explicitly to be said. But if this doctrine was first published by me, as I have till now been persuaded is the case, I will leave things as they are. The confidence I have in your liberal manner of thinking makes me hope that you will favour me soon with an answer, which i will be very gratefully received by*
> *Your obliged humble servant*
> *J INGEN-HOUSZ*

240 Priestley 1800

The version written by IngenHousz cited earlier differs substantially from this one reprinted by Priestley. So, either it is a draft copy which IngenHousz decided not to use completely, or Priestley left out half of it, viz. the part on the experiment from Paris.

Priestley's paper continues with saying that the copy of his answer was destroyed in the riot and makes the same evasive argument as read above in the original. As Gest says: "Priestley did not - in fact, could not - give Ingen-Housz a published reference to his own experiments in this connection." In his 1800 paper Priestley continued his tortuous argumentation by expanding the definition of 'publishing' to all occasions in which he wrote or spoke on topics with friends or acquaintances. He widened the scope of publication by including all anecdotal accounts in conversations or letters. In doing so he based his priority claim on this personal definition.

A reconstruction in detail

The priority question is more than a small child's competition on who is first, biggest or fastest. It can make or break reputations, it can open doors for further budgets and projects, it decides (often posthumously) on the status of a researcher. In the first place it is a natural drive for humans, at least for some of them, to try their best. In this particular case, the way Priestley, Senebier, Van Barneveld and IngenHousz were struggling with it is a nice example of how these things go. The strife for the ownership of a discovery can be a powerful emotional stimulus, which brings out the best and the worst in people. Even while most scientists know well enough that the attachment to one's name is only superfluous because once published scientific knowledge belongs to the whole of humanity. For IngenHousz this priority question was not a question of life and death. He did not crave for fame, just like he was not interested in titles or decorations. His main interest was honesty, as he deeply felt that without sincerity scientific knowledge is in danger.

In many of his personal notes, never published, IngenHousz probably tried to get clarity, to sharpen his mind and try to reconstruct what must have happened, and in what sequence. His story is history in the making. After all, one can begin to understand things by putting them on the right place in the string of events that unfolded in the past.

In a long personal note[241] IngenHousz' ideas about the attitude of Priestley are unfolded:

---

241 Breda Collection IngenHousz A8

on Dr Pr

The tribunal of the public and of posterity will never and can never cast a judgment about the priority of discoveries but by the priority of publications, for every body has it in his power to claim discoveries without any other proves than the mere assertion – private letters are of little more value than

[crossed out: After having received the letter of] I wrote a letter to mr Magellan conceived in such manner that if he had told me any thing of Dr Priestley having discovered that the sun's light on plants made them correct bad air or give out vital air, or if Dr Priest had realy wrote him any thing about it he must have told it me in his answer. I asked him whether he could recollect to whom he had communicated my discovery on the vegetables which sir joh Pr ... communicated to the company then present among which company mr Magellan was. I got for answer that knowing that what I than did I kept secret, he knew nothing of it and spook to nobody of it. if Magellan had known the discovery from Dr Pr it was very natural he would have written me that this was nothing news to him as he got the knowledge from Dr Priestley and was decided to communicate it me.

I do not want any personal authority of any one, as I have a public incontestable proov viz the priority, of publication by which alone the public can judge, the avowal of the Rev Dr himself in two letters is a quite useless testimony as neither he nor any body else has found it in his book except my other pretender Senebier. ..., viz the quoting and indicating the pages of the passages in the Phil Transactions, and in Dr Pr book, which do not exist either, and which Dr Priestley confesses himself in his two letters to me, not to exist, nor even to have conceived by imagination. this venerable divine this bouche de verité *not even ventured to quote passages in the books of others which do not exist, but even in his own book.*

... and he is confident that he deserves all the credit and believe of his readers, in the same confidens as he finds among his political and theological admirers full believe in every assertion, however extraordinary, as for instance the materialism of the allmost all christians and divene heathens believed immortal spirit our soul, of his friends attachement to the british institution even such an essential part of it, that its overturn could scarce be affectuated without overturning the state.

To understand his discussion with Magellan, it is helpful to read what IngenHousz noted elsewhere, a reconstruction of what happened on the day he first got to understand the beneficial activity of plants in sunlight. He was:

so pleased that he went to London to communicate it with sir Joh Pr, not suspecting that that gentleman who was struck by the novelty, would have communicated to any body else, but such was his astonishment of the fact, that he told me, he found it very difficult to believe it as a constant fact, and excited me to repeat it over and over and over and to communicate the result to him. the evening of the very day sir john had his meeting of philospohical friends, at wich meeting among some others Dr Watson of lincoln's cum fields, Sir William Watson and Mr Magellan. Sir Joh Pr in an ungarded moment told the gentlemen present that Dr Ing had been that day into town and communicated to him a very curious fact viz that leaves of plants give out deph air and correct bad air only in the sun. Mr Magellan was particularly attentive to the communication, as i heard a few days afterwards from sir Wil Watson, who told me that sir joh Pr had committed an imprudently in relating the discovery in the presence of a man who made it his business to communicate to his correspondents every philosophical news, and got for this reason the nick name philosophical spy. Sir William added that I may possibly loose the honour of the discovery by it and his advice was I should tell sir john of it. I went immediatly to sir john, and told him about it with some degree of warmth. he took it somewhat amis to being charged with imprudency (all sir john's friends knew that with the greatest mildness of temperament and the most amicable disposition, he possesed a great share of irritability, by which some of his friends were hurt

I was even in danger in forfighting his friendship by the step I took, and found the best means to prevent a breach of our long standing intimate friendship, by apologising and asking him excuse for my making use of improper expressions. he promised me to forget it, and we continued intimate friends. he aknowledged the danger in regard to that philosophical spy but added that he did not like to be entrusted with secrets, because sayed he, they often come to light by some other means.

I need not say that there is little doubt that Mr Mag wrote the news by the first post to his friend Dr Priestley. I think I may say that there is a certainty of it. if this is so, there is as little doubt but Dr Pr must have been struck more than sir john was, with the novelty, the more because at that time his volume was already published in which he gives a finishing stroke to his conclusion pag. [space left blank, clearly to be filled in later when the right page had been retraced] *where he says that he was convinced that the plants had no share as he had thought before, in the production of deph air, though very rightly, he added that by all that the former experiment, by which the plants had corrected bad air has not lost its force in demonstrating such power in plants, tho he did not know to what circumstance it was to be attributed.*

IngenHousz thought he understood what had happened the day he told John Pringle about his discovery. Magellan, infamous in that respect, had passed on this hot scientific news to Priestley, and later to Senebier as well, as we have seen earlier. In the way IngenHousz recounted this story, it is clear that he felt he was walking on eggs by counselling Pringle. He also did not understand how it was possible that Priestley was implying that he never performed experiments with whole plants. Priestley suggested as much in his letter of 24 November 1787 to IngenHousz:

> All the time I was making these experiments I wrote to my friends about them; particularly to Mr Magallan and desired him to communicate my observations to you as well as to others; but I believe you had not heard of them; so what you did with leaves was altogether independent of what I was doing with whole plants. The same summer and the same sun, operated for us both and you certainly published before me.

While admitting that IngenHousz published first, he implied that that publication was not saying anything about the thing in question, thus implying that IngenHousz had only observed a fraction of the truth. He, Priestley, in the mean time had been doing the right kind of experiments. The contrary is true: IngenHousz had been working with whole plants on a regular basis and refers explicitly to pages 192, 207, 209, 211, 219, 220, 222, 259, 264 and 265. He found it really impossible to explain:

> it is next to impossible that Dr Pr could say such unaccountable assertion to which the very titel of my book gives denial
> he can not have sayed such a thing without particular cause, which if not explained would seem unaccountable nay incredible.

In any case, Priestley had to act quickly. You can see IngenHousz think as he writes: he is trying to clarify for himself, trying to understand the incomprehensible behaviour of another.

> but why did he not publish it immediatly as his own discovery, as he did in a sly way the year after. this indeed he could have done without danger, as i would certainly have had charged him with such a shamefull plagiary, as at that time, sir joh pr mr watson and several others could have attested, that he took it probably from mr Magellan, besides, as he had so recently published a large volume on philos matter he could not so soon compose a whole volume and publish it in the same year, and he could not have guessed, no more than myself, that I should have published a book during that same year.
> it is very natural to concieve that having only understood that i made my experiments with leaves only he has put in his common place book this article as he understood from his friend Magellan...
> he had recourse only to his former notes, for if he had had recourse to my book,

the very title of it would have him ashamed to have charged me with having made my experiments with leaves only and not with plants. thus it happens that when a man digs a whole for another he hem self falls in it. nothing is more dangerous than an assertion, of which some art may be demonstrated as contradictory to an other, and it is not very rare that a man inflamed with passion what ever, be it love, jalousie, ambition, pride, anger, looses the faculty of reasoning and commits the most unaccountable imprudencies, which often admits no remedy but by committing a new inconsistency. Dr Pr whose inventive genius finds an easy manner of extricating ... will not find it difficult to explain this wonderful phenomenon, by saying, as he did with regard to the green matter (on which examination though very accurately made he adscribes the error to his eys being very weak, which I think he never sayd before) that the dimness of his eyes has made him see in the title of my book, the words leaves in stead of vegetables.

Some irony sneaks in. It seems as if IngenHousz was secretly amusing himself by putting all pieces of the puzzle together. Here too, his natural curiosity won over any proclivity for glory. In the margin, IngenHousz added another thought:

Dr Pr attempt indeed to console me in a reasonable way ....
it was easy enough to send to mr Magellan a copy of the letter to me 1787, but it might not to have been such an easy matter to have persuaded the same gentleman to keep among his papers the supposed communication if this letter should have been written. and for this reason I was not astonish'd that the supposed letter was not found among the correspondence of mr Magellan, the most trifling letters written to him by Dr Pr were found even such written so slightly that they were scarce legible and without date. if Dr Pr had ever written to mr Magellan the supposed letter it is more than probable that that gentleman would have presented such a precious letter with the utmost care, as it could serve to constate a memorable date of the discovery of his best friend.

He was a doctor after all, with a good faculty of observation and a lot of judgment of the human character. Magellan almost becomes a character from a Tjechov story.

... But as I deny with as much assurance the non existence of the supposed letter, Dr Priestley throws its existence on the world with the same boldness as he does all this for the constitution...
very probably he thinks him self to have send this supposed letter, though in reality he had only dream'd it, a dream about a thing so much wished for as the honour of a new discovery, may have made such strong impression in many minds, as never to become effaced from it.

IngenHousz was a fine psychologist and understood how Priestley was embarrassed and must have tried the most implausible tricks to turn the record to his advantage.

In the margin IngenHousz put a big X to mark another note of a totally different kind. He described John Pringle's state of alarm about the statement by Priestley that the Soul is just an imaginary concept and that after death nothing is left but inert dust. He and his friends assure Pringle that the soul is eternal and that they all, as good Christians, believe in spirituality and the immortality of the soul.

On another piece of paper he jotted down some more thoughts. This paper[242], in contrast to the former, can be dated. It must have been written after Priestley's house in Birmingham was burned in the summer of 1791, as IngenHousz talks about it. He also wrote about the turmoil and the disunity caused by Priestley's speeches. He thinks Priestley makes a mistake by thinking that the discord is caused by the authorities. To IngenHousz it is all the consequence of Priestley's way of thinking. He might have said he was a friend of the constitution and the state, but said at the same time that he wanted to overthrow the established church. As the latter is part of the state, he contradicted himself.

The force of words and thoughts can be just as devastating as that of guns and swords and he refers to Priestley's pamphlet of 1787 <u>A letter to the Right Honorable William Pitt</u>, sec ed, and cites:

> i shall inform them that the means we propose to employ are not force but persuasion of the gunpowder which we are to assiduously laying <u>grain by grain under the building of error and superstition*</u>, in the highest region which they inhabit, is not composed of saltpeter, charcoal and sulphur, but consist of arguments...
>
> *[under the page is this note] This is part of a sentence in a pamphlet of mine on the importance and extent of free inquiry, which sir William Dolben did me the honour to quote at large and discant upon it in his speech in the house.

On another folio[243] he made another effort to explain the behaviour of Priestley and Magellan. It may seem exaggerated to include these so extensively, but it is the only way to illustrate how IngenHousz was thinking about these things and how they occupied his mind. In this way he was preparing to defend himself to public opinion and scientific scrutiny. He wanted to stand firm and be able to defend himself with the proper and rightful arguments. He also has been talking about it with Dr Combe, as he did not want to get lost in his own thoughts and checked them against the critical perspective of others.

---

242 Breda Collection IngenHousz A8
243 Breda Collection IngenHousz A8

nb the letter of my self to mr Magellan were dated vienna the 30 october 1786,
1 may 1783, 13 october 1786

Dr Combe thinks that it very mysterious that Dr Pr should have sent to mr
Magellan a copy of a letter he wrote me 1787 and that the supposed letter
written by Dr Priestley to Magellan of 1779 should not have been preserved.
Dr Combe thinks that the letter of Dr Priestley to Magellaan 1779 never has
had any existence for if it had existed it would still be found among the rest of
Dr Priestley's letters to mr Magellan, the more so, as mr Magellan was very
carefull in preserving the letters of his correspondents, even to such a degree that
the most trifling letters and notes of Dr Pr written in the most slougish way,
even such as bore no date, are still in his collection. Dr Combe thinks that Dr
Priestley, knowing that he never wrote such a letter, had send to Magellan the
copy of his letter written to me 1787, as a substitute of an imaginary letter,
which should one day or another have the same effect as would have had the
letter written to Magellan in 1779 if it ever had been written.

Is it possible that a man so ambitious as Dr Pr is generally known to be should
have supported so tamely such a shameful incroachment on his right, as he
insinuates to have been committed by me, without any murmuring or without
appealing in his next work to the letter written to mr Magellan, which though
of no right before the tribunal of the public, as it was not publis'd, would have
had a good deal of right among his friends.

as Dr Pr did not give in any of his former books any the least hint of having
given such a communication to mr Magellan it is more than porbable that
he did not venture to doe it been afrayd that mr Magellaan would not have
countenanced such an assertion. and mr magellaan's answer to a letter I wrote
to him in 1788 (I doe not recollect the date as i left it at vienna) is next to a
demonstration that Dr Pr never made him such a communication.

if Dr IngenHousz had not by accident discovered what mr van Barneveld
has copyed out of his book the discovery of the sun's light which discovery mr
barneveld claims as far as 1778 in a paper print 1781 in the transactions of
the society of arts and science of utrecht, Dr Pr would have haboured in vain
by dating 1779 his having made the great step.

In this last paragraph IngenHousz ties the various tales of priority pretenders
together. And for a very good reason. As soon as anybody begins to tamper
with the exactness of dates or, for that matter with the facts, all construction
of knowledge about this world starts becoming shaky. Trustworthy knowledge
begins with the trustworthiness of the people that produce it.

One last citation from the many working notes from the IngenHousz' archive.
It repeats some of the arguments, but from a slightly different perspective.

It seems that Dr Pr was not so fond of claiming the discovery in the beginning,
when the french chymists had not yet applyed the action of the sun on vegetables
to their system as one of the means by which Nature decomposes water in an
insure [?] way as the plants absorbe thy hydrogene and let loose the oxygene.
that the Rev Dr seing more and more the importance of the discovery grew at
last impatient of seing that another became more and more universaly honored
with the discovery could no farther withstand the temtation to make the gigantic
steps that was stil to be made stepping over the head of Dr Ing with the greatest
intrepidity, by an assertion which he never has made before. he dares to make
any assertion or insinuation, though against all appearance of truth.
he says that he never did any thing to excite the peoples aversion against
him, though he teased continualy the people by preaching and writing against
their religion and against the constitution. he insinuates that the church and
government excite this hate against his person.

This note must have been made after Lavoisier's new system and nomenclature
had become accepted, at least in IngenHousz' mind. That renders the
discovery of photosynthesis even more valuable. It makes the world and its
interconnectedness somehow more understandable. IngenHousz suggested
that Priestley must have felt this too and that that might have been another
reason why he tried to preserve his claim on the priority of the discovery. This
also suggests that he thought Priestley was not so radically phlogistonist as
generally has been thought. This is an insight that has been formulated by recent
scholars too, especially after studying Priestley's 1800 paper on phlogiston. [244]
And not that he would have bothered in 1800, but many years later, his strategy
to hush up IngenHousz' claim to fame seems to have been effective, as we will
see further.

Opposition from Geneva

Another colleague and competitor to IngenHousz' name and fame was Jean
Senebier, who claimed that he was the first to understand the beneficial
processes in and by plants. IngenHousz was rather upset about this claim. Not
only because to him it was an unjust claim, but even more because it looked like
Senebier did not perform all necessary experiments and probably just copied
the work that suited him from IngenHousz' book. A major reason for thinking
so was that Senebier disputes the findings by IngenHousz that plants respire
in darkness - a phenomenon he should have been observed clearly if he had
just done the necessary experiment. While plagiarising IngenHousz' work,
Senebier only mentioned him as someone who inspired him while he himself

---

244 Schofield 1966

had had these ideas in the first place. IngenHousz read Senebier's *Memoires Physico-Chymiques*, published in Geneva in 1782, and scribbled abundant notes in its margins. This unique copy is being kept at the Archive of Breda, the Netherlands. One can imagine him furiously jotting down his thoughts in 1783:

> *Le premier article de l'ouvrage est rempli de tracts manifestes de vanité, d'envie, d'amour propre, de pretensions pour s'attirer autant qu'il prent de l'honneur des decouvertes de Mr IngenHousz.*

> The first chapter in this book is full of manifest traces of vanity, envy, egotism and pretensions to attire as much as possible from the honour of the discovery of Mr IngenHousz.

Curiously, IngenHousz wrote here about himself in the third person, taking an objective third-party perspective, probably because he made these notes as a rough draft for an official article to be published later. He further minutely analysed the rhetorical tricks Senebier used to make the world think he was the man who discovered photosynthesis, while resolutely indicating the moments where he must have been making mistakes, either in performing the experiments properly or in the interpretation of the results. To IngenHousz it was clear that Senebier had just copied his description of many experiments from his book of 1779 and not even bothered to reproduce them - for him a hallmark of proper scientific conduct. Senebier, according to IngenHousz:

> *il n'avoit que des ideés, comme il l'avoue pag. 3 lui même. Le date de ces prétendues lettres rend ces idées tres problematiques.*

> he did not have but some ideas, as he admits on page 3. The date of the assumed letters to Van Swinden & Volta[in which Senebier argues to have pronounced these ideas] makes these ideas very problematic.

Later in 1784 in a letter to Van Breda he wrote:

> *En observant donc de plus en plus, que la silence, que j'avois observé pendant si longtems, par pure politesse & ménagement pour un savant dont j'honore les talents avec toute la republique des lettres, ne servoit qu'à faire envisager mes decouvertes comme insoutenable; je me trouvai à la fin forcé d'exposer au public des preuves ulterieures de mon systeme. Il m'étoit pas possible d'y parvenir qu'en tachant de demontrer la foiblesse des arguments & des experiences, que Mr. Senebier a pu trouver bon de publier pour combattre les miennes.*

moral des observateurs pour graduer le degré de confiance qu'il doit donner aux observations qu'on lui raconte : il m'a donc paru convenable d'annoncer, que ce mémoire est moins le fruit de l'ouvrage publié par M. Ingenhous, que le résultat des idées que j'avois eues avant qu'il y pensât : j'avois communiqué ces idées par écrit à M. Bonnet, qui a souvent eu la bonté de recevoir mes confidences ; j'y avois même joint l'ébauche d'une théorie de la végétation, fondée sur ces principes, dans une lettre du 10 Mai 1779 : j'avois fait part de mes apperçus sur ce sujet, dans le même été, à MM. Van Swinden & Volta. Les nouvelles de la république des lettres & des arts, numéro XIII, pag. 119, parlent de tout ceci de la même manière, sous la date du 5 Janvier 1780. Enfin, j'aurois fait plutôt les expériences nombreuses relatives à ces idées, & je les aurois déjà publiées, si une maladie,

A 2

Handwritten comments by IngenHousz in Senebier's 1782
Memoires Physico-Chymiques

I have kept silent for a long time because I have highly respect Mr Senebier as savant and honoured member of the Republic of Letters. And I have kept silent for such a long time out of pure politeness, but this has only resulted in letting my discoveries appear as insustainable. I am now forced to reveal the ulterior proof for my system.

He would have preferred to leave it as it was. All the fuzz to him was a terrible waste of time and energy. And he respected Senebier as an honourable man, a worthy member of the Republic of Letters. But by leaving the things to take their course, he undermined his own position. Moreover, he could not explain his own findings and experiments without saying anything about Senebier's. IngenHousz prepares a "polemic article", but in the end did not submit it for publication. He didn't like the fiddling and the stress. He had too many other, more fruitful things to do and was plagued by his gallbladder stones. But by the early 1790s he was forced to prepare another article in defence of his position. Parts of the draft,[245] as far as they are readable and not rendered obscure by corrections and deletions, sounds like this:

> *Le celebre mr Senebier publia une theorie ingenieuse et très plausible de la production d'air vital par les plantes au soleil * plusieurs annéés apres que j'avois publié le fait que les vegetaux produisent de l'air vital au soleil, et dans l'obscurité de l'air fixe ou acide carbonique, et de l'air phlogistiqué ou azote. je prouvoi dans le premier volume imprimé à Londres en 1779 en langue Angloise que ce phenomène des vegetaux est constante dans l'etat naturel des plantes, c'est à dire, lorsqu'elles sont dans leur propre element l'air atmospherique. je developpoi cette verité plus au long dans l'édition française de cet ouvrage, que je publiai la premiere edition à Paris en 1780. avec un second volume en 1789. Quoique cette decouverte jettoit une nouvelle lumiere sur les loix que la nature a établie parmis tout le regne vegetal, en devoilant un double phenomène très important dont aucun ecrivain avant moi n'avoit fait la moindre mention, on n' a pas trouve cependant que j'ornoi un sujet aussi neuf avec des phrases bombastiques ou des exclamations, qui auroient pu servir à m'attirer des reproches de quelques vanité, dont il est bien difficile de ne pas en sentir au moins quelque peu d'atteinte, lorsqu' on se croit persuadé d'avoir reculé les bornes de nos connoissances.*
> *j'avoue franchement, que si le plaisir que je sentois de ne pas avoir fait des recherches tout a fait en vain etoit sans melange de vanité, j'avois bien de la peine à m'en defendre entierement lorsque je lisois l'ouvrage en trois volumes que mr Senebier publioit en 1782 +*
> *[...]*

---

245 Breda Collection IngenHousz A8

que je croyois infinniment plus importante, leur influence nocturne j'y trouvoit traité avec le .. mepris et des expressions les plus humilientes pour l'auteur.
[...]
Mais comme la vertu opprimée et reconnue sans tache acquiere un nouvel eclat de l'oppression même, ainsi son systeme blamé et avili par des savans du premier order ne peut manquer de reflechir sur son auteur un degré d'honneur egale, en raison inverse, à l'approbre dont les antagonistes vouloient le couvrir. Ce grand Physicien mr Senebier disoit a la page . du premier volume de son ouvrage cité, que mon système de l'influence nocturne des vegetaux sur l'air exposé à leur contact, le frappoit tellement par sa nouveauté et la singularité, qu'aucun objet avoit encore fait autant d'impression sur son esprit, jusqu'a même y sacrifier deux etés consecutives, en le soumettant a des epreuves innombrables et infinement diversifiées. Apres avoir pour ainsi dire epuisé toute la force de sa patience, de la curiosité et de son geni, dont la republique de lettre entiere reconnoit la penetration et le sagasité peu communes, il prononce avec le plus imposante solemnitude l'arret de condemnation contre ma doctrine comme insoutenable, erroneux et ême calumnatrice contre les sages et sublimes procedés de la Nature.
Mr Senebier se contentoit nullement du sacrifice de deux etés pour vanger le ciel outragié par ce systemè a ces yeux abominable n' a pas cassé de le persecuter de toutes ses forces depuis l'an 1779 jusqu' a l'anneé 1789 lorsq'il l'accabloit d'un ouvrage de pres de 500 pages rempli d'une vraie foule d'experiences toutes derogatoires à cette même doctrine
[..]. lui combattre ulterieurement dans des nouveaux ouvrages polemiques.
Qu'il me soit permis de faire remarquer le lecteur que mr senebier ne pouvoit, je crois, me disputer, que lui même croyoit, que si cette doctrine etoit fondeé sur les loix de la nature, elle auroit etée tres <u>importante</u>, tout à fait <u>neuve</u> et <u>singuliere</u>. mais sous la ferme persuasion que ma systeme etoit ... sur les loix de la nature, je me gardois (autant que la faiblesse attachée a notre nature pouvoit me le permettre) de m'attribuer toute l'honneur que ces expressions tranchantes m'attribuoient, car j'etois bien sur que ce savant celebre auroit reconnu le systeme aussi vrai qu'il le croyoit erroné, il l'auroit probablement traité avec autant d'indifference, qu'il l'attabloit de titres superlativement honorables.
[...]
Etant bien penetré de l'ancien proverbe Lans propria sordat, on n' as pas besoin de m' avertir que ce que je dis ici peut sortir de la bouche de qui que ce soit mais non pas de l'auteur.

It is a story that keeps popping up again and again in IngenHousz' life, right up to a last outpouring in his essay on the food of plants in 1796. It must have been a combination of things that made this dispute weigh so much on his mind and mood. It was not just a question of being the first or not to discover the wholesome influence of plants on the atmosphere. It was the fact that someone was using the wrong arguments, and on top of that manipulated the empirical evidence, that upset him. And it probably was just as much the attitude of Senebier, who was not playing the game of science like IngenHousz thought it should be played. He really became unhappy when people used the wrong arguments, or when they used rethorics and ideology instead of facts and evidence to make their point. Still, he was not inclined to turn this discussion into a public fight. That just was not his style. From what we can understand from his writings, he must have been too friendly to fight. But the stubborn attitude of Senebier finally forced him to react. He would have liked to publish his rectifications of Senebier's claims more quickly but because of the tardiness of his French publisher the first to appear of it in public was in the German edition of 1784 of the *Vermischte Schriften*. In this book he printed a 4-page letter from Senebier from 9 June 1784 followed by his own reply[246] that ran into 29 pages.

Senebier's letter was quite friendly though strategically a clever affair, expressing his respect for the brilliant mind of his fellow philosopher and agreeing with the facts which IngenHousz was making public. He reconstructed the chronology of things in such a way that he was publishing his results on the effect of fixed air on the production of dephlogisticated air in the summer of 1783 while IngenHousz was performing the experiments in Vienna that were to confirm his findings.

To IngenHousz it was the other way round. He was pleased to find the confirmation of his own findings in Senebier's book (sent to him by their mutual friend Landriani from Milan). What upset him most however, was the publication of a letter from Mr. Storr, professor in Tubingen, in the 1782 scientific magazine of Mr. Crell in Leipzig in which he announced the publication of a book by Senebier which mentioned that IngenHousz' book on vegetables had errors on every page. After all the polite and discreet actions he undertook to come to some mutual understanding, Senebier demonstrated resistance in his aggressive approach. IngenHousz recounted what happened at the time.

How he communicated via Lamberthengi and Landriani to Senebier that he was using inappropriate eudiometric equipment, which would explain his bad results. He did this consciously via mutual friends, because he did not want to hurt the name of a learned man with such a reputation by publishing it. "Discretion" and "delicatesse" were is key words. Senebier must have reacted favourably to his remarks and let him know (again via Lamberthengi) that he was willing to do a rerun of his experiments along IngenHousz' lines, if IngenHousz would publicly report on his methods. IngenHousz declined, as he described these methods already in his English and French books and again

246 IngenHousz 1784 p 477

insisted on not wanting to start a public discussion about something that could be handled perfectly well in private.

He suggested that Lamberthengi attend his experiments, not only as a witness, but also as someone who has seen IngenHousz perform these experiments in Vienna. He remembered that Lamberthengi had visited him in Vienna in the summer of 1783 and had been able to see with his own eyes how so many of the results published by Senebier in his *Memoires physico-chymiques* were erroneous. Plants in water with fixed air do not give constantly an abundance of dephlogisticated air. Under the eyes of his Italian guest, IngenHousz repeated the experiments he described in his book in 1779 and found the same results. Also when dissolving salts in the water he got results that contradicted those from Senebier.

L'eau feule, expofée à l'action du foleil dans mes récipiens, ne donne *opposition* pour l'ordinaire point d'air ; &, quelquefois, la quantité qu'elle en a fourni alors, n'égale pas la foixante & quatrième partie d'une de mes mefures, *Capacité de les récipiens* tandis que les récipiens que j'emploie, contiennent foixante & douze de ces mefures : j'ai eu pendant fix femaines un récipient plein d'eau, environné de mercure, où il n'a point paru d'air,

Page from IngenHousz' copy of Senebier's *Mémoires physico-chimiques* from 1782, in the margins of which he jotted down his remarks. Here about the useless type of low capacity eudiometer Senebier was using, explaining his contradictory results.

This being said, IngenHousz asked the reader to form his own opinion. He stayed polite and subdued, after Senebier started a violent and public attack. He was sad to see how Senebier persisted in slinging mud and held on to his conviction that the bad air IngenHousz found during the night was the result of fermenting, dead plants. While the experimental evidence pointed in the opposite direction. He was however pleased that Alessandro Volta and Professor Scarpa had seen IngenHousz' experiments in August 1784 with their own eyes and that they were totally convinced of their validity.

But keeping silent about the whole dispute did not serve his goals well because they turned his position untenable. He felt himself forced to deliver some public riposte. He sent an article to the *Journal de Physique* in November 1783, but learned in April of the year after that it got stuck in a long queue. It ended up in Molitor's first edition of the *Vermischte Schriften*. But in the meantime, Senebier's new book appeared in print and IngenHousz was glad that Senebier at least corrected some of his mistakes, clearly instructed by the information from Lamberthengi and Landriani. Nevertheless, he did not revoke the assertions against IngenHousz' discoveries. He still could not accept that plants in the dark produce mephitic air. IngenHousz found it hard to understand that a scientist of the calibre of Senebier is refusing to see the facts in the eyes. Worse still, he insisted on calling the discovery of the nightly influence of plants on the air an "incalumny against nature & the plants".

Finally the first of IngenHousz' memoires was published in May 1784 and the letter[247] from Senebier which he reproduced here was a reaction to that article. Not only did he send it to IngenHousz but he also published it - and because he did not mention where he would publish it, IngenHousz decided to publish it in his book anyhow. At first, IngenHousz was pleased with the reaction from Geneva. But on second thought, he was not so happy. It maybe sounded flattering, but because Senebier published it, it lost its amicable character and became a strategic tool in a public power play. Senebier misused the true nature of a private letter to reach his own goals. Praising IngenHousz was no more than a disguise for advancing his own hidden agenda. He only told part of the story and suggested that he already had doubts about his own results. These were only sophisticated ways of turning a story to his advantage. He admitted that he was glad to see that IngenHousz' results confirmed his. And by admitting some of his results were wrong, he implicitly admitted that IngenHousz' earlier results were right. If these results pleased him, he must have known about them as he had learned them from Landriani in the summer of 1783. But that's something he would not admit to in public. This is contrary to IngenHousz principles of building trustworthy knowledge. He always communicated all his knowledge and his doubts to anybody concerned.

Ceux qui me connoissent, savent, que ce n'a jamais été ma coutume de travailler en cachet. J'ai toutes les années plusieurs fois montré & expliqué publiquement toutes mes experiences sur les végétaux, même celles que je n'avois pas encore

247 IngenHousz 1784

publiées, à une compagnies de 100 & quelques fois plus de 200 personnes de distinction, à leur requisition. Je n'ai jamais refusé personne, qui s'est presenté pour y assister, & je n'ai non plus invité personne qui ne m'ait signifié son désir d'y venir. Les alies & admirateurs de Mr. Senebier m'ont souvent honoré de leur présence lorsqu'une telle compagnie s'assembloit chez moi.

Those who know me, understand that it has never been my habit to work in secret. All those years I have shown and explained several times in public all my experiments on vegetables, even those that were not published yet, to audiences of 100 & even 200 people of distinction. I never refused anybody who asked to assist & I never left anybody uninvited who expressed the desire to attend. The allies & admirers of Mr Senebier have often honoured me by being present in such company at my place.

And as he published his work on plants four years before Mr. Senebier published his book in which he attacked several of his discoveries, IngenHousz was consoled that it was not him who turned out to be the aggressor. And that in all the accusations of Senebier, he had not found a single one he had to take seriously. While he was pretty conscious that mistakes can be made and that he was not immune to them.

En fouillant sans relache dans les operations de la nature, je travaille moi même autant à rectifier ce que je pourrois avoir mal vu, qu'à d'couvrir des nouveautés. Si quelqu'un rectifie mes erreurs avant que de les avoir reconnues moi même, il a le droit à ma reconnoissance, & je lui en saurai toujours bon grès, pourvu qu'il me montre aucune aigreur par des expressions offensentes, qui ne servent qu'à déceler une ame jalouse & trop éprise d'elle même, & qu'il n'employe aucun detours ou sophisme pour s'appropier l'honneur de mes decouvertes.

While rummaging without cessation in the operations of nature, I worked as frequently to correct what I could have perceived wrongly, as to discover new things. If someone can correct my errors before I recognise them myself, he deserves my recognition & I will always be thankful, provided he does not show any bitterness in offending expressions, serving only to unveil a jealous mind, & provided that he does not use tricks or sophisms to appropriate the honour of my discoveries to himself.

It would prove difficult to find a better way to express IngenHousz' feelings and attitude than these words from that public reply to Senebier. It would be, however, of no avail. Senebier would keep on pestering him and he would be a nail to IngenHousz' coffin till the bitter end. So much so that in the second volume of the second edition of *Experiences sur les vegetaux* from 1789 he again went into all the details of this unsavoury soap story. First in the preface, at the end of which he apologised to his readers for so much quarrelling that might not be of interest to them. He referred those who were interested in the ins and outs of this never ending dispute to an afterword[248] in which he promised to put the whole case finally to rest. His reason why is simple enough: he thought it for the benefit of humanity that knowledge made progress and admitting when one was wrong proved essential for that progress. It was the only way to learn from one's mistakes and to get to know what is true. When the zeal for the Good Cause gets the upper hand, reason is easily repressed and enough examples are known of situations when that led to fanaticism and persecution "& tous les maux que l'histoire du monde nous a transmis". Senebier discredited him in front of the public tribunal, in an excess of fervour to defend the "outraged Heaven", this "system which is a calumny against nature, against the almighty and the sublime he created - a calumny that nature itself will revenge". IngenHousz dryly noted that he didn't know whether to laugh or cry on hearing these words. Although Senebier claimed this accusation of defamation was directed against IngenHousz' theory and not against his person, IngenHousz wondered what results these attacks would have had in other times and other places when men have been executed for lesser accusations.

And while IngenHousz would have preferred to let the case rest and forget about it, Senebier attacked him again in his book of 1788, having found new weapons and arguments to defeat this "hideous monster" that dishonoured the Providence. Even eight years were not enough to calm him down and IngenHousz was only able to observe how Senebier lost all sight on the real subject of the discussion (the experimental facts) in the heat of his combat. And these facts were what interested IngenHousz. He repeated that he performed experiments with plants under water saturated with fixed air and reported them in his 1779 book, but did not make much of it, because plants do not naturally grow in fixed air or acidic water. Of course, so many years later, he had gained new insights in the composition of the atmosphere and the role of fixed air in the nourishment of plants. In the mean time, Senebier had published his results on similar experiments - without mentioning IngenHousz - and got contradictory results because he used inaccurate equipment. And on top of that, IngenHousz could not find any single experiment in Senebier's three volumes published since then in which Senebier put plants in a jar in the dark. IngenHousz recounts again the whole story with Landriani and Lamberthengi and what happened in the communication between Milan, Geneva and Vienna in 1783. And he closed this slightly bitter final chapter of his book with the words:

> Je finis cette tracasserie insipide, dans la ferme espérance que je n'y reviendrai plus. Je respecte trop mon lecteur pour l'en occuper de nouveau.

---

248 IngenHousz 1789 p 452

Again, IngenHousz apologised to his readers for having bothered them with this tasteless and unproductive hassle. But it left him with mixed feelings, hating polemics on the one hand and loving sincerity and justice on the other.

A Haarlem shuffle

IngenHousz had met Van Marum for the first time in 1789 in Haarlem, just before he left for England for the last time. They had become friends and kept up a lively correspondence on subjects both scientific and personal until IngenHousz' death in 1799. Undoubtedly the quarrels with Priestley and Senebier, both about the priority question and the interpretation of scientific facts, have been a returning topic. In a letter[249] of 1792 he asks Van Marum to consider all the facts.

> *Londres*
> *14 sept 1792*
> ...
> *Dans mon traité sur la matiere verte de Dr Priestley que vous trouverrez dans un de ces volumes, divisé en 12 sections, vous verrez que le Dr Priestley a cherché, par un tour jesuitique, à faire passer sa matiere verte (c'est a dire celle qu'il a decrite dans son 4ieme volume, et dont il enseigne dans le volume suivante la generation acceleré par la putrefaction des substances vegetales et animals) pour une plante à fin de pouvoir dire, que c'est lui non pas moi) qui a le premier trouvé <u>que les plantes fournissont par l'assistance de la lumiere solaire de l'air vital</u>. Il n'a jamais prononcé une telle phrase que long tems apres la publication de mon ouvrage. Si vous imitez exactement ce que j'ai fait, et ce que Priestley lui meme enseigne de faire pour produire cette matiere verte, vous ne pouvez douter, que son origine est animale.*
> *Il est bien vrai qu'en parlant vaguement il existent dans la nature assez de matieres vertes vegetales. Toute la nature vegetale est verte. En vous demandant à lire avec attention cette dissertation de moi, je vous enjoigne une tache assez difficil et peut etre annuiante, mais pour un homme de votre sagacité rien ne doit couter, s'il veut examiner les loix de la nature. Pour moi je defie qui que se soit de trouver dan mes 4 volumes une seule experience que je m'attribue comme auteur, controuvée en fausse, au lieu que la plus part de celle de mon <u>zoilus</u>, le Predicant declamateur Senebier sont imaginaires, pour ne pas dire mensongeres, comme vous verrez, dans ma preface de 2 volume sur les vegetaux et le postscriptum. Ces deux pieces vous divertirons par la singularité de l'attaque. L'academie R de Sc de Paris a nommé 6 commissaires pour examiner mes experiences, et au lieu de revoquer en doute l'influence méphitique nocturne*

---

249 Haarlem Noord Hollands Archief

des plantes sur l'air en contact avec celles, ils y attachent une theorie suivant leurs nouveau systeme. Mr Senebier qui m'a menacé d'une novelle attaque pag 100 de son derniere attaque imprimé en 1788, s'en repentie à present et me cede le champ de bataille, a ce qu'un de ses meilleurs amis m'a dit.

...

IngenHousz describes how Priestley *post hoc* claims to have said that the green matter was of vegetable nature in order to be able to pretend that it was him who was the first to discover that plants produce vital air with the help of the sun. While Priestley never wrote anything like it in any of his works but a long time after IngenHousz published these findings. Similarly he debunks Senebier's claims as imaginary, as he already demonstrated in the preface and the postscript of his second volume (of the French edition). In the mean time a commission of six savants appointed by the French Academy of Sciences has been scrutinising IngenHousz' experiments. They did confirm that plants mephitise the air in the dark and found it compatible with their new system.

The letter wasn't just a complaint. He described also new data on the electric eel: "il parait que le cerveau de les animaux est une espece de machine electrique". This conclusion was based on experiments in which muscles contract after an electric shock, done by an Italian anatomist whose name he forgot to write down. Politics was also never far away in IngenHousz' letters, especially not in his latter years, when he became more and more concerned about the ways of the world. He was afraid that the chaos and turmoil in France might infect other countries: "J'espere que le delire feroce et presqu'universelle au France ne se communiquera pas à cette isle, qui se remplit d'une maniere alarmante de democrates et aristocrates francais fugitifs." He concluded by saying that he had written this letter in a hurry because he would be leaving for Lansdowne the next day.

As far as is known Van Marum never took any clear standpoint in the whole priority dispute. He probably did or could not do so because he kept a friendly correspondence with Senebier too.[250] He would visit Senebier in Switzerland later, in 1802.

The Geneva connection

His politeness had worked against IngenHousz, as can be learned from how he has been referenced through the years. In 1797 Théodore de Saussure published an essai in the *Annales de Chimie* in Paris on the question if carbonic acid was essential to the growth of plants.[251] De Saussure was only thirtythree then and was born and bred in the world and the words of Lavoisier's new chemistry. It offered him a great ability to work with the new concepts and De Saussure was

250 Wiechmann 1987
251 De Saussure 1797

to deliver some strong scientific stuff of his own. He shifted the limits of plant physiology as he started measuring the weight of the air contained with the plants in the vessels, as well as the weigth of the plants themselves. So he was able to quantify the growth of plants and determine where the carbon went.

1. *Que les plantes, comme les animaux, forment continuellement de l'acide carbonique en végétant dans l'air atmosphérique, soit au soleil, soit a l'ombre:*
2. *Qu'elles forment, comme les animaux, cet acide carbonique avec l'oxygène de l'atmosphère, et que quand on ne s'apercoit pas de la production de l'acide, c'est lorsqu'il est décomposé à mesure, qu'il est formé:*
3. *Que la présence, ou plutôt l'élaboration de l'acide carbonique, est nécessaire à la végétation au soleil:*
4. *Que la lumière favorise la végétation, en tant qu'elle contribue à la décomposition de l'acide carbonique:*
5. *Que la plus forte dose d'acide carbonique qui favorise la végétation au soleil, lui nuit à l'obscurité.* [252]

1. Plants, just like animals, produce carbon dioxide.
2. Plants, just like animals, produce this carbon dioxide with the oxygen from the atmosphere.
3. The metabolisation of carbon dioxide by plants needs the light of the sun.
4. Light favours plants as it contributes to the decomposition of carbon dioxide.
5. The highest concentration of carbon dioxide that favours plants in the sun, harms them in the dark.

He brought the theories of IngenHousz and Lavoisier together in one quantified whole, based on accurate measurements of weigths and volumes. Thanks to these exact measurements he was able to conclude that plants accumulate biomass through the absorption of carbon from the carbon dioxide in the air. But one thing is missing in his essay: the fact that only the green parts of plants do their oxygen-producing work in the sun. Not a word on what we would now see as a core element in the process of photosynthesis. Still, he mentioned in a footnote that Ingenhoutz [sic] has done similar experiments and drawn the same conclusions on the enrichment of soils with acids as he himself presented at the Société d'Histoire Naturelle de Genève on 16 January 1779. He referred to the *Journal Britannique* no 38, Juillet 1797, where these results from IngenHousz had been published. That was the only reference to IngenHousz, while he more or less repeated what IngenHousz said in his essay on plants in 1796. But that was hardly available outside England, except in a

---

252 De Saussure 1804 p 56

Dutch translation and most unlikely that De Saussure had seen it. But it is not likely that De Saussure deliberatly kept silent about IngenHousz. As Wiesner has analysed,[253] De Saussure was not keen on studying the available literature.

In 1804 De Saussure published his seminal work[254] on the metabolism of plants. He had continued his investigative work on the carbon metabolism of plants and by then he was also able to confirm that water was playing an important role in the metabolism of plants, as a fundamental ingredient in their nourishment. In the mean time, in 1800, Senebier's five volumes' *Physiologie des plantes* had appeared in which the role of carbonic acid was extensively described in volume three. Senebier did do his utmost to minimalise or even undermine the historical role of IngenHousz, while repeating his old errors such as the carbon dioxide that he thought was transported by rain water. So it should be no wonder that in De Saussures' book IngenHousz is hardly mentioned. Most honours went to his friend and fellow Geneva citizen Senebier, still alive at that time, who was meant to have discovered that green plants decompose carbon dioxide, keep the carbon and expel the oxygen.[255] Furthermore he even ascertained on page 54 that it was Senebier who "has seen that those parts of the plants that do not contain the green colour, such as the wood, the roots and major parts of the flowers etc; do not expel oxygen". Next, Hassenfratz humus-theory is critically taken apart without even one reference to IngenHousz who had described this eight years earlier in just as much detail. De Saussure did not study the history of his scientific discipline, he just recounted the story as far as he knew it, not reading the relevant works or going back to the original sources, relying on Senebier's falsifying accounts. Only a few years after IngenHousz' death, Senebier's continued and clever rhetorics had done their work. When Liebig took up the research in plant physiology, almost fourty years later, he would follow De Saussure's lead and did not give IngenHousz the attention he deserved while Priestley and Senebier got all the honours. This was the reason why Wiesner wrote in 1905:

> *Die historische Inkorrektheit des Saussure'schen Hauptwerkes, freilich auch in bezug auf Ingen-Housz vielfach irreführende Darstellung Senebier's sind des wahre Grund, warum Ingen-Housz für lange Zeit fast des Vergessheit anheimfiel und dass seine grossen Entdeckungen, welche ja doch die grundlage der späteren Arbeiten Saussure's bildeten, lange Zeit hindurch und zum Teil auch jetzt noch Senebier zugeschrieben werden.*[256]

The historical incorrectness in the main work of de Saussure, mainly as a consequence of the misleading representations by Senebier of the contributions of Ingen-Housz, is the true explanation for the long period

---

253 Wiesner 1905
254 De Saussure 1804
255 De Saussure 1804 p 39
256 Wiesner 1905 p 158

during which Ingen-Housz' name has been forgotten and for the fact that his great discoveries, that even form the basis for De Saussure's later work, have for long and partly still today been attributed to Senebier.

Wiesner's analysis has not been able to prevent IngenHousz' name vanishing in the mists of time. His reconstruction of the IngenHousz' story has been equally forgotten. No wonder, as even the preface of the great Alexander von Humboldt to the German translation of IngenHousz' *Uber die Ernährung der Pflanzen und die fruchtbarkeir des Bodems* from 1798 did not turn the tide. It begins with:

Herr Ingen-Housz gehört zu der kleinen Zahl arbeitender Physiker, welche das fruchtbare Talent besitzen, nicht nur einzelne Gegenstände mit bewunderenswürdiger Anstrengung zu verfolgen, sondern auch jede neue Erscheinung, statt isoliert aufzustellen, harmonisch mit der ältern zu verbinden. Seine Schriften lehren, dass er den grossen Zweck aller naturforshung, dieses Zusammenwirkung der Kräfte zu untersuchen, nie aus dem Auge verliert...[257]

Mr Ingen-Housz is one of that small group of physicians who have the fertile talent not only to investigate a subject with admirable intensity, but also to discern every new phenomenon and to connect it harmonically with the rest. His published works teach that the biggest goal for all science is never to lose the investigation of the cooperation of all these forces from sight...

The Dutch pretender

And even in his homeland Holland, somebody tried to claim the discovery of photosynthesis or to "rob him from the sunlight" as he wrote poetically "dat ik my van het sonneligt niet kan of mag laten beroven" in his 22 October 1785 letter to Van Breda. Willem van Barneveld manipulated the dates of handing in his publication on the purifying powers of plants at the secretary of the *Verhandelingen van het Provinciaal Utrechtsch Genootschap*, in order to claim primacy. He too, didn't mention IngenHousz, whose experiments he copied. As Smit already noted,[258] IngenHousz was living and working in a period during which numerous new discoveries were made and in which some investigators did not shun at all to claim priority for this or that discovery. And IngenHousz knew that too, as he wrote to Van Breda on 11 May 1785:

---

257 quoted by Wiesner 1905
258 Smit 1980

Men heeft er fontana al publiek van beschuldigt zyne uytvindingen te datummen op tyden wanneer hy er waarschynlyk nog niet op dagt, en van ontdekkingen aan te kondigen die hy niet heeft gemaakt. Ik heb reden om te denken dat priestley er ook niet vry van is, terwyl de H. Franklin, de laatste keer als ik in Parys was, een brief bequam van Priestley, ontdekkingen bevattende, welke brief omtrent 5 maanden voor den ontfangst gedatumd was. Ik wil hierdoor maar alleen beweren dat het niet onmogelyk is dat Senebier al vroegtydig de wind van myn ontdekking, misschien wat verward, heeft gevat. Hoe dit ook zy, hy had de couragie niet van in zyn boek zig op iets anders als ideén te beroepen., zig waarschynlyk niet betrouwende van enige proeven te spreken, omdat zyne vrienden hem zoude schuldig konnen kennen van onwaarheden te schryven.

Some have been accusing fontana in public of putting dates on his inventions when he probably had not even thought about it and of announcing discoveries which he did not make. I have also reasons for thinking that priestley is not free of this weakness, while Mr Franklin, when I was last time in Paris, received a letter from Priestley, containing discoveries, dated about five months prior of the date of reception. Therewith I just want to argue that it might not be impossible that early on Senebier can have picked up rumours about my discovery, however muddled. Whatever happened, he did not have the courage to call upon anything else but ideas, probably not trusting to refer to any experiments, because his friends would be able to know him to write untruths.

Local heroes seemed to have overtaken IngenHousz from the left and from the right, highlighting one of IngenHousz' vulnerabilities. He never stayed long enough in one country to build up popularity. Moreover, he was no societies-man. He preferred to work on his own and keep some friendships alive with people that really interested him, such as Franklin. As he was not associated with learned societies, nobody there was inclined to defend his honours. He was a member of the Royal Society in London, but did not have the occasion very often to appear there live, as he was travelling most of the time. He left few traces in the minds of people apart from those of some good friends and almost no traces at all in the public records. Although his books sold well, they were no bestsellers, as he was catering for a rather selected audience. His final essay was hidden away in an official government report. His complete oeuvre was spread over four languages in many different versions. No wonder that the historians would end up having a difficult time in giving him a place in history.

A story retold

In 1809, John Murray, lecturer in Edinburgh in chemistry, medicine and pharmacy, gave his account of the events.[259] By that time, the whole story was told in terms of oxygen and carbon dioxide. All references to the phlogiston-era were gone. He stated that Priestley was the first to find that plants generate almost pure oxygen when exposed to the sun, and that he thought the green matter was of a vegetable nature. IngenHousz, so he mentioned, prosecuted this investigation to a greater extent. According to Murray's (rather muddled) account, IngenHousz merely determined that it was some organic structure in the plants' leaves that was responsible for the production of oxygen. Murray rightly acknowledged that IngenHousz was the one who determined that plants in the dark consume oxygen and produce carbonic acid. In his account of the details of the research of Priestley and IngenHousz it sounds as if both have been doing similar things in parallel. Discussing the fact that plants use carbon dioxide (carbonic acid) as a basic ingredient to produce oxygen, he left the honour to Senebier for discovering this. Although IngenHousz is mentioned as the one who did the crucial experiments, the lecture by Murray gave no clear account, most probably because "this subject has been more lately investigated, and several series of experiments performed, but with results which are still discordant". Murray did not have a fitting framework as yet to explain the many phenomena that were observed in plants, water and sunlight.

Half a century later Chevreul gave his account of the story to date in the *Journal des Savants*.[260] He was very clear in his judgement: Priestley definitely saw something very important but he could not explain it because of his way of working. He performed experiments without submitting them to verification: "il faisait donc des expériences sans contrôle et conséquemment sans se conformer à l'esprit de *la méthode expérimentale*, telle que nous l'avons définie". IngenHousz, according to Chevreul, could explain perfectly well what was so contradictory in Priestley's results. Thanks to his meticulous and systematic way of working, he could demonstrate three principal facts: leaves produce a gas, which is exceptionally good at making candles burn and which is only produced under the influence of the light of the sun. "La découverte du troisième fait est la gloire d'Ingen-Housz". Priestley in the mean time was not as indifferent to glory as he pretended to be, which is clearly proven, according to Chevreul, by his behaviour towards IngenHousz after he did his big discovery and published it in 1779. Chevreul was delighted about IngenHousz' methods and findings, and went into some detail on the experiments by which IngenHousz could demonstrate his three findings: "ils sont un exemple de la manière de procéder à la recherche de la verité dans les sciences expérimentales, lorsqu'on est animé

259 Murray 1809
260 Chevreul 1856

de l'esprit de *la méthode a posteriori*". He specifically highlights how IngenHousz used rubber (caoutchouc) bottles and tubes, just standard procedure in the laboratories of his time.

Senebier gets a critical treatment in Chevreul's account.[261] Indeed it is Senebier who wrote - in 1800 - that one had to admit that carbonic acid dissolved in water, is taken up by plant leaves and decomposed while the carbon is deposited in the plants and the oxygen expelled. Before admitting that, he did have a long way to go. This is understandable as Senebier (and Priestley), from the 1780s on, tried to explain the interactions between plants and the atmosphere within the framework of the phlogiston theory. They could not explain the origin of carbon in plants, as this element did not exist in their view on the world. It was waiting for Lavoisier's theory, which offered Senebier the right conceptual frame that could explain the link between the fixed air taken up by the plants and the vital air expelled. Still, he framed the role of carbon in the physiology of plants in a rather weird way, speculating as he had always had done, that this element was deposited in a network of yellow 'mailles'. Senebier did "commis la faute d'y attacher des explications vagues, sans parler d'erreurs évidentes qu'il a commises".

However, Chevreul concluded that this case is "un des exemples les plus remarquables de la manière dont l'esprit de plusieurs procède pour découvrir la vérité dans les sciences d'observation et d'expérience". Finding out how nature functions is not and cannot be the work of one man on his own, but will always be the result of many men and women who all have their input, sometimes contradictory, other times polemic, to result in trustworthy knowledge.

Reputations in the making

However, measuring reputations, certainly after two centuries, is a perilous undertaking. With no popularity polls being done at that time, we will have to do with its closest derivative, which is available from the francophone perspective at least. The *Biographie Universelle* sums up in alphabetical order "all men, ancient or modern, who distinguished themselves through their writings, actions, talents, virtues or crimes". It was a gigantic enterprise set up by Michaud in Paris, expanding to 90 volumes. In the 21st volume,[262] printed in 1818, IngenHousz gets a full page biography, describing mainly his medical career. His photosynthesis-research is treated in one line: "On lui doit ... l'importante découverte que les végétaux vivants exposés à la lumière et répandent dans l'atmosphère le gaz oxygène". Without a single word to describe the concept behind this natural phenomenon, this one sentence is the best the writer could

261 Chevreul 1857
262 Michaud 1818

do at that moment. In 1823 this biographical encyclopaedia[263] arrives at the letter P. Priestley is described in five full pages. Three of these pages cover his theological and political work. The fact that he was proclaimed a French citizen and had always been a fierce proponent of the French Revolution, gets prominent treatment. One page describes his character as being, sweet, tolerant (even for people with opposing viewpoints) and absolutely not jealous. Concerning his merits in natural philosophy, the biographer remarks that Priestley could come to his original findings in chemistry due to his ignorance, but he nevertheless underlines Priestley's great contributions to the discovery of different airs (especially "l'air déphlogistiqué") laying the fundaments of the system build later by Lavoisier. Two pages further John Pringle's entry covers two pages.

In the 42nd volume,[264] published in 1825 Senebier's entry covers three pages and a half. His life as a man of letters is described extensively and with great respect. His translations of the works of Spallanzani and his large book on the art of observation are pointed at. Some sentences describe his controversy with IngenHousz:

> Il démontra que elle [la lumière solaire] agissait puissamment sur la décomposition de l'acide carbonique par les végétaux; mais il soutint l'opinion contraire à celle d'Ingenhousz, sur la nature de l'air qui échappe des feuilles pendant la nuit. Ingenhousz en avait exagéré la qualité délétère; Senebier en exagéra la pureté. Tous les deux se trompaient; mais dans la dispute qui s'éleva sur ce point, Senebier conserva l'avantage de la modération, et fut le premier à reconnaitre son erreur.

The editor takes to a diplomatic stance in which IngenHousz and Senebier were both supposed to be exaggerating. Both were wrong, so he claims, but in the ensuing dispute Senebier is described as being the most moderate and the first the recognise his mistake. It may be clear to whom the most sympathy goes in this version of the story.

The *Biographie Universelle* got a second edition, growing to 45 volumes, of which the first was published in 1843. The 20th volume[265] came out in 1858 and still covered IngenHousz in one page, repeating the same story as in the earlier edition. Volume 34 sees the entry on Priestley[266], slightly reduced to three pages and a half, due to smaller type. This story is essentially the same as in the earlier edition: he is described as one of the towering heroes of revolutionary thought, both in natural as in political philosophy. In Volume 39[267] it is Senebier's turn to gets exactly the same treatment as he did some twenty years earlier. A last

---

263 Michaud 1823
264 Michaud 1825
265 Michaud 1858
266 Michaud 1860
267 Michaud 1860b

man to have a quick look at in Volume 40[268] is Spallanzani. Surprisingly he gets almost five full pages, as much as the three other men together. From the French perspective on that era of innovation and scientific research, he must have been a heavy weight. In Italy and in the world of Enlightenment experts his name is well known. He was one of the great proto-biologists and took a prominent place in the professional network of IngenHousz. He followed up on IngenHousz' experiments on the electric ray, visited Vienna in 1786 where he was cordially welcomed by Joseph II and spend many months on the study of respiration: plenty of cross-over points with IngenHousz' work. No mention however of his possible intermediary links in the controversy between IngenHousz and Senebier.

It is difficult to weigh the impact or the importance of these entries. In any case, Priestley was not going to be worse for it, as his fame would last two centuries. The Photosynthesis Bicentennial Symposium took place in November 1971. The Proceedings of the National Academy of Science that covered the symposium opens with the words of chairman Kenneth Thimman:

> In early August 1771 Joseph Priestley, chemist from Birmingham, England, performed his famous experiment with the mouse and the mint plant. This experiment provided the beginnings of our understanding of that remarkable process whereby the organic matter of our biosphere is produced and our atmosphere continuously purified.[269]

He dryly notes in the following sentences that it is perhaps not widely known that almost twenty years earlier Joseph Bonnet observed that leaves when immersed in water and exposed to sunlight produced bubbles of gas. Thimann struggled as much with the same priority problem as all the others and it is a question of historical interpretation to choose one or the other. One year later a two-volume book[270] with a overview of photosynthesis research appeared with the title *Photosynthesis, two centuries after its discovery by Joseph Priestley*. The message was clear. One year later, Hill attributed the discovery of photosynthesis to Priestley and concluded that "the 'photo' part of the photosynthesis was discovered independently by the two men". By that time it was the fact that Sachs had written in 1875 that IngenHousz had observed more and more correctly than Priestley had already been forgotten.[271]

Nash would redress the balance in 1952 when he wrote, based on extensive study of some of the original materials, that "There can be therefore be no question that the effective discovery of the joint action of light and vegetation

---

268 Michaud 1860c
269 Thiman 1971
270 Forti 1972
271 Sachs 1875

in maintaining the atmosphere belongs to Jan Ingen-Housz"[272] although he admits that "... reliable monographic literature dealing with the problems under discussion is conspicuous only by its absence". He pointed out that IngenHousz observation of the plants' respiration in the dark was a discovery of great importance. In saying so, Nash joined the small group of historians who gave IngenHousz the place he deserved. Plant physiologist Howard Reed had published a reprint of IngenHousz' 1779 publication with some comments,[273] while Rabinovitch had written in 1945 that "Ingen-Housz' main achievements were the discovery of the importance of light for the 'dephlogistication' of air by plants, and the proof that the plants improve the air not merely by absorbing its 'mephitic' constituents (as suggested by Priestley), but that they also actively produce 'vital air' (oxygen)."[274] Still, in 2009, Matthews proposes to take Priestley's discovery of photosynthesis as a prime example of a scientific breakthrough to use in the secondary school curriculum.[275] He provides three educational and social considerations. The Priestley & Photosynthesis-story offers plenty of material to teach three important topics: the pressing environmental problem of 'good air', the 'nature of science' and a reappraisal of the Enlightenment values and tradition. All three are perfectly valuable subjects too often missing in school curricula, but one may wonder if IngenHousz would not be more accurate point of entrance if one wants to talk about photosynthesis.

Priestley's fame seems to be overpowering. In February 2004, an article in *National Geographic* magazine[276] covered the carbon cycle on earth. Carbon is an essential atom on the earth, as it forms the backbone of many of the molecules that make up living organisms. It is also a key ingredient of the greenhouse warming of the atmosphere. And it is photosynthesis that keeps this carbon carrousel going. A picture showed a mouse in a sealed glass jar, alive and kicking because of a mint plant standing alongside in the jar: without the plant, the mouse would die, as Joseph Priestley had demonstrated and described in 1771. The article praised Priestley as "an English minister who grasped the key processes of the natural carbon cycle". As we have seen, Priestley indeed did many fascinating experiments, concentrating on the quality of the air produced under various circumstances. But he combined these observations and experimentations with "maddening failures in interpretation"[277] as has already been critically remarked half a century ago. Priestley realised there was an interdependence of plants and animals, but he did not grasp the fundamental mechanism behind this phenomenon. In the *National Geographic* account this is only

---

272 Nash 1952
273 Reed 1949
274 Rabinovitch 1945
275 Matthews 2009
276 Appenzeller 2004
277 Nash 1952

a small but incorrect historical anecdote in a comprehensive and well researched article. It again demonstrated how the popular history of photosynthesis (if such a thing exists at all) boiled down to a distorted account of the facts. Somebody else did the crucial experiments and formulated the process by which sunlight, water, oxygen and carbon make up the wonderful proces of photosynthesis, but his name was lost in the 250 years that passed since then.

Still, it shows how, this many years later, Priestley's fame is capable of overshadowing that of his contemporaries. Priestley's manipulative silence about what IngenHousz had done, combined with his prolific writership, have no doubt contributed to the fate of IngenHousz' name in history. Still, that can't explain all. The twists and turns in the history of science are driven by more than just some rivalry between two men who were totally different in their ambitions, style and background.

Ingredients for an amnesiac cocktail

The fate of IngenHousz and his lack of fame is clearly the result of a mixture of factors. The battle between the four contenders to the title of 'discoverer of photosynthesis' has drawn attention away from what really happened. IngenHousz himself was reacting to things happening in far away places, in a period where communication was becoming increasingly difficult. This resulted in the three other men involved being able to do more or less what they wanted, claiming in their own countries and in their own writings whatever they liked. Every one of them followed his own strategy. Priestley simply hushed IngenHousz up. Senebier depicted him as unreliable and blasphemous. Van Barneveld cheated with the delivery dates of his paper. IngenHousz was on the road most of the time, never spending a very long time in one place. He was a 'man of the world', but that resulted in not being a local hero. He was not present to take care of his personal public relations and so could be easily sidetracked by the local VIP.

Talking about public relations, IngenHousz did not become member of the many societies all over Europe that asked him to become member. He was not a member of the freemasons. He was a critical commentator on popular things such as charlatans. He refrained from entering the public debate as long as possible. His networking was based on personal friendships, which seem to have been based on genuine interest in other people on the same wavelength. IngenHousz was not a man of the rhetorical argument either. His two main opponents were priests, well versed in influencing people through their words and argumentations that did not necessarily have to be based on facts, but on the interpretation of texts.

Priestley was a clergyman and a revolutionary. While IngenHousz was humble and non-polemic, Priestley was a revolutionary and outspoken public person. And he was a very prolific writer, with 50 works on theology, 13 on education and history, 18 on political and metaphysical subjects and 12 books and 50 papers on physics, chemistry and physiology to his name. He overshadowed any other writer by the sheer volume of his writing. Senebier was a catholic priest from Geneva, a librarian and man of letters, verbose and argumentative, trained to reach his goals by rethorics. His books blossomed into volumes of never-ending argumentation. IngenHousz in contrast, seems to be rather the man who is more interested in obtaining proper results than in discussing the theological implications of his findings. Moreover, IngenHousz was an independent man of means. After his successful inoculation of the Imperial family, he received an annual stipend that made him virtually and literally, spiritually and materially, independent. He did not belong to a school or ideology. This coin has a flip side: one doesn't belong to a fraction that can stand behind you, or that can push you upfront as their spokesman.

The picture is even more complex. IngenHousz did not really belong to one profession. The first years of his professional life he was a doctor, but afterwards he was involved in investigating physical, chemical, electrical and agricultural problems. At that time these were not yet part to separate disciplines. Later it would prove difficult to put IngenHousz in a pigeonhole. Making things even more difficult, IngenHousz not only travelled a lot but also wrote in at least four different languages. He published not only exactly translated editions, but new versions that grew in length and breadth. In the end, although he did not publish much, it did not add up to an easily overseeable oeuvre.

And last but not least, everybody seems to find it difficult to remember this strangly unfamiliar name, being a word with unusual sound and form. Another, very prosaic but nonetheless quite realistic reason why IngenHousz' name has been forgotten is that he left no children who could keep the name of their father high. It is only recently, 200 years later, that some great-great-nephews dusted off the name of the man they fondly call "Ome Jan" (Uncle John). So, if no nation, no society nor your family stands up for your name, how could it survive the erosion of history?

rewind

Memory and priority: who was first?
and why has he been forgotten?

The reason why so few people know who discovered photosynthesis and that even fewer people seem to know the name IngenHousz was one of the motivations for start this research. It soon became clear that most of those who believe they know who discovered photosynthesis think it is Priestley.

The reconstruction of the history of the first half-century of photosynthesis research and of how this history has been written, may explain these facts, at least partly. The historians over the last two hundred years are not completely to be blamed for this biased view on things. The main protagonists in this story each contributed to the final result, consciously or unconsciously. The characters of Priestley, Senebier, IngenHousz and others are very different and may explain why they behaved the way they did. It shows science is also a rat race for fame and glory in which not everybody is playing the game along the same rules. Some are even bending the rules to serve their purpose.
Next to all the theories, experiments, technology, peer review, logic and rationality, this is also a most natural ingredient of the game called science. Science is a human pursuit and priority questions like these show how human it is.

Still, it is the scientific method that makes it possible for the knowledge that finally results from all these efforts to transcend the personal perspectives and behaviours of individual scientists. Everything that we now know about photosynthesis is the result of centuries of dedicated hard work of many people whose names nor motives, mistakes nor misdemeanors make any difference in the end. Trustworthy knowledge is the property of humanity. One can only wish or hope that the historical knowledge about how we got there is just as trustworthy.

ESSAI
SUR
L'ART D'OBSERVER
ET DE FAIRE
DES EXPÉRIENCES ;
SECONDE ÉDITION,
Considérablement changée et augmentée.

Par JEAN SENEBIER, Membre associé de
l'Institut national , de diverses Académies
et Sociétés savantes , et Bibliothécaire de
Genève.

TOME III.

A GENEVE,
Chez J. J. PASCHOUD, Libraire.

AN X. (1802.)

ESSAI
SUR
L'ART D'OBSERVER
ET DE
FAIRE DES EXPÉRIENCES.

CINQUIÈME PARTIE.
L'art de faire les expériences.

CHAPITRE I.
De l'art de faire les expériences.

On n'a pas examiné long-temps en quoi
consiste l'art d'observer , sans s'appercevoir
qu'il serait insuffisant pour interpréter la nature dans toutes les circonstances, et qu'il
faut le joindre à l'art de faire des expériences.
J'ai résolu de m'en occuper encore , parce que
ces deux arts ont plusieurs choses communes ,
Tome III.   A

'Science' as they knew it in the eighteenth century. How to observe, how to perform experiments and how to assemble that into reliable knowledge, as described in Senebier's magnum opus.

Science as IngenHousz knew it
& science as we know it

'Science' as we know it today did not exist in the eighteenth century. IngenHousz referred to himself and his colleagues as 'natural philosophers'. That was the common term for all people (most often men) who investigated nature in all its aspects, doing natural philosophy. 'Science' as a word existed and was often used by IngenHousz & co. 'Scientist' however did not. That word would only be used for the first time at the third meeting of the British Association for the Advancement of Science in Birmingham in 1833 - as an alternative to 'natural philosopher' or 'savant' and as rhyming counterweight to 'artist'.[278] It was the culmination of an evolution that started in the last decades of the eighteenth century. Slowly the divide between philosophy and natural philosophy had begun widening, under the influence of the increasing use of mathematics, ever more accurate instruments and rigorous experimentation. Natural philosophers saw themselves more and more as 'experimental philosophers' and started making claims that their knowledge was a particularly privileged form of knowledge that avoided deference to textual authorities or sophisticated wordplay. The territory covered by these philosphers grew larger while some areas of investigation started claiming a separate identity[279]. Some disciplines such as chemistry and physics began to be seen as more or less delineated fields of research. Hearing the words 'chymist' or 'physicist', people would know what was being talked about, just as much as an 'electrician'. Electricity was a physical phenomenon, people investigating it would be electricians, and it was not as yet seen as a subfield in physics. Chemistry too had subspecialties that were not yet recognised as such, with vague grey areas such as pneumatic or pharmacological medicine. The properties of metals or magnetism, they were all the wonderful and amazing elements of nature. In that sense 'natural philosopher' was and is an appropriate term for people who thought about the world and tried to understand how it worked. While they were as well thinking about how they could get to know these things. That sounds like the mission statement of a modern philosopher of science. So maybe, IngenHousz' practices can teach us a thing or two about what science was and maybe still is.

---

278 Holmes 2009
279 Gascoigne 2003

More than a doctor

IngenHousz can rightly be called a 'physician', in every sense of the word, both old and new. He was a medical doctor in the first place, what we now would call a 'physician'. As for most doctors, he was primarily interested in people, sick people, and in ways to make them better. He knew that medicine in his time only had limited means to do anything useful and he hoped it was only at the beginning of its possibilities. Too many patients still died in the hands of the surgeons or quacks, lacking suitable techniques, drugs or therapies. He noted with horror how children died in their first years, how women didn't survive childbirth, how infectious diseases rampaged through the world and how chronic ailments such as gout, gall stones or gravel[280] made lives (including his own and many of the other protagonists in this story) really miserable. Finding better drugs and trying them out in order to find the most effective formulation or dosage was one of those goals that form a connecting theme through his life. His pragmaticism was very likely the result of his medical training. In the end, as a doctor, you want something that works, because in medicine the verdict is rather simple. Either the patient lives, or not. One's remedy helps, or not. In contrast, in natural philosophy all sorts of theories about the structure of nature or wonderful explanations for phenomena had been around for ages. Without implications for real life, any nice and convincing story could survive for a long time. Aristotelian biology survived until the renaissance anatomists started looking inside the body. For IngenHousz the bottom line was always Checkpoint Reality.

But he was also a 'physicist'. Interested in all things physical, he was investigating the natural world, as any natural philosopher would do. This has to be taken literally, as he was really interested in almost all aspects of the physical world. Chemistry can be seen as an extension to his medical practice. Chemistry led to electricity and magnetism. And these linked up to conductors, lightning, microscopy &c. When the botanist Julius Wiesner wrote his monumental but long forgotten book on IngenHousz in 1904, he asked his physicist-colleague Ernst Mach to summarise IngenHousz' importance to physical science. This is what Mach wrote:

> Ingen-Housz kennt die Physik seiner zeit sehr gut und interessiert sich lebhaft für alle Entdeckungen auf diesem Gebiete. Mit besonderen Interesse erfolgt er die Franklin'ische Entdeckung, dass man Wolken die Elektrizität abzapfen und dieselben zum Laden von Flaschen verwenden kann. Die Erfindüng des Blitzableiters (für desssen praktischen Anwendung er, wie wir gesehen, so viel getan) scheint ihm so beteutend, dass er meint: 'Künftige Jahrhunderte

---

280 Keeman 2006

*werden dieses, orin wir leben, bewunderen und immer eingestehen, dass
keines der verflossenen an grossen und nützlichen Entdeckungen so fruchtbar
gewezen ist.' Ingen-Housz' Abhandlung über Elektrizität, insbesondere über
den Elektrophor, gibt eine gute sachverständige Darstellung der Franklin'sche
'unitarischen' Theorie. Ingen-Housz ist auch ein tüchtiger Experimentator.
Seine mechanischen Konstruktionen, seine Versuche über künstliche
Magnete, über zweckmässige und leicht Aufhängungen der Kompasnadeln,
seine Improvisation der Scheibenelektrisiermaschine durch einen gläseren
'Kredenzteller', welche alle ähnlichen Konstruktionen veranlasst haben dürfte,
seine Verwendung der Deckgläschen bei der mikroskopischen Beobachtung
beweisen dies hinlänglich. Durch jede neue Entdeckung wird Ingen-Housz zu
eigener Arbeit angeregt. Auf Fontana's beobachtung der Luftabsorption durch
frisch ausgeglühte Kohle gründet er die Konstruktion einer neuen Luftpumpe.[281]
Als Physikalische Leistung ist heute noch am populärsten der Ingen-Housz'sche
Versuch zur Bestimmung der Geschwindigkeit der Wärmefortpflanzung. Wie
Ingen-Housz selbst sagt, hat er den Gedanken zur Ausführung des Versuches
von Franklin übernommen. Merkwurdig ist, dass Ingen-Housz an einen
Parallelismus der Wärme- und Elektrizitätsleitung schon denkt, während man
diese Einsicht für eine moderne Errungschaft hält.[282]*

Ingen-Housz knows the physics of his time very well and has a lively interest
in the new developments in this field. He follows with special interest the
discoveries of Franklin, how one can tap electricity from the clouds and store it
in bottles. The discovery of the lightning rod (for the practical development he
did a lot of work) was to him so significant that he wrote: "In the future they will
look back at this time in admiration and think that none of these discoveries was
as useful as this one." Ingen-Housz' dissertation on electricity, and especially on
the electrophore, gives a good and informed explanation of Franklin's 'unitarian'
theory. Ingen-Housz is also a thorough experimentalist. His mechanical
constructions, his research into artificial magnets, his work on the functional
and light suspension of magnetic needles, his improvisational electrical machine
with glass discs which may have led to all comparable constructions, his use of
cover glasses for microscopes, all of these provide sufficient proof. Ingen-Housz
was inspired by every new discovery to continue his research work. Based on
Fontana's observation of the absorption of air by freshly extinguished coals, he
built a new airpump. One of his physical achievements that are still very popular
is the investigation of the conductivity of heat. He was inspired by Franklin's ideas
to perform these experiments, as Ingen-Housz himself testifies. It is remarkable
that Ingen-Housz already thinks about a parallel between the conductivity of
heat and of electricity, as these insights are taken to be a modern attainment.

---

281 IngenHousz 1784 p 431
282 Wiesner 1905 p 189

As in his other activities, IngenHousz demonstrated his spirited individuality in the field of physics that much is made clear in this account of a top-ranking physicist. He linked one subject to the other, found inspiration in every new discovery, saw no distinction between applied or theoretical physics and tried to combine all this in a grand overview.

He was not a professional chemist either. But neither can he be seen as a dilettante. His methodology was rigorous. The way in which he was performing series and series of experiments in which he methodically eliminated irrelevant parameters, his use of controls, his concern about measuring errors, they all compensated for his lack of specialised knowledge. On top of that, these were combined in a clever and clear mind with a realistic idea about what scientific knowledge could achieve. He thought and worked in the mindset of the natural philosphy of the eighteenth century, which was nicely phrased by Buffon and quoted by Felice Fontana, IngenHousz' friend, in his preface to his book[283] on grain rust: "It is by well reasoned and prolonged experiments that one forces nature to disclose her secrets..."

The hunger for knowledge

IngenHousz gathered knowledge about the world and he did that in a very specific manner. He wanted that knowledge to be trustworthy. He wanted it to be founded on evidence, not on revelation. The former is what everybody can experience and what can be reproduced by anybody willing to put in the right effort. The latter is what was based on hearsay, sacred or secret texts, intuition or opinions, however wellmeant these might be. The Royal Society had as motto "Nullius in Verba", or "On the words of no one". It was the philosophical organisation's mission to promote experiments and careful observations rather than unsubstantiated and untestable opinions, old or new. IngenHousz must have felt at home in a close circle of friends and peers that were strongly committed to this approach. IngenHousz did not systematically elaborate theoretical considerations on science, but his method shines through in what he did and how he did it. He placed himself in a long tradition of knowledge building, which is as old as mankind itself.

It is something men and women have been doing since the beginning of humanity. We were not even the first living organisms on the earth to understand the world. Knowledge is a prerequisite to survive. Lions or bacteria 'know' the world too, crows and apes are even using instruments, but humans were the first to start developing knowledge on a grand scale and do it consciously (supported by a brain that was big enough to handle the complexity). They started to see the relationships between things. They discovered causality. Hit a man with an

---

283 Fontana 1767

axe hard enough and he will not stand up again. That's knowledge about the world, and the beginning of emergency medicine (or its reverse). Every day the sun comes up in the east and she does that every day a little bit earlier, until in the middle of a vast period of 356 days, she starts coming later every day. That's the beginning of astronomy. They did start counting before, but now they started to use big numbers and handled them cleverly, a logical thing to do if you want to put things into categories and want to know how many of each you can find. Gradually all sorts of knowledge emerged, and certainly not all knowledge could or can be called science. Priests and shamans, fishermen or pigeon lovers know things about the world, or believe they know. (One of the reasons why their knowledge is not that trustworthy is that they tend to forget what did not work and only remember the 'facts' that support their beliefs. In contrast, scientists learn from what does not work.) Darwin though learned a lot from pigeon lovers.[284] Not from their stories about why and how you can get a pigeon to get home first; these were (and still are) just-so stories, which cannot stand critical scrutiny. Darwin learned from what they did to breed ever-faster homing pigeons through selection, thereby offering a format for his theory of natural selection.

Science gathers knowledge that wants to be reliable, reproducible and systematic. People soon learned that not all people that tell you how the world works are as trustworthy. And some of them told really nice stories, which people tended to believe. Better a nice story about the origins of thunder and lighting than no story at all, even it was a scary story. People wanted to know why things were as they were. But as soon as you start looking around, you realize there are many things out there. And not just objects. When you look out of the window and try to describe what you see, in great detail, you will soon realise there is rather more to it than meets the eye. Moreover, these things change, as well as their relationships. For a detailed description you have to bring time into the picture. One sees the weather change, one sees seasons come and go. One sees animals do things to each other and to plants. One sees people go to work. One sees some of them do some hefty things with each other. And nine months later a small sized subject of the species falls out of the big one with the breasts. Try to capture all of this and even the biggest and fastest hard disk won't do. There is just too much happening out there and it's happening too fast. If you really want to understand at least some of what is happening out there, you will have to make choices. Limit yourself to what you think is relevant. Divide the jobs. One of you watches the falling objects, another one concentrates on the living things on two legs. First you have to define what a living thing is, or what constitutes falling, but that's part of the job. As long as somebody brings in the food and drinks to keep you going, and you have the instruments essential to do the good observations, you can keep on watching and describing.

---

284 Desmond 1991

But one day you realize there is something in between you and that outer world. The window gives you a nice view but also distorts what you see. From the room next door, they see it slightly different. And then there is the other side of the building…, as if it weren't complicated enough without it. Your knowledge turns out to be not that reliable; your perspective has been biased all the time. Some of the colours of what you see may be different from what you think you see. And you want to know the reality out there as it really is. And even worse, comparing the measurements from different observers, you realize that there are some serious differences. Your measurements are not all that reliable. Not all samples that have been taken are comparable or relevant to the whole population. You learn to eliminate so called variables that are not relevant. For some plants to bloom, it's not the temperature that matters but the length of the day. You learn to eliminate variables that disturb your observations. The weighing of very small objects turns out to be disturbed by vibrations of passing trucks, so you make vibration-free tables or forbid trucks to pass while you are measuring. Slowly but surely, in a collective effort that spans generations of philosophers, natural philosophers and scientists, we got an inkling of an idea about what we could consider to be trustworthy knowledge. After all, science begins with knowing e.g. how to cook an egg and to be able to do it each time in such a way that the yolk is fluent and the white just not solid, and in such a way that you can tell it to others so they can do it too.

Science in the making

That's how science came into existence. Nothing spectacular, nothing fancy. Just clear thinking and critical evaluation of what you measure and even more critical of what you think that might mean. That's not to say it is easy. It took humanity some 10, 000 years (since the first Neolithic settlements formed the cradle of civilization) to come where we are now. And still there is more that we don't know than what we do know. This however is a crucial part of what we call science: knowing that there is still a lot one doesn't know. Look out of the window again: the complexity out there is so big that not one theory can grasp it all. (Although some dream of the Theory of It All, but until further confirmation that belongs to the realm of religion.) And on top of that, one of the most complex things to study stands at this side of the window: it's inside your head and we call it the brain. Thinking about knowledge implies that we have to know how the brain works.

Science has apprehended that it's learning by doing and trial-and-error that make up the game. Permanent doubt is necessary. Everything one finds turns out to be just a stepping stone to the next question. And surprisingly, all that has been found turns out to be fitting together. From the behaviour of subatomic particles to the behaviour of *Homo sapiens*, from the falling of objects to the movement of populations, we are able to construct a more or less coherent whole of knowledge (admittedly containing still quite some black holes). Some people call it rational, others naturalistic. It is a gigantic body of facts and concepts and models and theories by which we can explain in a fairly reliable way how the world came to be, how it fits together and sometimes how we can try to make it a better place. That's far from claiming to know the Truth, if ever we would be able to define such a thing. Science is rather sceptical towards those who claim to know the truth, especially if spelled with capital T, just because centuries of measuring, experimenting, thinking and finding oneself wrong have thought us that truth is an elusive concept prone to abuse by absolutists. The more so because we learned (the hard way) that humans, even with years of academic study under their belt, are prone to false beliefs and illusions. The human mind is in some respects a weak instrument for investigations and very good at concocting convincing stories that seem to deliver explanations. The human mind is prone to bias and often follows weird heuristics, which can lead to mistaken opinions and wrong beliefs, even in the most rationally trained people.[285]

That's probably why some people claim they can cure cancer by laying their hand on the patient's head. Others say their carpets can fly. And still others believe that they can read your character and fate from the stars at your birthday. These kind of beliefs are rife among the general public and even scientists aren't always immune from such pseudo-scientific beliefs.[286] 'Pseudo-scientific' because they superficially may look like science, but turn out to miss some of the essential ingredients of what makes up science.

Producing knowledge

IngenHousz was trying to assemble a reliable and useful picture of the world and therefore earns a prominent place in the abbreviated reconstruction above of the slow emergence of science as a special and systematised version of everyday trustworthy knowledge. Looking at IngenHousz' publications, at his correspondence and the discussions he had with fellow natural philosophers it becomes clear that he worked at the front line of the birth of modern science. While Hales, Bonnet or Priestley were occupied with collecting qualitative information about the world, IngenHousz, just like Senebier, Lavoisier or De Saussure, collected qualitative data. IngenHousz measured and measured again,

---

285 Kahneman 1982
286 Piattelli-Palmarini 1994

calibrating his instruments, trying to measure as exact as possible. This was part of his methodological approach. Priestley was experimenting haphazardly, setting up experiments inspired by his creativity and the inspiration of the day.[287] He explicitly described them as such to show that discoveries were largely accidental because that was to teach philosophers that their actions were caused by divine providence, not a result of their own will. Priestley's methodology was definitely guided by religious inspiration; the adulation of human genius would thus give way to worship of the divinity of nature. IngenHousz was not troubled by such divine regulatory principles and built long series of experiments, eliminating irrelevant variables one after the other, zooming in on the crucial aspects of the process he was studying. He had a hypothesis and wanted to confirm or refute it. His observations were laden with theory. His reasoning went from deduction to induction and back, slowly building more insight in the transformational power of plants. Along the way, vague concepts such as 'airs' and 'earths' were replaced by more accurate concepts such as 'gases' and 'elements'. IngenHousz follows one of his intellectual masters, l'Abbé Fontana who wrote in 1785:

> Dans le siècle éclairé où nous vivons, ...on ne s'attache à présent qu'aux faits, & toute doctrine, qui n'a pas pour fondement des expériences réelles, n'est regardée que comme une pure hypothèse.[288]

In this enlightened century in which we live, ... we hold on to the facts. And all doctrine, which has no foundation in real experiences can only be considered as pure hypothesis.

Experimental data, acquired through methodical inquiry, were very important for IngenHousz. He spent many pages in his books and endless paragraphs in his letters to describe his measuring apparatus and how he used it, almost like a modern day protocol. He delineated what other experimenters do wrong and how that influences their results. He calculated the error margins and demonstrated how big the error can become if one handled the eudiometer inappropriately. (It is very likely he was inspired by the work of people such as Herschel. These astronomers and mathematicians calculated the error margins on the lenses of their telescopes. IngenHousz frequently visited Herschel, who discovered Uranus in 1781[289], and might well have been inspired by his approach to nature, importing new techniques into biology.) As described in chapter 5, IngenHousz gave lengthy descriptions of the eudiometer, included

---

287 Conant 1950
288 Fontana 1785
289 Lemonick 2009

drawings and extensive information on how to assemble the apparatus down
to the addresses where to buy the right and reliable materials such as glass
tubes and copper fittings. He wanted to give the reader a complete picture of
his procedures and materials so that anybody interested should to be able to
faithfully reproduce the experiments. In his letter of 24 November 1786 to Van
Breda he stressed the importance of being careful, both in doing experiments
and in publishing their results:

> *Men kan niet te voorzigtig zyn in experimenten. Ik denk nog al vry gelukkig
> geweest te zyn in het punt van voorzigtigheid omtrent het uytgeven van
> experimenten. Ik weet nog niet dat Priestley, Senebier, of ymand anders iets
> tegen myne experimenten met fundament heeft in kunnen brengen. Indien u
> iets diergelyks zoud gewaarworden hebben in 't lezen of experimenteren, zou
> het my genoege zyn er van onderrigt te worden.*

> One cannot be too cautious with experiments. I think I have been rather
> lucky with my cautiousness concerning the publishing of experiments.
> I don't know yet if Priestley, Senebier, or anybody else has been able to
> formulate any fundamental criticism against my experiments. If you
> would observe anything of that kind while reading or doing experiments,
> I would be pleased if you would inform me.

Finding hypotheses refuted was one of his greatest delights, as he realised that
that was the only way to learn how nature really works. That is undoubtedly
the reason why in all his writings he tried to be as clear as possible, stressing
the importance of straightforward specifications so that others could easily
reproduce his experiments. This is essential to get any findings confirmed
by independent third parties. It is the only way to unmask mistakes, to learn
from error and to progress to knowledge that is, yes, trustworthy. It is no
coincidence that the following citation from Horatius is the motto to his book
on experiments on plants:

> *Si quid novisti rectius istis,*
> *Candidus imperti; si non, his utere mecum.*
> Horatius, I.6.67

> If you know a better guideline than this one,
> share it with me; if not, then follow this one with me.

Epistemic rules of thumb

IngenHousz opened his article on the electrophorus in his 1782 *Vermischte Schriften* with a farewell to all speculation and a plea for an empirical basis for all knowledge, in this epistemological consideration:

> *Seitdem das eitle Wortgeprängte, welches alles zusammen dem geiste nicht eine einzige Kenntnis beybrachte, aus der Physik verbannt ist, und man den Argumenten oder vielmehr den Sophismen Untersuchungen der Werke der Natur entgegengehalten hat, sind die Wissenschaften auf einen Grad gestiegen, welchen man nicht vermuthete, dass sie ihn erreichen könnten; und nachdem endlich die Wuth, Systeme zu schmieden, der allgemein empfundenen Nothwendigkeit, heutigen Tages alles unser Wissen auf dem sicheren Grunde, durch Erfahrungen, zu befestigen, gewichen ist, ist man überzeugt, dass der Gebrauch unseres geistes, wofern er nicht durch das Licht geleitet wird, welches Thatsachen und ächte Beobachtungen anstecken, oft zu nichts diene, als uns in Irrthum zu stürzen. Der reissende Fortgang, den die neueren Naturkündiger fast in allen Theilen der Physik gemacht haben, ist der Beweis von dem, was ich soeben behauptet habe.*[290]

Since all idle usage of words, that does not bring any knowledge to the curious mind, has been banned from physics and all sophisms and arguments in the investigation of the works of nature has been left behind, the sciences have been upgraded to a level that could not have been imagined before. And after finally the rage to construct (hypothetical) systems has given way to the generally experienced necessity to base nowadays all our knowledge on firm ground through experience, people are convinced that the use of our mind, in so far as it is not guided by the light which initiates facts and actual observations, often serves to nothing but to plunge us into fallacy. The boisterous progress that has been made in all fields of physical investigation is the proof of what I have just said.

Here speaks a 'modern scientist'. It confirms the picture that has been growing throughout these pages of a man who believes in honesty, sincerity, empirical evidence, progress, curiosity, objectivity and responsibility. These attitudes help if someone is looking for trustworthy knowledge. If a claim is made by a person who cannot be trusted, how likely is the claim to be true? If a claim is not based on evidence, how much is it worth? Is this claim based on the opinion of one person, or on the objective observation of many? If you don't feel responsible for the results of your research, how valuable could they then be? What drives scientists but curiosity, if not fame or money?

---

290 IngenHousz 1784

These scienctific attitudes give an indication of the epistemic guidelines that lead to trustworthy knowledge. These are not a question of ethical imperatives. They are the simple pragmatic consequences if what you would call trustworthy. It is an easy question that is valuable in- and outside of science: when can a claim be trusted? The trustworthiness of a claim may depend on a wide variety of things. The person who makes the claim. The journal in which the claim is made. The amount of evidence for the claim. The way the claim fits or does not fit with what is already know. The way it may conflict with assumptions about nature. If possible, one would like to check it in person, if that's not possible some of the above become even more important. In short: when do you trust someone's recipe for boiling an egg? Here is how IngenHousz himself describes it, on a loose note on a little piece of paper, not referenced to another source, his thoughts on trustworthy knowledge:

A true lover of natural knowledge never looks with indifference on any phenomenon which seems somewhat extraordinary though he does not understand at first sight neither the cause by which it is produced, nor the utility which might be derived from it. he delights first of all to trace the cause without any regard to the benefit, which he may derive from it either for himself or for mankind; But men, whose pursuits are limited only to enjoyment of these senses which we have in common with the whole animal creation, who look with a cold indifference on every thing that is unfit to draw money from *or to flatter their sexuality (fruges consumere nati) have allmost a general custom to ask immedialy, by being shown for the first time a new discovery of experiment, what is the utility to be derived from it (cui bono?) whereas the true Philosopher or even a man of good sense, is immediately struck by the fact, and can scarce withstand his curiosity to know by what means the phenomenon is produced, without any regard to its utility.*

There are indeed very few new experiments or discoveries, which are capable of being turned to any immediate advantage except perhaps that of surprise or admiration, and in the discoverer a kind of delight, unknown to any but true rational minds, mixed with a kind satisfaction bordering on a degree of irresistible pride inseparable from the consciousness of having enlarged the boundaries of human knowledge. [291]

This is a plea for fundamental research, the urge to know what things really are and how they really work. Without being driven by profit or gain, fame or power, the satisfaction of really understanding some natural phenomenon is to pride oneself on the little contribution one has made to mankind. Elsewhere he pronounced more pragmatic considerations on the business of gathering knowledge. In the foreword to *Experiments upon plants* he wrote:

... experiments may indeed be very useful when a sufficient number is collected to draw some conclusions from them; but without pursuing methodically the same object, discoveries are to be expected only by mere chance, and are sometimes overlooked. [292]

One measurement, one figure, one observation is not yet a hard fact that can serve as evidence. Such a single result is rendered untrustworthy through the complexity of a world in which so many variables are involved. Approaching the truth can be achieved through repetition, as IngenHousz understood. In many repeated observations by several experimenters the subjectivity of each is cancelled out for the objective core of truthfulness to be left over. Some claim that objectivity as scientific virtue only came into existence in the 1850s.[293] That may be true for the word, but not for the attitude, as the works of IngenHousz and Senebier demonstrate. Indeed, Senebier understood just as well, although the general public maybe did not yet do so. (And still seems not to at the beginning of the 21st century.) In a review[294] of Senebier's book of 1783, the reviewer opened with describing how Senebier studied leaves of plants in a multitude of circumstances. "Il revient vingt fois pour chaque vérification; rien ne paroit omis pour constater des faits. L'auteur se croit même obligé de

291 Breda Collection IngenHousz A8
292 IngenHousz 1779 p xliv
293 Daston 2007
294 Anonymus 1784 p 75

demander grace pour sa prolixité." Senebier repeated each experiment twenty times and excuses himself for so much prolixity. But that's science too: rigorous repetition, endlessly reproducing almost identical experiments. Most of the work scientists do is nothing spectacular, though that does not mean it has to be boring. Some people just love to do that kind of thing.

Still Senebier's reviewer had some critique too:

> *On reprochera peut-être à l'auteur des répétitions, des expressions obscures: nous nous contenterons de lui observer qu'il a presque entièrement oublié les dates des ses experiences. Il parle peu du thermometre, presque point du barometre, jamais de l'hygrometre ni des vents. Cependant l'état de l'atmosphere influe beaucoup sur toutes ces recherches.*

> One could reproach the author for repeating himself and using obscure expressions, but we are content with observing that he almost never mentions any dates of the experiments. He does not talk much about the use of thermometer, never about the hygrometer, nor about winds. At the same time the state of the atmosphere has a lot of inluence on this kind of experiments.

Apart from the fact that the author repeats himself and uses obscure expressions, he does not give information that can be crucial in these kinds of investigations. To be able to evaluate the empirical results one has to have at one's dispositon all relevant data.

Another unavoidable consequence of looking for trustworthy knowledge is that one soon realises it is hard to get at on one's own. Sooner or later, one wants to know whether what one has observed has also been seen by others. Was it a hallucination, a perceptional bias, or have your colleagues seen it too? IngenHousz knows all too well that one can easily be mislead, that measuring methods can give systematic deviations, that you can keep concentrating too long on the irrelevant parameter. That is why the critical contributions from others are so welcomed, and explicitly so by IngenHousz and Senebier in many of their writings.

IngenHousz' claims have been checked and double-checked, not only by himself, often in front of an audience, or with the help of other scientists. There was also a group of like-minded people, mainly in Vienna and the German speaking part of Europe, who reproduced his experiments, gained expertise in using the eudiometer and communicated about IngenHousz' findings. Molitor, also a medical doctor, was one of them, just like Scherer and Pickel. A famous defender of IngenHousz was also Jozef Franz Freiherr van Jacquin, IngenHousz' nephew and later Professor in Chemistry and Botany at the Vienna University and Director of the Botanical Gardens. Both Molitor and Scherer not only translated IngenHousz' works in German, they also reproduced his experiments, added comments to his books and articles and took up his defence against his adversaries and critics. They often were more combative than IngenHousz himself, who often tried to calm them down. A close ally and collaborator, although they did not see each other very often, was Jacob Van Breda, his translator in Delft and intellectual right hand, as we have seen already. One of IngenHousz' letters to Van Breda was deemed important enough to be published in one of his books with collected work.[295] They wrote regularly about the technique of using the eudiometer and other instruments, about results and their interpretations, and about what others where doing. Priestley, Spallanzani, Lavoisier and especially Franklin were some of IngenHousz' scientific peers that were valued very highly by the travelling scientist. (Unfortunatly most of the letters between Franklin and Ingenhousz have been lost.[296] The correspondence with the others still deserves to be explored thoroughly.) He may not have been member of many academic societies but he did have a network of professional friends and acquaintances.

Eighteenth century philosophy of science

These eighteenth century natural philosophers were true philosophers, who spent quite some time thinking about their methods and mission. They were scientists and philosophers of science at the same time, more so than their twentieth century counterparts can ever dream to be. Of course, they were not yet restrained by the academical partitioning of the departmental boundaries which we know now. Joseph Priestley wrote at the beginning of his book on nature and air in 1781:

> *The greater is the circle of light, the greater is the boundary of the darkness by which it is confined. But, notwithstanding, this, the more light we get, the more thankful we ought to be, for by this means we have greater range for satisfactory contemplation. In time the bounds of light will be farther extended; and from infinity of the divine nature, and the divine works, we may promise ourselves an endless progress in our investigations of them: a prospect truly sublime and glorious.* [297]

The more we know, the more we realise what we don't yet know. Science is a way of living with the unknown and with the permanent question mark. As we have seen in IngenHousz' writings, he was also not just doing experiments and verifying hypotheses but also thinking about the sense of doing so and

---

295 IngenHousz 1785b
296 Smyth 1907
297 Priestley 1781 p ix

what it could mean in the wider scope of things. Senebier, his great opponent, went even further. He developed into something of a epistemologist avant-la-lettre. His scientific work is only a small part of his huge oeuvre that consisted mainly in historical, theological and epistemological works: "His pen was never dry."[298] He published in 1802 *Essay sur l'art d'observer et de faire des experièences*. It was only the end-point of a long development that started in 1769 when he responded to a question from the Academy of Sciences in Haarlem (Holland) "Qu'est-ce qui est requis dans l'art d'observer? Et jusques-où cet art contribue-t-il à perfectionner l'entendement?" Published in 1772, the essay Senebier wrote was only the priming layer of a book he published in 1775 titled *L'art d'observer*. This book was completely reworked to his final and complete view on epistemology of which the last part was published in 1805. It is a huge book in three volumes which runs into almost thousand pages. This book cannot have been read by IngenHousz, as he had been dead for six years already. There is no evidence if he ever saw or read the other epistemological works of Senebier. One wonders if he would have liked it if he had done so.

Senebier claimed the existence of a precise method to study nature. He wanted to legitimize the rules of this method based on the difference between the "art d'appercevoir" as opposed to the "art de penser". With the distinction between observing and thinking he hoped to be able to give a fundament for the practice of science. Moreover, he was quite ambitious as he believed that this method, which he specifically aimed at young and budding naturalists, was applicable to any subject of enquiry, be it the "beaux arts, des belles-lettres, comme ceux des arts mécaniques". In his very detailed step-by-step analysis of the scientific enterprise he distinguished the moment before, during and after the observation or experiment. He underlined the fundamental difference between an observation and an experiment. In the first nature presents itself as it is, in the second it is the experimenter who uses methods to perceive these same objects in a more distinct way, by putting them in circumstances in which nature would not put them. This all sounds a little like we know Senebier by now, skilled in theoretical musings, bad in practical logic. Especially so because in the concrete rules he formulated and in the advice he gave, these seem to be equally applicable to both observing and experimenting.

Being a good observer requires some behavioural characteristics such as attention, stubbornness and patience while one should expand one's erudition. One should also only study that which has a use, an aspect which fits nicely within the culture of the Genevan protestant ethics. Then he treated the proper execution of an observation: repetition, variation, using a witness and the use of instruments. Next came the rules of proper thinking: analysis and synthesis, the use of hypotheses and the formulation of discriminating hypotheses, the three

---

298 Huta 1998

rules of Newton (simplicity, uniformity and analogy). Finally he gave some tips on how to publish: take care in being persuasive, use the right rhetoric, use illustrations and be clear in your definitions and descriptions.

Hidden in the hundreds of pages of a verbose and literary flood, are some remarkably modern thoughts. Many scientists in 2010 would still recognize themselves in Senebier's descriptions. When one publishes a scientific book, so warns Senebier, it reports the reconstruction of the reality in the laboratory, which masks the doubts, hesitations and failed experiments. He further illustrates his methodical rules with a multitude of examples from the history of science, in order to try to find the secret behind great discoveries. This can happen when one new fact puts a panoply of known facts in a new perspective, or when one is able to eliminate an error that has distorted the knowledge of a phenomenon. It looks like the modern idea of falsification and critical rationalism was already alive and kicking in 1800, only to become one of the hallmarks of the scientific method later on. This should not come as a surprise. Anybody in the search for trustworthy knowledge would adopt it as one of those rules of thumb that help one get reliable knowledge. Of course one has to start rethinking the claim that all animals have two hearts, if one can find only one in every beast one dissects.

Senebier knew also very well how the investigative mind works. He warned for

> le désir de voir c'une manière fait voir précisément ce qu'on souhaite, et l'illusion est telle qu'on voit précisément, comme on veut voir.[299]

One should beware of the desire to see what one would like to see, because that may deceive one in seeing exactly that. Senebier was well aware that even our simplest observations are coloured by our expectations. This is an echo of what Adam Smith already wrote in his *The theory of moral sentiments* in 1759 where he warned for the dangers of "self-deceit, this fatal weakness of mankind".[300] Anglofile as IngenHousz was, and having studied in Edinburgh, this could well have been a book that he had read. In the eighteenth century it was therefore a well-established insight that all our experiments are biased by our theoretical perspectives and expectations. We cannot see something 'as if it was for the first time' with a maiden vision. Here he showed himself to be far less naive than his great predecessor Priestley who wrote frequently about his "unexpected discoveries" and that "more is owing to what we call chance, ... than to any proper design, or pre-conceived theory in this business. ... I wish my reader to be not quite tired with the frequent repetition of the word surprize ... ".[301] Surprise is his key word to demonstrate in his narrative that he was not expecting anything prior to any experiment.

---

299 Senebier 1805 p 55
300 Schliesser 2005
301 Priestley 1775b

Priestley as a genius in the 1770s played a major role in Senebier's narrative, thirty years after the facts. He referred to Priestley as a prime example of someone who was so eager to see phlogiston as the basic chemical building block that he could not see how water was decomposed in oxygen and hydrogen. Here, Senebier gave the living proof of the danger for which he warned his readers. However, Senebier seemed to be oblivious to his own blindness: he never came to accept that plants produce carbon dioxide at night because it did not fit into his teleological frame of mind.

Another proof of Senebier's uncritical attitude is that he rewrote the history of science in his own way. Only decennia after the facts he described how Priestley, a genius he admires, discovered hydrogen and nitrous gas and how he made oxygen. Priestley never used those words and Senebier appeared to have forgotten that names and concepts have a history too, which cannot be discarded. Accuracy was not always his biggest quality. Nor was completeness, because he names almost everybody in the history of chemistry and plant physiology, but IngenHousz' name is nowhere to be found. But that should not come as a surprise by now. Anyway, this all shows that methodological considerations were very much part of the game of performing science at that time. IngenHousz worked in an intellectual climate in which thinking about the methods of how to gather knowledge was a very normal thing to do. It was a tradition that went back to Bacon and Galileo, while men such as Newton or Boyle introduced their books with prefaces on methodology. Three quotes may illustrate how epistemology - thinking about how we think or how we come to know things - was part and parcel of the everyday concerns of the natural philosphers.

> I seem to have been only like a little boy playing on the seashore, and diverting myself in now and then finding a smoother pebble or a prettier shell than ordinary, whilst the great ocean of truth lay all undiscovered before me.
> Isaac Newton, *Memoirs of Newton*, Vol 2 Ch 27, ed. David Brewster, 1855

> There are three principal means of acquiring knowledge... observation of nature, reflection, and experimentation. Observation collects facts; reflection recombines them; experimentation verifies the result of that combination.
> Denis Diderot, *Pensées sur l'interprétation de la nature*, 1753

> The love of discovery, and of intellectual acquirement, may indeed be called an almost instinctive faculty of the mind; but still a combination of circumstances is required to call it into useful exertion. The truly insulated individual can effect little or nothing by its unassisted efforts. It is from minds nourishing their strenght in solitude, exerting that strenght in society, that the most important truths have proceeded. Humphry Davy, extract from unpublished lecture[302]

302 Davy 2001

It is the tradition in which Diderot wrote his *L'Interprétation de la nature* and Deslandes his *Discours sur la meilleur manière de faire les expériences.* Deslandes' essay was published as the introduction to Petrus Van Musschenbroeck's *Cours de physique expérimentale et mathématique.* Musschenbroek was the professor who taught the young IngenHousz when he studied in Leyden. So, epistemology was not an unknown subject at the time of the Enlightenment.[303] The only difference between Senebier and his fellow natural philosophers was that he went into great detail to describe the experimental method they were all using. Another difference is where Senebier mixes his epistemological insights with teleological preoccupations. Both IngenHousz and Senebier were religious. But the former was catholic, the latter protestant. That may explain how differently they combine their religious attitude with their scientific work.

Science & the Supreme Being

IngenHousz did not hide his religious thoughts. They were as natural and evident to him as the clothes he was wearing. His God was part of his world and could flow freely into his scientific work. In the conclusion of Part I of his 1779 book he included this digression:

> *Though we are too much accustomed to look upon the most obvious operations of nature with a kind of unconcern and indifference, such as, for instance, the vegetation of plants; yet we cannot look with so much indifference upon the final causes of those every where obvious scenes when we discover them; for they do not so much affect the organs of our sight and other external senses, as they do our understanding, our reason, our judgment; by which only we are superior to all other living animals. The consideration of final causes gives us to understand that this great universe is not the offspring of chance, not coëval with the beginning of time, or of an eternal origin, but that it has been made by an Omnipotent Being, who, by giving it existence, has, at the same time, endowed it with the most wonderful qualities and powers, continually in action, and tending it with an astonishing harmony to one general end, the preservation of the whole.*
>
> *An upright mind, averse to that manner of living which induces many to wish, rather than really to believe, that this world is not superintended by an intelligent being, takes delight in finding out those deep insights, which, by their obvious tendency to promote the preservation of the whole, inspire him with that awful reverence we owe to the Supreme Cause of every thing, and fill him with that consoling expectation that only being upon earth capable of true reason and of tracing the existence of God in his wonderful works, and of*

---

303 Huta 1998

contemplating him in adoration, may expect not to be entirely annihilated after his body is returned into dust, out of which it took its origin.
But to come back from this digression to the purpose, let us consider how much the real facts drawn from nature itself are concordant with the theory deduced from my experiments.[304]

The biological world was considered a very ordered place in the eighteenth century. Plants and animals were classified in the Linnean classification. Physiology was envisioned in mechanistic of hydrolic terms. The generation of new animals and plants proceeded from preformed seeds and germs that had existed since creation. And all this order arose from God. In the course of the eighteenth century this strictly delineated and mechanistic worldview was disintegrating. Micro-organisms had become a challenge to any naturalist watching them through the miscroscope. The borderlines between animal and vegetable kingdom were blurring, as we can see in the discussion about Priestley's green matter. The mystery of the real nature of this organism haunted Priestley, IngenHousz and Senebier until the end. It produced dephlogisticated air just like plants, but it moved, just like animals. It appeared out of nothing, especially in quiet water with fermenting material in it. It did not appear however in boiled water.[305] The ongoing discussion on these plant-like insects or insect-like plants added fuel to the ongoing debate on the distinctions between life forms and on the way they came into being. Did they propagate from unseen eggs or seeds, or did they arise by some sort of spontaneous generation? Could new organisms be produced from dead matter?

Some natural philosophers moved towards vitalism, explaining the difference between living and dead matter through a "live force", opposed by others who preferred a strict materialism. Various biological theories had important implications for the existence and the role of God and for the moral basis of society.[306] As one can deduct from Buffon's claim in 1747 that:

enfin le vivant en l'animé, au lieu d'être un degré métaphysique des êtres, est une propriété physique de la matière.[307]

in the end, living and animation, instead of being a metaphysical degree of being, is a physical property of matter.

Long before Darwin, religion kept an anxious eye on the developments in the biological sciences. It even used biology to its own ends. Teleology was a distinct intellectual discipline specialised in proving the existence of God through

304 IngenHousz 1779 p 139
305 IngenHousz 1789b
306 Roe 2003
307 Buffon 2007 p 143

the final causes of everything, and as such it was some sort of theology and a scientific way of shaping your religious beliefs. Senebier wrote six versions of an *Essai de Téleologie* between 1773 and 1803. This essay was never published but has been extensively studied by Huta. Her study makes clear how Senebier's epistemological interests were primarily driven by his teleological preoccupations to discover the final causes. To Senebier teleology was a practical science, not a theoretical pursuit. Every observation or experiment was motivated by just one aim: to elucidate the final goal of nature. In his essay on the art of the observation he referred often to notions of the plan of the universe and the order in that universe. Understanding that order through the scientific method leads one straight to God: that was Senebier's main thesis. He wrote to Van Swinden on 22 May 1784:

> ... *toutes mes expériences me semble rayonner tellement vers le même Centre qu'elles ont l'air de se servir mutuellement de preuves.*

> ... all my experiences seem to radiate in such a way from the same Centre that they give the impression to serve mutually as proofs.

Senebier was on the same track as Priestley, who had always maintained that the accidental nature of his discoveries increased the sublime feelings provoked by the contemplation of the progress of science. The experimental study of nature was at the same time an act of religious devotion.

To IngenHousz, God seems to be rather only the eternal cosmic motor behind everything in the universe, some kind of static mover. His main fear was that by dumping this idea and the interlocked idea of an eternal soul the whole of civilisation would collapse. His religious thought seems to have been mainly a pragmatic one, as one can see in his letter to Van Breda on 12 October 1794:

> *als men de onsterfelykheid van de ziel lochend valt alles in duygen. Den theismus, materialismus en atheismus hebben de zelfde effect op de menschenlyke samenleving. zy veranderd de menschen in ergere schepselen dan wilde dieren, als het gemeen volk ze aanneem, men ziet er nu reeds een staaltje van in Fr. ... Dwaasheid is meer aanstekelyk dan wysheid.*

> if one denies the immortality of the soul, all falls apart. Theism, materialism and atheism have all the same effect on human society. They change men into creatures that are worse than wild animals. When the people take up these ideas, you can see what happens then, a nice example you can see now in France... stupidity seems to be more attractive than wisdom.

For IngenHousz there was no contradiction between believing in God and trying to understand the world that was supposed to be created by that God. He only believed that this God installed some order in the world, an order in which mankind could find its place. Denying the existence of that God implied denying that order and throwing open the door wide to all sorts of disarray and chaos. IngenHousz was not a revolutionary, so much is clear. But he was not a zealot either. He probably cannot be counted as a participant to the "radical enlightenment", as Jonathan Israel calls it.[308] But it is difficult to say much more about it, based on the available testimony. As far a we can tell he did not go to church regularly, at least he did not talk about it anywhere. But it might just as well have been such a normal habit that he never thought about mentioning it. But God was omnipresent in his life and thoughts, although presumably in a different way from Senebier. IngenHousz was probably more inclined to follow Linneaus who elevated the scientist as much as he elevated the creator: "If the Maker has furnished this globe ... with the most admirable proofs of his wisdom and power; if this splendid theatre would be adorned in vain without a spectator; and if man ... is alone capable of considering the wonderful economy of the whole; it follows that man is made for the purpose of studying the Creator's works."[309] Religion was not simply retreating while science was advancing. The same scientific discovery could have both sacred and secular readings. Men such as Priestley and Senebier were men of God and believed strongly in revelation. (While for Priestley science and enlightened religion were on the same side against popular superstition.) IngenHousz believed more in the book of nature than in the book of the bible. One thing is sure: the relationship between seeing and believing, between science and religion, was just as precarious and complex in 1790 as in 1990.

Science as a public good

The Enlightenment has also been called the "Republic of Letters" and that has to be taken literally. These natural philosophers wrote a lot, and not just personal letters to each other, but also published widely. Their books were often translated in many languages. Publication was as important then as it is now. It just took more time and effort. And it was not easy.

It is not widely known that in France, from about the sixteenth century to the Revolution, censorship of speech and thought was official policy, with a "thought police" deciding on what was to be published and what was not.[310] There was no freedom of press. Strict surveillance of printers and booksellers was in place and put up rather effective mechanisms for controlling the dissemination of ideas.

308 Israel 2001
309 quoted by Brooke 2003
310 Darnton 1989

Before publication, every manuscript had to be inspected for content. This censorship was applied through a selective policy rewarding publishers who, in return for their cooperation with the established order, enjoyed the advantages of a monopoly. After publication, the police actively controlled all books on the market. The results of this system can be seen on the last pages of every book published by IngenHousz and so many others in France, where two pages reproduce the official approval of his majesty's censor. But when books slipped through the nets of censorship, different measures were taken. As Darnton describes it: "When the public hangman lacerated and burnt forbidden books in the courtyard of the Palais de Justice in Paris, he paid tribute to the power of the printed word." But he often destroyed fake books, keeping the originals for the magistrates, who tried to keep such actions limited and gave them a low profile, as they knew that a book burning was the best sales promotion for a book. In the mean time hundreds of booksellers and printers spend their days in the Bastille.

It may be no surprise then that so many writers of the Enlightenment published their books in Holland, where the freedom was much greater. It also explains why publishing one's ideas was a tricky business, often involving clandestine operations, smuggling books into a town like Paris and publishing under a different name. Voltaire was only one of the many authors struggling with this autocratic system.[311] The absolutist state and its rulers clearly realised the danger of the printed word as a principal vehicle of knowledge and thought, the medium of all political and religious discussion and the instrument to express whatever one got in one's mind. Science was probably not seen as the most politically dangerous activity, but it was science that came up with all kinds of unorthodox ideas, so it most certainly had be kept in check. Science had yet to arise from the fundaments of natural philosophy and 'philosophy' was a label that got the book censors alerted.

When IngenHousz struggled for years to get his books published in Paris, the Dutch and German versions had already long sold out. This was undoubtedly the result of the administrative hurdles to get something printed. Only when he finally went personally to Paris, could he get the presses rolling again, probably because his own presence could get the grinding machinery going. Which is not to say that things were better elsewhere. IngenHousz ventilated his anger in a letter to Van Breda on 23 March 1790

> In Engeland moet alles gepubliceerd worden in de Transac. Philos. En zolang het daar niet instaat, wordt er over gezwegen. Dat bevordert de uitwisseling en discussie van nieuwe ideeën niet. Het past ook niet om over andermans bevindingen openbaar te praten en men krijgt dus geen totaalbeeld van de nieuwe ontwikkelingen.

---

311 Pearson 2005

In England everything has to be published in the Philosophical Transactions [of the Royal Society]. As long as it is not published there, nobody talks about it. This is not advantageous for the exchange and the discussion of new ideas. It is not being appreciated to talk about other person's findings and so one can never get a good view on all the new developments.

Monopolies on knowledge are counterproductive. Free speech limited by a state censorship is unproductive, but a *de facto* control through a historically grown monopoly can be almost as inhibiting. If this solitary scientific publishing platform is only supplemented by a biased public press it becomes hard for interested readers to get a balanced picture of the state of knowledge. So much can be learned from the foreword of IngenHousz' German book. Its translator, Molitor, writes in his introduction "Einige Bemerkungen über des Einfluss des Plfanzen aud das Tierenreich" to *Vermischte Schriften* in 1782:

> *Ingen-Housz musste durch Briefe aus England erfahren, das die Schärfe dieser kritik (es handelt sich um die obengenannten Rezension in der Critical Review), welche seine Lehre intergräbt, in diesem Lande einen solchen Eindruck gemacht hat, dass, wofern er sich zu verteidigen ausser stand wäre, eine zweite Auflage seines Werks über die Pflanzen einer zo günstigen Aufname wie die erste, sich nicht zu gewärtigen hätte.*
>
> *[...]*
>
> *Man erlaube mir hier anzumerken dass mehrere Journalisten des festen Landes, deren Schuldigkeit es wäre, die Werke selbst zu lesen, ehe sie einen Auszug davon liefern und ihr Urteil fällen, sich begnügen, die englischen rezensionen wörtlich abzuschreiben.* [312]

Ingen-Housz had to learn from letters from England that the fierceness of this criticism (it concerns a review in the Critical Review) that undermined his theory, had made such an impression in those countries, that he, in so far he was unable to defend himself, could not expect the second edition of his work on plants to have had equally good sales as the first one.
[...]
The reader should allow me to remark here that several journalists on the continent whose duty it is to read these works for themselves before summarizing them and passing a judgment on them, were very happy with copying the English reviews word for word.

Molitor's complaint sounds all too familiar. Two hundred years later, scientists complain about how their work and its results are reported in the media as journalists copy stories uncritically, do not go back to the original sources and lend

---

312 IngenHousz 1782b

airplay to superfluous anecdotes and gossip. Then like now, a writer of scientific essays had to walk the thin line between popularising one's findings to the point of ridicule and keeping it serious and incomprehensible for the greater masses.

Writing is not the only way of bringing science to a wider public. Part of the practice of science in the eighteenth century was also the public demonstration. As we have seen before, IngenHousz was getting lots of visitors in his laboratory in Vienna, and often these were high ranking people. Showing interest in the world of the natural philosophers was a prestige thing among the nobility and the powerful. But as IngenHousz documented, he often gave demonstrations in front of big audiences. And he entertained his friends and their guests with scientific experiments. He always carried a flask of pure oxygen with him in which he burnt a steel spring with great sparkles. Science and scientific knowledge had become a common good and IngenHousz' practices fit in nicely with science as a public culture as described by Golinski.[313] It shows how science and its developments cannot be seen as separate from its social context. The society in which IngenHousz lived and worked applauded progress in whatever guise it manifested itself. The first sparks that jumped off the Leyden jars and the first balloons that took to the heavens were magnificent demonstrations of a new and exciting view on the world, a natural culture medium for the birth of the Romantic era in which the crossovers between science and poetry or litterature were a most natural thing.[314]

The public manifestation of scientific knowledge produced by instruments such as air pumps, Leyden jars, eudiometers, burning steel wires, electric generators or air pistols was a powerful stimulus to spread the ideas of the Enlightenment. This spreading enlightenment would expose the illegitimacy of repressing and corrupt religious and political powers, so these men of science hoped. As Condorcet would write in the *Gentleman's Magazine* in 1790 that Franklin's science reached beyond truth, as it was the essential condition for freedom: "An ignorant people is always enslaved."

A way of reasoning

At the core of all these developments was the practical scientific work. Theories were developed through hard logical thinking and laborious experimental work. How did the reasoning of IngenHousz and his contemporaries develop, how did they develop their conceptual schemes in which they could put their findings in order to understand them?

One of the few historical accounts of the discovery of photosynthesis is the detailed reconstruction and thorough analysis by Leonard Nash, as Case Five in

---

313 Golinski 1992
314 Holmes 2009

the *Harvard Case Histories in Experimental Science*. He follows the path from Van Helmont over Bonnet and Hales to Priestley, IngenHousz and Senebier to end with De Saussure. He describes how they developed their thoughts, building on each other's findings, theories and techniques. He tries to reconstruct the way these men have been thinking, how they reached conclusions based on the facts they observed and what assumptions they made to knit everything together. He illustrates how science does not follow a clear-cut straight logical path from neutrally observed facts to unavoidable conclusions, a causal explanation of those facts. In reality, the building of a scientific explanatory theory follows a tortuous road, led and often misled by assumptions, intuitions or theoretical constructs. Investigators are often puzzled by contradictory or unexpected results, which would not end up appearing that contradictory or unexpected if they had known more about the real relationships hidden in the phenomena they studied.

Nash unravels Priestley's, IngenHousz' and Senebier's thinking in particular detail with regards to the way they thought about the air dissolved in the water and how that air was influencing their experimental results. Remember that they used water as the medium in which their plants were floating. It was known that water contained some air, although little to nothing was known how that was happening exactly. That's why they had to guess at what was really happening when the experiments with plants suspended in water saturated with fixed air gave inconsistent results.

> *Ingen-Housz goes on to suggest that this penetration of fixed air into the leaves may have upset their internal "vital motion" - the vital motion that he supposed to be responsible for the transmutation through which dephlogisticated air was produced by leaves immersed in water. The postulation of a derangement of the vital motion was, of course, completely ad hoc: Ingen-Housz was simply looking for some way of reconciling his transmutation hypothesis with the apparently contradictory observation. However, this speculative postulate was not very wide of the mark. We now believe that the air-producing activity of certain species of leaves is paralyzed when they are exposed to excessive concentrations of fixed air. This paralysis - not unreminiscent of Ingen-Housz' idea of a "derangement of the vital motion" - may have been responsible for Ingen-Housz' failure to obtain the large volume of dephlogisticated air that he expected.* [315]

Nash' analysis neatly illustrates how science advances in a continuous pendulum movement between inductive and deductive reasoning. IngenHousz saw something happening which he could not immediatley explain. The one thing he knew was that plants produce at least one sort of gas in sunlight and another in the dark. He had however made only a few of these experiments specifically with fixed air, because he thought this would test the plants' actions

---

315 Nash 1952

in a situation which was not relevant to their natural environment. And water with lots of fixed air in it was especially different from the natural atmosphere in which only such a small amount of fixed air is present that they could hardly measure it at the time. On the basis of these few experiments he induced a general principle that the concentration of carbon dioxide influences their photosynthetic productivity (to rephrase it in twentieth century terms). So he speculated that the amount of fixed air could block or harm this gas-producing mechanism. He needed to formulate such a general principle to be able to formulate a hypothesis. This hypothesis was then the next thing to try and verify. Which is what he would do in the years following the publication of his first book and certainly after Senebier came forward with his general principle that plants need fixed air to produce vital air, in the form of fixed air in the water that surrounds them or in the form of acids added to that water. This ping-pong between induction and deduction does not only play in the mind of one investigator, but also between the minds of all researchers involved in the same research. Their (combined and interchanging) development goes from the particular to the general and back. After postulating a general principle they take that back to the laboratory bench to check it, but on other experiments and new facts. That is how scientific knowledge bootstraps itself towards a better understanding of the world. Science is a continuous recycling from observation to hypothesis and back on ever-higher levels of theoretical understanding.

In his account of the interaction between IngenHousz and Senebier, Nash appears a trifle non-critical. He describes how "Senebier conscientiously repeated practically all of IngenHousz' experiments and confirmed the broad outline of his results, though there were some differences in detail". These so-called details were rather fundamental differences, which were the basis of the life-long quarrel between the two men. Senebier flatly denied that leaves vitiate the atmosphere in the dark and gave empirical evidence that submerged leaves liberate dephlogisticated air at the expense of fixed air in the water surrounding them. This led him to a different conceptual appreciation of the phenomenon at hand. To Senebier the vitiated air that was produced in the dark was the result of the fermentation of rotting plants. Healthy plants could not, as we have seen him declare time and again, do something harmful to the atmosphere. If anything like that happened, it had to be the result of an unnatural process, the consequence of an unnatural state of the plants' leaves. Nash rightly remarks that both investigators had the same facts in front of them, still drew different conclusions: "How much more than the simple facts goes into the making of a scientific opinion!" Neither IngenHousz nor Senebier could rely on trustworthy quantitative measurements on the quantity of vitiated air plants produced. It may well be that IngenHousz exaggerated when he claimed that the bad air from plants at night was one of the most poisonous airs ever known. While Senebier could not fit the same results in his own conceptual framework which in his mind created an opposite tendency to undervalue it. Senebier developed a train of thought in which this sounded perfectly reasonable: if plants produce pure air when exposed to sunlight and the volume of this air seemed to be roughly proportional to the duration and the intensity of the amount of light falling on the leaves, then, what could be more reasonable than to suppose that when no light fell on them no gas would be produced. If any gas was produced in the dark, so Senebier reasoned, this must be the product of side effects peculiar to the unnatural conditions in the experiment. Senebier had been fixated on the role of fixed air, right from the beginning of his research. As described earlier, IngenHousz had done some experiments with plants submerged in water to which extra fixed air was added. But because he got inconsistent results, he had not followed this line of thought.

From imagination to evidence

Something interesting is at work here. Senebier came to a completely different conclusion than IngenHousz, based on the same facts. Nash quotes Senebier in saying that "imagination is passé in science, corporeal phenomena can be phatomed only by a profound study of the facts".[316] Then as now, so continues Nash, it was always necessary to face the small but treacherous question of how to interpret the facts. In his interpretation Senebier undoubtedly followed what to him was the most rational explanation of the nocturnal gas production. Yet, as IngenHousz pointed out there was nothing irrational in interpreting this production as a natural function of plants. He saw in the first place the parallel between respiration of animals and the expulsion of gas by plants. As he remarked in his second French edition in 1789 that nobody seemed to think it strange that animals did not dispose of their foodstuffs in the same condition as they were ingested, just as they vitiate the air they breath. He summarised that we had only learned that animals vitiate the atmosphere day and night and that plants do the same with one exception of the time during which they are exposed to the sun. The alteration made by animals to the air in which they live is by no means due to a state of sickness or fermentation in their economy. So let us not interpret the impression made of plants on the atmosphere as a pathological effect, but as a phenomenon owing to their true nature.

Could it be said that IngenHousz was reasoning more rationally than Senebier? They were both filling in the gaps in their knowledge by educated guesswork. One can be more sympathetic towards the one or the other in where they found the inspiration for their imaginative leaps, but neither can be discounted for not making their final picture being plausible.

---

316 Nash 1952 p 73

In the meantime, Lavoisier slowly but certainly felt at ease in his new system. By 1786 he knew how molecules were assemblies of atoms: "Fixed air is composed of 28 parts carbon and 72 parts oxygen and that's why he called it carbonic acid; and secondly water is the result of the combination of 15 parts inflammable air or hydrogen and 85 parts oxygen." The experiments and theories of Priestley, IngenHousz and Senebier had been of interest to him all along. Knowing that water can be decomposed in oxygen and hydrogen, he wondered if plants would not be doing the same. He had analysed the composition of sugars and concluded that it was a combination of big quantities of hydrogen and oxygen united with carbon. That led him to the following hypothesis:

> *Pour se faire une idée de ce qui se passe dans cette grande opération, que la nature semblait avoir, jusqu'ici, environnée d'un voile épais, il faut savoir qu'il ne peut y avoir de végétation sans eau et sans acide carbonique; ces deux substances se décomposent mutuellement dans l'acte de végétation [...] l'hydrogène quitte l'oxygène pour s'unir au charbon, pour former les huiles, les résines, et pour constituer le végétal; en même temps, l'oxygène de l'eau et de l'acide carbonique se dégage en abondance, comme l'ont observé MM. Priestley, Ingenhousz et Sennebier, et il se combine avec la lumière pour former du gaz oxygène.*
>
> *Je ne fais qu'annoncer cette théorie, dont je ne suis pas encore en état de développer les preuves, et qui, d'ailleurs, ne présente pas encore à mes yeux des résultats évidents; ce ne sera que l'année prochaine que je pourrai répéter les premières expériences que j'ai faites à ce sujet, les rapprocher de celles de MM. Priestley, Ingenhousz et Sennebier, et en ajouter quelques autres que je médite.*[317]

To get some idea of what happens in this great operation, which until now has been shielded by nature in a thick mist, one needs to know that one can't have vegetation without water nor without carbon dioxide. These two substances decompose both in the process of vegetation [...] hydrogen leaves the oxygen to unite with the carbon to form oils, resins, and to constitute the plant. And at the same time the oxygen from the water and from the carbon dioxide frees itself abundantly, as observed by Messrs Priestley, Ingenhousz and Sennebier [sic], and combines with the light to form oxygen gas.

I do no more than announcing this theory, while I am not able yet to produce any firm proof for it and which proofs do not as yet present any evident results. It will be next year before I can repeat my experiments, which I already performed on this subject, to compare them with those of Mr. Priestley, Ingenhousz and Sennebier and to add some others on which I am currently thinking.

---

317 Lavoisier 1786

These fragile suppositions were going to inspire IngenHousz in his later years, and were also the theoretical substrate in which young scientists such as De Saussure grew up. He made it clear in the opening sentences of his 1804 book how knowing more (and valuable hypotheses are part of that knowledge) leaves less room for imagination. Growing knowledge fills in the gaps of our ignorance that until then were filled in by speculation and fantasy.

> *J'aborde les questions qui peuvent être décidées par l'expérience, et j'abandonne celles qui ne peuvent donner lieu qu'à des conjectures.*
> *[...]*
> *La solution de ces questions exige souvent des données que nous n'avons point; elle requiert des procédés exacts pour l'analyse des plantes, et une connaissance parfaite de leur organisation.*
> *[...]*
> *Les fonctions de l'eau et des gaz dans la nutrition des végétaux, les changements qu'ils font subir à leur atmosphère, sont les sujets que j'ai le plus approfondis. Les observations de Priestley, de Senebier, d'Ingenhoutz, ont ouvert la carrière que j'ai parcourue; mais elles n'ont point atteint le but que je me suis proposé. Si l'imagination a quelquefois rempli les vides qu'elles ont laissés, c'est par des conjectures dont l'obscurité et l'opposition ont toujours montré l'incertitude.*[318]

I attack the problems that can be decided by experiment, and I abandon those that can give rise to mere conjectures.

...

The solution of these problems often involves data that we lack completely; exact procedures for the analysis of plants, and perfect acquaintance with their organisation, are required

...

The function of water and gases in the nutrition of plants, the changes that the latter produce in their atmosphere - these are subjects that I have investigated most. The observations of Priestley, Senebier, and IngenHoutz have opened the road that I have crossed, but they have not all attained the goal that I set myself. If, in several instances, imagination has filled the gaps that these observations have left, it has been by conjectures the obscurity and opposition of which have always shown them to be uncertain.

It may be clear that De Saussure was at a point in time when new possibilities were opening up. He knew more, had better (quantitative) techniques and felt he could move forward, thanks to the men he duly honoured as the ones who opened up the path. He could do what his predecessors could not. They observed that plants thrived or that they wilted, and that the air had ameliorated or not.

---

318 De Saussure 1804 p iii

He measured the gain of weight of the plants and analysed their biomass into its different components. He followed the balancesheet reasoning with which Lavoisier gave chemistry its modern method. The total amount of substance at the beginning of the reaction is the same as at the end of the reaction. De Saussure had applied the same method to plant physiology and could so determine the water was not just a carrier for nutrients, but a fundamental nutrient by itself.

Conceptual schemes and practical perspectives

Like Priestley before him, Senebier worked with or within a conceptual scheme in which the amelioration of the atmosphere was due to the removal of phlogiston from the system through phlogistication of fixed air, rather than an addition to it of dephlogisticated air. This was not much more than a new version, a little more formalised and refined, from Priestley's depuration hypothesis.

Of course, the whole phlogiston business was soon to be overcome by the new theory of Lavoisier. Air phlogisticated by respiration or combustion would be nothing more than the oxygen in the common air being replaced by an equivalent portion of carbonic acid. Even when Senebier had become informed about Lavoisier's new system he was not that impressed by it. He did not even consider subjecting his ideas to a simple test in the new conceptual framework. Just like IngenHousz had not performed the simple test to see how soluble dephlogisticated air really behaved in water. (He thought that plants produce less measurable vital air when submerged in distilled water, which thus contains no more air, because the vital air they produce is first of all taken up by the water itself, before accumulating above it so that he could measure it.) This was a test Senebier performed successfully, putting him on a completely different track than IngenHousz, which led to the importance of fixed air in the water. However, the fact that both of them did not check some of their basic assumptions shows how key assumptions are often not recognised as such. A hypothesis can be so attractive or convincingly fitting into the used framework that it becomes highly unimaginable that this main assumption could be mistaken.

This shows how facts only will never do. Assumptions play a role in interpreting what one observes. IngenHousz started to control carefully the illumination of the plants and this delivered results that were unprecedented in reproducibility and meaningfulness. It was a gigantic step forward. But that was not all. In his interpretation of his experiments he could easily get lost, misled by the complexities of the experimental systems he and his peers were setting up. A complexity of which he was well aware. One of the first things he had to decide on was where this dephlogisticated air came from. Was it an effect of purification, at the surface of the plant, powered by the sun? Or was it a process of creation inside the leaves? This is something he tried to find out. When one tests the air that is expelled from pump water by boiling this water, that air is less good than common air. The air produced by plants on the contrary is much better than common air. This led IngenHousz to discard the first possibility and adopt the second: plants actively produce vital air. In such a way he slowly built his theory, a process in which he would sometimes stray away from the real state of affairs. And often, his limited insights would leave him in the dark. It took him a long time to realise how really important the fixed air (carbonic acid) was, about which Senebier continuously boasted.

IngenHousz was quite sure that plants use carbonic acid as an essential nutrient. But he was at a loss to explain where they got it from, as there is so little of that element in the atmosphere. It was De Saussure who could measure the amount of carbon in a plant in order to demonstrate that the relatively little carbon in the air is in absolute terms sufficient to feed plants. With his more precise quantitative methods De Saussure could also measure the exact amount of carbon dioxide in the air and he found out that plant growth is enhanced by higher carbon dioxide concentrations, of up to 8%. At even higher concentrations, growth was harmed. Life has some minimum and maximum thresholds in between which it can survive, as IngenHousz already suspected on the basis of his rather crude results in measuring the effect of fixed air. De Saussure however, would get stuck at almost the same spot as IngenHousz. How to reconcile that plants equally absorb oxygen and emit carbonic acid (in the dark) and absorb carbonic acid and emit oxygen (in the light)? He remained unable to trace the oxygen and carbon and hydrogen molecules from different sources flowing into different metabolic pathways. He could not even image such a process existed.

Still, De Saussure could start in a landscape of which a partial map already existed. He knew the various substances that make up common air, he could measure them. In his evaluation of De Saussure, Nash draws some conclusions that are unfairly rigid and severe, especially for IngenHousz. Nash clearly did not read all IngenHousz' writings (much of which was not yet available in the 1950s) and feels that De Saussure was finally able to contradict some of the earlier findings. De Saussure is supposed to have contradicted IngenHousz in his opinion that plants die in vacuum. IngenHousz did claim that they die in a sustained vacuum, but that they can survive (also in water from which all gases were removed by boiling) because they seem to be able to produce their own fixed air. De Saussure, of course, could confirm what IngenHousz wondered about: plants produce carbon dioxide from their own substance, while they metamorphose part of their carbon dioxide into oxygen when illuminated. When De Saussure wrote that "My investigations lead me to show how water and air contribute more to the formation of the dry matter of plants growing in fertile soil than does the humus matter that they absorb, in aqueous solution, through their roots", he reframed IngenHousz' words in a much more transparent framework.

The continuity between De Saussure and his predecessors was even greater than Nash could or would presume. It shows how the development of new concepts, hypotheses and theories is a very slow and gradual process in which practical things such as a very precise balance and grand things such as theological assumptions can and will influence the proceedings.

intermezzo

Cognitive models

Genuine reasoning can take place through constructing and manipulating models.[319] This helps to make sense of what scientists have been doing through the ages. The traditional philosophical accounts see reasoning as carrying out logical operations on propositional representations. Only recently has some more attention been given to the power of models in trying to understand the world. We have seen above how IngenHousz, Priestley, Senebier and De Saussure struggled with their concepts. They tried to fill in the gaps in their knowledge by creatively drawing a picture of the world. They actively recombined the elements they could see and those they could not see in novel recombined settings. Phlogiston was one element in that model, to be replaced by oxygen later on, by which their model became more powerful. The eudiometer was to them what the telescope was to the Herschels or the microscope to Malphigi. These instruments enriched their practices and took their thoughts to new depths. They could 'see' better. They saw more, to understand better.

It is tempting here to draw some parallel with art and play. Children search novel recombinations while playing, their brains being primed for survival and learning. Artists conserved this childlike freedom to explore new possibilities and ever-novel recombinations.[320] That's why the comparisons between science and art are sometimes relevant. On the cognitive-emotional level of creativity, fuelled by curiosity, rewarded by recognition.[321] Or as Adam Smith in the introduction to his *The history of astronomy*, posthumously collected in 1795) wrote: "Wonder, Surprise, and Admiration, are words which, though often confounded, denote in our language sentiments that are indeed allied, but that are in some respects different also, and distinct from one another. What is new and singular, exites that sentiment which, in strict propriety, is called Wonder; what is expected, Surprise; and what is great or beautiful,

319 Nersessian 2008
320 Ellenbroek 2006
321 Magiels 2002

Admiration."[322] In the eighteenth century it was openly acknowledged that this was one of the main drives of natural philosophers. The gap between science and art was not as wide and deep as it has become in the twentieth century. People were interested into science as they were into music, art, literature, for the pleasure to do it, and with the intention to carry out amusing and entertaining experiments. The "amateurs" of the Enlightenment "played" science in a relation of sane and harmonic equilibrium with their other interests and activities.[323] In many other aspects, however, science and art differ as much as water and fire, as in their degree of objectivity or reproducability. Not recognising these differences explains why many flirtations between science and art run into postmodern dead-end streets.[324] Not recognising the similarities has alienated many from science, which some have come to see as the impersonal, deadly rational, poetry-killing machine. Acknowledging the passions and the struggles and creative moments of individual scientists might help to strenghten the understanding between scientists and others.[325]

A revolution that wasn't

At this point in the story of IngenHousz and his contemporaries, it is about time to look at something that did not exist in their time, and did not exist yet at the time Nash wrote his analysis. The conceptual beast of the 'paradigm shift' was only unleashed in 1962. That's when Thomas Kuhn wrote his renouned book *The structure of scientific revolutions*. (Who must however have known Nash's work as he was explicitly mentioned in the foreword by James Conant,[326] the editor of the series in which Nash's account was published.) Since then paradigms have continued to shift, both within science (what Kuhn was writing about in the first place) and all over society. It has become a household word to describe anything that changes (theories, styles, viewpoints etc.) and changes abruptly, irreparably and in such a way that before and after cannot be compared anymore. It has become cliché, even so cliché that the web magazine *Wired* elected it in summer 2009 to be thrown down a black hole together with the 'silver bullets' and the 'holy grails' of this world.

Wired was not the first to criticise Kuhn's concept of a paradigm and its shift. It has been under heavy flack from many sides and has been a point of controversy in the philosophy of science for all those years. This is not the place to rehash all those comments and critiques, but something can and has to be

322 Schliesser 2005
323 Piccolino 2007
324 Obrist 1999
325 Lightman 2002
326 Conant 1950

said about it, because the overthrow of the phlogiston-theory by Lavoisier's 'new system' is a major element in the saga of the discoveries of IngenHousz, Senebier and Priestley. More so because it was one of Kuhn's 'paradigmatic' examples of scientific revolutions. Kuhn argued that the new system in such a revolution was incompatible with the old in some crucial ways, that there is not a fully logical way to reason from the one to the other, or back. That's also why there is, according to Kuhn, no possibility to make a truly logical or rational choice between them. If one reads Lavoisier himself however, he described his system as a completely new edifice, rather than a reconstruction of the old. He did not attempt to refute the phlogiston system, but simply offered a new theory in which he believed he could explain the same phenomena in a simpler, more natural way with fewer internal contradictions - which sounds rather like a rational choice for something better. Priestley on the other hand perceived Lavoisier's revolution in chemistry primarily as an assertion of power over the philosophical community. We have already seen how his ethics on experimentation were profoundly influenced by his ideological view on the world. His resistance against oxygène was equally deeply coloured by his idea about knowledge. He rejected the imposition of discipline upon practitioners (as e.g. the precise use of the eudiometer). Because that would impair their independent powers of reasoning. The proposed transformation of chemistry was based on rigorous and complex measurement techniques and methods that could accurately be reproduced. This is why Priestley associated Lavoisier's chemistry not with the revolution in France but with dogmatism and despotism. His criteria for 'bad science' did not have much to do with the essence of the methods or the possibility of error, but with a deeply ingrained disgust for everything authoritarian. Laudable as that may be, it did not produce great scientific insights and prevented Priestley and many of his followers from taking the step towards a better understanding of the chemical structure of the world. They refrained from using the relevant rational arguments and thus cognitively blocked their way into the new theory.

In Kuhn's approach a scientist switching from the one to the other was supposed to be going through a Gestalt-switch, an idea he probably got from Ludwig Fleck or at least borrowed from Norwood Hanson. But no sudden switch of that kind can be found in Lavoisier's work, according to people who know the man better.[327] Lavoisier worked and thought for years in an incoherent world in which he jumped from one point of view to the other. It was a halfway position with its inbuilt inconsistencies. The worlds that are, in Kuhn's terms, completely different could be better described as one hybrid playground. Kuhn's image of a Gestalt-switch may well have been highly suggestive but is difficult to hold up in face of what Lavoisier was doing. He did not switch from phlogiston to oxygen like switching from the image of the

---

327 Holmes 1985

rabbit to that of the old lady. The conceptual complexity of the new chemistry cannot be compressed into one image with two meanings. Lavoisier worked on his system for many years and developed it in parts, putting the puzzle together in different episodes. Lavoisier was not even the initiator of a new chemical nomenclature but captured a nomenclature for his purpose that was already taking shape.[328] Even in his famous statement of February 1773 predicting a revolution in physics and chemistry, Lavoisier may not have viewed himself as an initiator of an upcoming revolution but as a participant in a revolution that had already begun.[329] As Frederic Holmes refutes: "The tendency to seek a single "key" to the Chemical Revolution, whatever it might be, oversimplifies a multidimensional historical proces." [330]

The switch is more like a "passage" as Holmes calls it, a long and winding road, with many unexpected vistas and stretches through impenetrable underwood. A revolution it definitely was, looking back at that turbulent half century between 1750 and 1800, in many different ways. And something changed in the conceptual framework of chemistry that was a fundamental step towards a better understanding of the chemical structure of this world. But the structure of that revolution was more that of an evolution, a gradual growth of new ideas. Some of these were to survive in the new theory; others were discarded on the way. That is what we saw happen while IngenHousz and Senebier were developing their experiments and thoughts on plants. That is what they did when they connected these with Lavoisier's system.

Paradigm lost

When looking at the men who lived in that same period and who needed a theory of chemistry to understand what they observed in the plants they were studying, the word "passage" is quite appropriate. Priestley, IngenHousz and Senebier began their research in the phlogiston-era. Priestley's work in chemistry evolved totally around that concept. He was (together with Scheele) the first to isolate and described oxygen (vital air). In explaining how this air helped candles to burn and mice to survive he used the concept of phlogiston. In that system the air produced by plants was dephlogisticated air. He would never in his long live show the slightest public or unambiguous sign of dropping his belief in the existence of phlogiston, as his last letter to IngenHousz testifies. Senebier and IngenHousz entered the new field of chemistry a little later and they did so only because they needed it to investigate plants. Chemistry was not their profession but their research tool. When they started to perform their first

---

328 Holmes 2000
329 Holmes 1998
330 Holmes 2000

experiments Lavoisier still had to formulate his new theory. This was to appear during the 1780s in several steps, the first prudent the later ones bold and triumphant. That story has been told elsewhere in great detail.[331] IngenHousz and Senebier must have closely followed these developments, sometimes even firsthand. That is what one can see reflected in their writings.

In 1779, IngenHousz was a true phlogistonian in the wake of Priestley, one of his friends and colleagues and the main defender of this theory. When he described the results of his experiments, he did so in a phlogiston-world. The following winter he spent in Paris, where Lavoisier had done his crucial experiment producing oxygen in 1777 (inspired by a demonstration by Priestley when visiting the French capital as the scientific and philosophical escort to Lord Shelburne[332]). The first tentative ideas about a new chemistry must have been in circulation. And IngenHousz must have been aware of these, as he was one of the people frequenting the 'chemical dinners' at the Lavoisiers' house where good food and wine, in exquisite intellectual company, were combined with philosophical demonstrations and scientific discussions.[333] Still, it would take Lavoisier until 1789 to develop it into a consistent theory that would overtake the old phlogiston story. By that time, in a footnote in the second volume of the French edition of his book, IngenHousz mentions Lavoisier's system, as promising but still requiring more confirmation. In a letter of May 1792, he mourned his friend Lavoisier who had been beheaded by the revolution, not giving IngenHousz much confidence in what was called 'democracy'. In a letter from 21 November 1794, written from London to Van Breda, he endorsed the new terminology fully, proving himself to have become a 'new chemist', applying the new knowledge to clinical purposes.

> ... *verscheide lugten gebruyken om verscheide ziekten te genezen die tot hier toe genoegzaam als ongeneesbaar gehoude wierde... het levensmakende principum of oxygene bedraagt 27/100 van de gewone lugt, de overige 73/100 is azote of dodelyke lugt...*

> ... various airs can be used to treat different diseases which until now have been uncureable... the life-giving principle of oxygen is 27/100 of the common air, while the remaining 73/100 is azote of deadly air...

During the course of less than fifteen years, IngenHousz changed his chemical perspective on the world. This is a far cry from the idea that old paradigms only disappear with the death of the last pensioned professors believing in them. It is also different from the idea that the switch from one paradigm to the other

---

331 Bensaude-Vincent 1993
332 Jackson 2005
333 Bensaude-Vincent 1993 p 89

is based on irrational, argumentative elements. IngenHousz shows in his move from phlogiston to oxygen that it can be perfectly rational for a man to change his point of view. The new system offered him a better toolbox to handle the problem facing him. Senebier made the same step, albeit a little quicker. In the texts of both men, one can see how they use both terminologies together, side by side or separately over a period of more than ten years. Senebier wrote in his book of 1788: "I propose nothing that is unique or unheard of in chemistry when I say that sunlight decomposes the fixed air contained in the leaf. I am no less consistent with the principles of sound chemistry when I say that the carbonaceous substance or phlogiston is combined in the plant with the resins." Senebier, like IngenHousz, used the two terminologies together. As if they were translating the concepts in their minds, or on behalf of their reader, to make it easier to understand what they were talking about. Others would make the switch even later, or not at all. Priestley stuck to phlogiston until his death; Jean Baptiste de Lamarck would end his opposition to the Lavoisierian system in 1802. Another great biologist, Lazzaro Spallanzani, was very quick in recognizing the value of the new system to understand processes of decomposition and recomposition in living organsisms.[334]

There is another consistency that spans this period of 'paradigmatic change'. All men involved in this story used the same instrument. In 1805 De Saussure was still using eudiometers to measure the amount of oxygen in his air samples. The apparatus was slightly improved compared to the primitive tube Priestley first used, but the principle was the same. On the instrumental level, things stayed pretty much the same. At that level everybody concerned talked about the same things. That is probably why Priestley and other phlogistinists were well equipped to understand what those fashionable chemists from France claimed. So much so that they could think about what their opponents' next argument would be. This tale confirms Galison's remark: "Missing from the positivist account is the arena of argument situated between perception and the establishment of "facts". In the laboratory lies all the interest, for that is where experimentalists muster arguments, rearrange equipment, test apparatus, and modify interpretive skills."[335] It explodes the monolithic image of scientific practice. The significance of experience is not so "local" as to be commensurable only with the theory or the group of theories that support its results. Both theorists and experimentalists have breaks in their respective traditions, as Galison puts it, but they are not typically simultaneous. The history of plant physiology and chemistry in the late eighteenth century shows how Kuhn's version of the history of science leaves room for improvement. The so-called incommensurability between two paradigms dissapears as an artifact when

---

334 Bandinelli 2007
335 Galison 1987

one looks at the process, not just at the theories as end products. The history of plant physiology and chemistry in the late eighteenth century shows how Kuhn's version of the history of science leaves room for improvement. The so-called incommensurability between two paradigms dissappears as an artifact when one looks at the process, not just at the theories as end products.[336]

The concept of the paradigm has shifted itself. In many ways Nash's account of the research on plants and the atmosphere is very inspiring as it was written before Kuhn's conceptual thunderstorm hit the philosophy of science. He apologised himself for having omitted mention of comparatively large numbers of less significant experiments and less talented investigators:

> Note carefully the multitude of the inspirations, fruitful and sterile; the multitude of experiments, well or ill conceived and executed; the multitude of the "trivial" points, the hidden assumptions, the uncontrolled variables, the slight misapprehensions, that ultimately made all the difference between success and failure. And then consider the meagreness of the results brought forth by all this work: highly significant though they were, they constitute no more than the cornerstone of the still uncompleted comprehensive conceptual scheme of photosynthetic processes. Here it is that we gain some sense of the travail, the waste, the reverses that invariably accompany scientific research. Its triumphs are known to all.[337]

Nash illustrates how people try to understand what they see happen in front of their own eyes. And how difficult it is to do that properly, in order to get reliable knowledge. He shows how time- and energy consuming it is when people try to come to intellectual grips with a complex world that is out of reach of the theoretical framework they have at their disposal. So they start again and again, with trial and error, and with all the creativity they can generate. They cast a critical eye on their methods, aware that they can't possibly know everything. As the writings of IngenHousz, Senebier and De Saussure amply illustrate. Scientists were no know-it-alls with the hubris to think they could understand everything. Just like they still aren't today.

The conceptual evolution in the minds of IngenHousz and Senebier shows how scientific evolutions have to be seen more as organic changes in complex systems, rather than as cataclysmic breaks in theoretical thought. How biology might help in clarifying the workings of science is the subject of the next chapter.

---

336 Nersessian 1984
337 Nash 1952 p 120

rewind

Science as IngenHousz knew it & science as we know it

In the work of the eighteenth century natural philosophers, methodology and epistemology went hand in hand with experiments, hypotheses and theories. Science and philosophy of science were to them a most natural whole. Some wrote complete books about knowledge and how to acquire it; others expanded regularly on methodology in their scientific works. And it shows that their methods and the way they thought about them were not that different from our contemporary thinking. Trustworthy knowledge is something of all times because the criteria to judge something as trustworthy have always been the same, independent of what or how much you know.

Other 'modern' aspects of science pop up in the works and writings of the explorers of that bygone age. The relationship between science and religion was a hot issue. Some natural philosophers kept the two almost in two different spheres of their lives, as if Gould already had told them about the different "magisteria".[338] Others don't consider them conflicts of interest. Still others built their science on religious fundaments. The relationship between science and society was just as precarious as it still is now. Knowledge can be just as threatening to some as it is attractive to others. Reliable knowledge can be used to improve the fate of many or to increase the power and prestige of a few. Communicating will in any case be crucial and the way they communicated their knowledge was just as important in 1775 as in 1995.

The story of photosynthesis research in the second half of the eighteenth century is a nice case of the development of new concepts and models. Based on a small collection of 'facts' several people tried to understand a specific aspect of reality. With the same facts as starting point, they ended up with different explanations. Twenty years later these would merge in one theory that was accepted by all people concerned. Not in the least because everybody by that time had accepted the new framework and terminology of chemistry. The theoretical and experimental developments in plant physiology surfed comfortably on the waves of change in chemistry. This illustrates how the Kuhnian concept of paradigmatic change is too meagre to explain the complex changes that take place when scientists change their mind and start to use new tools, alongside old tools, to gain a better understanding of reality. IngenHousz and Senebier are two clear examples of men who changed 'paradigms' during the course of around ten years and felt perfectly at home in this slowly shifting conceptual playground.

---

338 Gould 1999

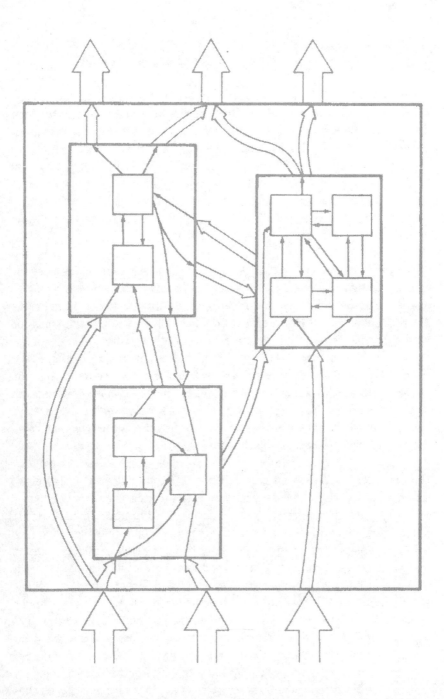

Complex systemes can be seen as processes and structures, nested in a hierarchy of subsystems, connected and interrelated by inputs and outputs. This web-like multilayered model may help to clarify complex phenomena that arise from the interaction of simpler components.

A case of complexity:
ecological questions
for the philosophy & history of science

The disappearance of IngenHousz' name and fame can - as has been analysed above - be attributed to the interplay of many variables in his life: from the setting in which he worked, to the particularities of the experiments he was performing and to the theory he was trying to expand, from his colleagues and competitors, friends and foes, to the situation in science and society in Western Europe in the second half of the 18th century.

This complex story has come to life through a chronological and diachronological reconstruction of his life and works. That was a work of history of science. As this story also contained a treasure of elements of interest to the philosophy of science it was very tempting to use it as an occasion to revisit some famous topics in that philosophical discipline. Not in the least because it is interwoven by the phlogiston-oxygen controversy, one of the paradigmatic examples of a paradigm shift in the philosophy of science. IngenHousz turned up, after all those years, at the crossroads of the philosophy and the history of science. It illustrates nicely what Imre Lakatos said: "Philosophy of science without history is empty; history of science without philosophy of science is blind."[339]

The IngenHousz case seems to offer the possibility of doing what Robert Kohler was dreaming of when he pleaded for a generalist vision in the history of science[340]. It brings in new data, in a story that can speak to all, to both specialists as well as casual readers. It may pay attention in a generalist way to the process of trustworthy knowledge in the making. As Kohler says:

> *Science has features that transcend the particularities of subject, period and locality, simply because knowledge is by definition communal and because there are many ways in making it so. Processes are always fewer and simpler than products.*

339 Lakatos 1981
340 Kohler 2005

It has the same ring to it as McEvoy's comment that "postmodernist historiography lost sight of the patterns of historical change in a bewildering array of individual instances and specific situations."[341] He admits that many of the necessary elements and levels necessary to understand a complex system such as the Chemical Revolution, have already been identified and characterized in scholarly literature, and he names: empirical objects and information, theoretical strategies and constructs, experimental techniques and practices, methodological and epistemological principles, political formations, linguistic conventions, pedagogical and professional organizations, and social, cultural, and economic institutions, values and regularities. And he is struggling to find a framework in which to bring together these various and very different dimensions or levels.

It also echoes what Anne Fagot-Largeault said during her key-note lecture on "Styles in the philosophy of science" at the first European EPSA conference in 2007. Being specialised in the extremely complex discipline of psychiatry, she held a plea for diversity in the philosophy of science. Just as there are different scientific approaches to different problems or aspects of reality, she argued, do we need more innovative styles in philosophy of science. She named some. In a "formal style" one is interested in the logic and the concepts, not in the contents or the results. In the "historical style" history and philosophy meet often in the debate of those who see continuity in the development of science and those who see discontinuity. In the style of the "philosophy of nature", science is just part of a grander scheme and becomes in its most ambitious form a kind of speculative philosophy. Following Suppes, she sees room for "emergent styles", with input from climate science, neuro-cognitive science, medicine and technology. Knowing that rationality is a mix of knowledge, belief and trust, while depending on the collective functioning of the group, her plea is to find ways of looking at science in its overall complexity. Maybe IngenHousz, this multi-faceted figure from the eighteenth century, is able to inspire to look afresh at scientific and philosophical practices of our own and of all times.

Towards an eco-philosophy of science

As Kohler mentioned, science is no monolithic block of knowledge but a continuous process. That process is driven by individuals with their passions and ambitions, or lack of them, and by their curiosity and sometimes their strive for status. Still, building trustworthy knowledge can't be done in isolation. It is necessarily a group activity, because that's the only way objective knowledge that transcends the individual standpoint can be obtained. Sooner or later, peer-review filters out the idiosyncratic influences of individual researchers

---

341 McEvoy 2001

and, in the worst cases, the moments of fraud and dishonesty. All this activity accumulates around certain subjects crystallized into scientific disciplines that follow pathways steered by models, concepts and - to use the big word - paradigms. Science is a group of people formed around a common subject with a certain method. They do this within the context of a society which finances all this work and approves or disapproves it in the next round of subsidies, or through some wealthy benefactors as was often the case in the past. From this bird's eye view on science, watching long enough, one can observe how this very complex system of people and their activities evolves, its form and content changing through time, while remaining recognisably the same.

It reminds me of how I learned to look at the world while studying ecology. Maybe, science could be looked at successfully as an ecosystem. According to the definition by Francis Evans from 1956:

> *In its fundamental aspects, an ecosystem involves the circulation, transformation, and accumulation of energy and matter through the medium of living things and their activities.*[342]

Ecology studies an ecosystem as the interplay between four factors: individual organisms (a squirrel, a fungus or a beech), groups of animals and plants (the population of all squirrels or beech trees), their environment (the geological, climatological and geographical surroundings in which all these populations live, say, the forest) and the flows of energy and information that link all components together and define their interactions and dynamic equilibrium (the nuts the squirrels eat, the squirrels that are eaten, whose leftovers decay through bacterial and fungal action and are recycled into a new little beech). Each can be studied in its own right, but we will never get the whole picture if all these detailed findings are not put together. It will be necessary, even inevitable, that somebody studies the activities of some squirrels in great detail. And the fate of the forest will be mirrored by the lives of the squirrels. But that doesn't mean to say that you understand the forest if you understand the squirrel. At least, that's what I learned when studying ecology in the 1970s.[343]

Looking at the whole

In an 'ecological' approach to the philosophy of science this could be translated as the interplay (through their common ground in theory and practice) within society (which decides where the money or the attention goes) of a group (science is defined by its creative and critical co-operation within and over

342 Evans 1956
343 Odum 1971

disciplines) of individuals (each with their own idiosyncrasies). (See table 1) Some philosophers of science will have to look into the details of the logical format of theories, others will have to reconstruct the historical chronology of events, only together will they be able to shed some light on the complexity of this ambitious human enterprise called science. And when we look at this science as primarily a process, we are inevitably talking history. Looking in this way at photosynthesis research in the second half of the eighteenth century can help to explain how this chapter in the history of scientific enquiry is representative for what science, as a method for acquiring trustworthy knowledge can do. It may also explain why few people know about the discovery of the most important biochemical process on earth. And why - though all ingredients seem to be present to make IngenHousz a household name in the long and glorious history of scientific discovery - his name doesn't figure in the history or the biology books.

individual	science as inspiration	the lonesome investigator, with all his or her idiosyncrasies
population	science as teamwork	people together, in a lab, in a department, within a discipline
interactions	science as theory & practice	the content that binds them, the flows of information and energy that link them up
biotope	science as politics	within the wider world of knowledge and within society

table 1
science as an ecosystem: interaction on and in between four levels

There are quite a few philosophers and historians who have been detecting some 'missing ingredients' in the traditional kitchen of 'classical' philosophy of science. Kuhn was the first to feel his approach was too narrow. In the preface to his *Scientific Revolutions* he wrote that "Aside from some occasional asides I have said nothing about the role of technological advance or of external social, economic, and intellectual conditions in the development of the sciences."
In his analysis, development is driven by the internal rules and expectations of science, by the logical internal structure of that paradigmatic business.

He was trained in physics, not in biology: otherwise he would have appreciated his own remark more, as it was pointing in the direction of science as a complex, multi-facetted activity.

The clearest and most encompassing critique on the old ways of doing philosophy of science comes from Susan Haack. She characterises these established philosophical approaches to science[344] as the battle between the "Old Deferentialism" and the "New Cynicism". The adherents of the former were convinced that formal logic would be able to formulate the core of scientific epistemology. It would give rise to Logical Positivism or the Logik der Forschung, to the battleground between deductivists and inductivists, flowing over in the works of Kuhn, Lakatos, Feyerabend and their followers. In following Feyerabend and other postmodern sources of inspiration, some did put all focus on power, rhetoric and politics and became cynical about the possibility of any true knowledge. Both ends of this spectrum, the deference and the cynicism, reflect the deep cultural currents of admiration for and uneasiness about science of which they are manifestations. That is how Haack describes the dichotomy between the philosophers of science on the one hand, studying the rationality and logic and rigour in science as its essence, and on the other hand the sociologists of science who study science as an amalgam of socially constructed stories and perspectives that are all true in their own way. Haack shows these two viewpoints on science as - at best - the extreme results of exclusive focus on partial aspects of what real science is.

> *Science ... is a thoroughly human enterprise, messy, fallible, and fumbling; and rather than using a uniquely rational method unavailable to other enquirers, it is continuous with the most ordinary of empirical enquiry, 'nothing more than a refinement of our everyday thinking', as Einstein once put it. There is no distinctive, timeless 'scientific method', only the modes of inference and procedures common to all serious enquiry, and the multifarious 'helps' the sciences have gradually devised to refine our natural human cognitive capacities: to amplify the senses, stretch the imagination, extend reasoning power, and sustain respect for evidence.*

As a scientist I had been struggling to recognize in the picture of so many philosophers of science the scientists of flesh and blood who I know in the libraries, laboratories and fields all over the world. The "one-dimensional accounts in dull academic prose of so many academic philosophical analyses" did not seem to fit the vibrating reality out there. With Haack there is someone who comes close to the complexity of the real thing. She describes how evidence is both context-dependent as well as logical. They are, as she calls it, 'worldly', depending on the scientist's interaction with the world and on the relation of

---

344 Haack 2003

scientific language with things and kinds in the world. She shows how scientific knowledge forms a continuum with everyday knowledge. And she illustrates her story with examples that show how not all theories are supported by good evidence nor that all scientist are allways objective, unbiased, perfectly logical, &c. She seems to be able to bridge the divide between philosophers and historians of science, who have been living on hostile footing for a long time. The former are looking for a general theory that can explain all aspects and the essence of science, the latter find every event in the course of history unique and only understandable in terms of its particular context. These two approaches seem to be incompatible or irreconcilable. Unless you step back for the wider view and take the complexity of reality, and of science, for what it is: complexity. Haack uses the metaphor in an old story about an elephant to make this clear. That story has been my guiding metaphor since I studied ecology in the seventies. I was glad to encounter my elephant again in good company.

The case of the elephant

The story goes that six blind men were asked to determine what an elephant looked like. Every one of them touched a different part of the animal's body. One felt a pillar, another a rope. The third felt a tree branch, the fourth a hand fan. Another says he touched a wall, the last saids he was holding a solid pipe. A wise man explains that all of them are right. The reason every one of them told it differently is because every one touched a different part of the elephant. The elephant has all the features mentioned, but to understand what the elephant is, all the different perspectives have to be integrated. This story has roots in Jain, Hindu and Buddhist traditions and got a place in Western thought by the children's poem by 19th century poet John Godfrey Saxe:

> It was six men of Indostan,
> to learning much inclined,
> who went to see the elephant
> (Though all of them were blind),
> that each by observation,
> might satisfy his mind.
> ....
> And so these men of Indostan
> Disputed loud and long,
> Each in his own opinion
> Exceeding stiff and strong,
> Though each was partly in the right,
> And all were in the wrong!

Being partly right may mean one is totally wrong. In the philosophy of science something similar happens. Many approaches co-exist. Many perspectives are defended, others sometimes furiously attacked. Understandably so, as none of them is able to catch the essence of what science really is. There is always some escape in another's black hole. Susan Haack demonstrates this convincingly in her book. Her analysis of the problem gives science its proper place on a continuous spectrum between scientism and cynicism. She rightly recognizes the human side of science, both in its individuality and its group-effects, embedded in a wider context of culture and society. She picks up all the ingredients of the ecological view, which I introduced before.

She is not the only one to discern this weakness in the recent attempts to explain science as a phenomenon in all its complexity. Johnson in his book on Priestley comes to the same conclusion. He describes Priestley as a core figure in "a story of science, faith, revolution and the birth of America".[345]

> One of things that makes the story of Priestley and his peers so fascinating to us now is that they were active participants in revolutions in multiple fields: in politics, physics, education, and religion. And so part of my intent with this book is to grapple with the question of why these revolutions happen when they do, and why some rare individuals end up having a hand in many of them simultaneously. My assumption is that this question can not be answered on a single scale of experience, that a purely biographical approach, centred on the individual life of the Great man and his fellow travellers, will not do justice; nor will a collectivist account that explains intellectual change in terms of broad social movements. My approach, instead, is to cross multiple scales and disciplines - just as Priestley and his fellow travellers did in their own careers. ... To answer the question of why some ideas change the world, you have to borrow tools from chemistry, social history, media theory, ecosystem science, geology.

As may have become clear by now, the story of IngenHousz is just as complex and multi-layered. And Steven Johnson's proposal may sound all too natural to anyone who read this story of sunlight and insight up to this point. He finds the mainstream approaches to explaining 'science' too meagre, uninspiring and insufficient. His account of Priestley's life is fascinating and entertaining, though here and there he is a little too eager to link Priestley to the current relationship between science, politics and religion in the United States. But he comes up with the billion dollar question:

> Is there a better organising principle, a better metaphor for making sense of the conceptual revolution like those that Priestley helped bring about? One

---

345 Johnson 2008

might be the twentieth-century concept that neither Priestley nor Marx had
available to them, and which was still a new idea for Thomas Kuhn in 1962:
the ecosystem. Ecosystem theory has changed our view of the planet in countless
ways, but as an intellectual model it has one defining characteristic: it is a
"long zoom" science, one that jumps from scale to scale, and from discipline to
discipline, to explain its object of study: from the microbiology of bacteria, to the
cross-species flux of nutrient cycling, to the global patterns of weather systems,
all the way out to the physics that explains how solar energy collides with the
earth's atmosphere.

That is exactly where IngenHousz comes in, at the interface between the sun
and the ecosystem of the earth. And Johnson's sketch echoes perfectly well
the framework I set up above. It captures the four-part diagram in a different
perspective. His 'long zoom' view can be understood like a system in which each
level of complexity gives rise to a next level, to finally fold back on itself.

neurochemistry of emotion and cognition
⇓ ⇑
the story of an individual's life and works
⇓ ⇑
the social networks in which he or she functions
⇓ ⇑
the information flows connecting them, in permanent interaction
⇓ ⇑
the technology (soft and hard ware) carrying and shaping these flows
⇓ ⇑
the scientific conceptual frameworks within which all this works
⇓ ⇑
the political or ideological regimes they are embedded in
⇓ ⇑
the local and global economical modes that define them
⇓ ⇑
the light of the sun driving nature and human society as part of it

One could link the last back to the first, as all the neurotransmitters in our
bodies are proteins manufactured from plant stuff we have been eating, which is
recycled sunlight. It is just one of the results of this immense web of relationships
on which IngenHousz unknowingly did put his finger.
Johnson's proposal reminds us of another recent attempt to come to grips
with the astonishing complexity of the human activity called "science". In her
recent book[346] Nancy Nersessian makes an effort to understand how creative
thinking leads to conceptual innovation. In her preface, she circumscribes

---

346 Nersessian 2008

the problem. Any attempt to understand creativity by focussing on the act instead of the process would be like attempting to understand a rainbow by looking at what is happening inside one drop of water. Creative thought in science is a multidimensional process. That's why it has been addressed in history, philosophy and the cognitive sciences. The latter show how scientific practices align with those non-scientists use to solve problems and make sense of the world. Nersessian sees a continuum between mundane problem solving strategies and scientific efforts to do the same, albeit in a more articulate explicit manner. This corresponds very much with the way I described earlier how the efforts of science - in order to gather trustworthy knowledge - are an extended version of the everyday necessity to understand the world. In order to comprehend how concepts are constructed and how they are used and how they develop, she chooses to combine historical and cognitive approaches. She sees conceptual change as a problem-solving process. These processes are "extended in time, dynamic in nature, and embedded in social, cultural, and material contexts". (This harks back to the contribution in the 1987 book[347] she edited, in which Marjorie Grene was pointing at a new, developing perspective on the philosophy of science in which an *ecological* realism (her emphasis) was needed to understand the process of science in which scientists as real live people are concerned about real puzzles about the how's and why's of the world.) Scientific practices, so Nersessian continues, can be investigated at different levels of analysis: "at the level of researchers as individual, embodied, social, tool-using agents; at the level of groups of such researchers; at the level of the material and conceptual artefacts comprising the context of activities, such as laboratory research; and as various combinations of these". She explicitly refers to her "environmental perspectives" from 2005 where she argued that cognition is embodied, acculturated, distributed and situated. Reasoning is more than applied logic, scientific knowledge is more than theories and experiments, science is more than scientists rationally gathering knowledge.

Ecological tools

Ecology could be a workable metaphor to chart the complex, multidimensional process that is science, as it has been hinted at by various philosophers of science. Which perspectives can ecology offer on the biotope of science, which handles could it offer to get a hold on science? Which links can be forged between ecology and philosophy? We are entering uncharted territory here, as ecology has not been featuring significantly in philosophy (of science). The philosophy of biology has been concerned mainly with evolution. Textbooks on the philosophy of biology are taken up for ninety percent by reflections on Darwin's

---

347 Grene1987

theory and its scientific, epistemological, religious, political and philosophical implications. Looking at the non-biological world through an ecological lens therefore is new. The hints given above by some thinkers and writers are not much more than a suggestion of what could be. More food for thought can be found in what some philosophers have done over the last decennia. They stressed aspects of science that were in their opinion underrated and which all fit in with the ecological approach that will be unfolded. We will encounter Hull, Wimsatt, Nickels, Nersessian, and others in the paragraphs below.

But it would be useful first to see what ecology does, what it observes and measures, what results it delivers, how it tries to understand the world. Let's look at an example with not too many species, in a quite desolate place at one of the far ends of this world: Svalbard.[348] It is an archipelago in the Arctic Sea halfway between Norway and the North Pole. For the few species that live there it is an extraordinary cradle of life, ruled by water, light and temperature. The warm and nutrient rich Gulfstream arrives here, keeps the waters relatively ice-free and nurtures a massive plankton bloom in spring. The plankton lures whales and great schools of cod and capelin, which serve dinner to seals and seabirds. The seals keep the ice bears' menus filled. These energy-rich waters draw in massive amounts of seabirds, some three million of them, of just about 28 species, only one of them living permanently on the ground, the ptarmigan. And it is the geology that makes this work. Svalbard's coastline consists of near-vertical cliffs, containing millions of outcroppings big enough for a nest, too small for a predator like a fox. Foxes, just like other animals that stay on through winter, spend their entire summer eating, 24 hours a day, day in day out, fattening up for winter. In this ecosystem sky, sea and shore are intimately linked. One of the cycles goes as follows: dovekies (a kind of auk) dive for copepods and nest on the rocky slopes. The guano and the carcasses that the flocks deposit on land fertilize a mossy garden, ideal lurking ground for the arctic fox. This hunter preys on puffins and other birds and eggs. Nutrients left over from all these life forms are swept off the shore or drop from the sky into the ocean, nourishing vibrant colonies of anemones and corals.

This abbreviated overview nicely pictures how living and non-living interact, how energy is continuously cycled through different trophic layers of the system. And how material factors, such as the hardness of the rock that built the cliffs millions of years ago is a determining factor in the numbers and the kinds of birds nesting on them now. To study this system one needs specialists in different areas. One has to count the amount of species and the number of individuals in each species. One has to measure how much of one species is eaten by another. One has to measure the temperature and the nutrient load of the water arriving. One has to study the way the seasons influence behaviour. One has to keep this up for many years to see what changes if one of the variables changes. What are

---

348 Barcott 2009

the crucial variables for the population of the ice bears or the puffins? What if the Gulfstream becomes colder? What is the effect of the whaling on the whale population that was once so abundant that in 1612 whalers described how their ships parted the whales like pack ice? How to build a model in which you can simulate this system, or parts of it? What can we learn from this to manage such a system, in interaction with human activities such as whaling, fishing, mining or tourism? Changes at macro level such as climate fluctuations interact with micro level changes, such as the available amount of plankton. Changes take place over the time span of a year, through the seasons, but can also be shorter (migrant birds arrive after the snow has melted and depart before winter falls again) or longer (warming of the climate might cause longer snow free periods).

All these phenomena may seem very unlike those philosophical problems of understanding science. But if you see how Gunderson and Holling[349] describe the essence of ecology, it starts to look different: "The emergence of novelty that creates unpredictable opportunity is at the heart of sustainable development". In other words: how does science succeed time and again in finding new creative solution to ever changing problems of understanding nature? Science in 2010 is different from the science done by IngenHousz and his contemporaries, but it is still recognisably science. The system and the contents of science has changed beyond anything IngenHousz could have imagined, though he would still have felt himself at home in any plant physiology lab at the beginning of the twenty-first century. Science has proven to be a human system that is resilient to change, and has reinvented itself in response to a changing environment. Adaptive cycles are nested in a hierarchy across space and time, as Gunderson says. His conceptual tools seem to be readily applicable to a new view on science in which history and philosophy meet. They can help to explain how in nature adaptive systems can generate, sometimes for brief moments, novel recombinations that are tested during longer periods of capital accumulation and storage. Windows of experimentation, I am still using Gunderson's words, open briefly but do not trigger cascading instabilities of the whole because of the stabilizing nature of nested hierarchies. Larger and slower components of the hierarchy provide the memory of the past and of the distant to allow recovery of smaller and faster adaptive cycles. In ecosystems, for example, seed banks in soil, biotic heritages and distant pioneer species are all critical accumulations from the past that are available for present renewal.

Stability & change

It would be hard to deny that science is in constant flux. Scientific knowledge changes continuously as the accuracy improves, new experiments are developed,

---

349 Gunderson 2002

bold hypotheses are formulated and theories are reshaped. This stands in contrast to other forms of knowledge that have been static and immutable since they were first recorded (see: the books of the Bible, Koran, Mormon or Torah, the theories of Hahnemann or astrology). Many questions in the philosophy and history of science have at their core the attempt to understand this change. How and why does an individual change ideas or come up with innovative solutions to a problem? When is one idea better than another? How do groups of people change their idea about something? How are new insights incorporated in a new theory? And how new is that theory? What's the difference between one theory and the next? How does society react towards or interact with new scientific developments? The concepts from ecological science with which to think about change, and its counterpart stability, may help to clarify these questions in one coherent framework. This is not to say that ecologists pretend to understand their ecosystems completely. The framework presented here should be understood as work in progress.

According to Holling and Gunderson[350] change in ecosystems is (1) neither continuous and gradual nor consistently chaotic. Change is episodic with periods of slow accumulation of natural capital such as biomass, physical structures or nutrients, punctuated by sudden releases and reorganisations of these biotic legacies as the result of internal or external disturbances. Rare events such as hurricanes or earthquakes or the arrival of an invading species can unpredictably (re)shape the existing structure, at critical times, or at locations of increased vulnerability. (2) Productivity and textures are patchy and discontinuous on every scales, from the leaf to the landscape to the planet. There are several different ranges of scales, every one with different attributes of architectural patchiness and texture. The world is lumpy, just look at how daisies cluster in a meadow or how people are distributed over a train platform. (3) Ecosystems do not have a single equilibrium with homeostatic controls that help to remain near it. Stabilizing and destabilizing forces are both part of the game. It is a game of multiple equilibria pushed around on the rhythm of both fast and slow evolving variables. (4) Ecosystems are moving targets, in which multiple futures are uncertain and unpredictable. Policies and management that apply fixed rules to achieve constant yields (such as a fixed carrying capacity for cattle or wildlife, or fixed sustainable yield of fish or wood) lead to systems that increasingly lose resilience. That way one ends up with a system that suddenly may break down in the face of disturbances that it could have absorbed previously.

Even without being a thoroughbred ecologist, some of the above will be recognisable in the human knowledge production system called science. The growth of knowledge - in this case about photosynthesis - follows a process with characteristics similar to those of an ecosystem. (1) From the time of Van Helmont the science of plant physiology changed very slowly, not much happened for

---

350 Gunderson 2002b

almost a century. Hales' pneumatic through was a little technical step forward, but did not lead to great new insights. These came with Priestley who observed plants producing good, breathable air. That was the catalyst of some tumultuous years in which Priestley, but mainly IngenHousz, closely followed by Senebier and Van Barneveld and various others, produced fundamental insights in what green plant parts do with the air, during the day and at night. Knowledge became restructured around new insights. Both the eudiometer and the new chemistry were external factors that did reshape the theory about a specific part of nature. The fact that these experiments were best done during summer time could be called an internal variable, timing the moments of data collection and publication. Phlogiston was an element that increased the theory's vulnerability, as it was the source of an accumulating amount of anomalies. (2) The gathering of knowledge was not uniform, but patchy and discontinuous. The idea that fixed air played a role was first seen by IngenHousz but discarded, to be picked up again only many years later, influenced by Senebier's findings. Some people were working with the eudiometer, but not everybody was using it to measure gas production by plants. Some used a particular type of eudiometer, others another. Groups with different practices were spread over Europe. Groups were clustered, with central points in England, Paris, Geneva and Utrecht. (3) IngenHousz' thoughts shifted over the years, just like Senebier's. They were in different ways and at different speeds influenced by Lavoisier's ideas. While their theory about plant gases was first centred on the phlogiston concept (a point of equilibrium) but it gradually shifted to a new equilibrium around the oxygen concept. This happened over a period of ten years, while the slow growth of the concept of photosynthesis took more than two centuries to come to full maturity. (4) It was unpredictable what would happen after IngenHousz wrote his book in 1779. History, although it is suitable to reconstruct the past, is hardly of any use to predict the future. Nobody could foresee that De Saussure would push the science of plant physiology onto a higher level by the beginning of the new century. The phlogiston theory by that time had collapsed (although Priestley did not want to hear about that). That concept was beyond repair. The new system was so powerful that it completely overwhelmed the old, almost as an exotic invading species, like red squirrels pushing the grey ones out of existence.

Dimensions and scales of change

Traditionally ecosystems were seen as systems that went through a succession of stages, controlled by two functions: exploitation with a rapid colonisation of recently disturbed areas; and conservation in which slow accumulation and storage of energy and material takes place. For example, a woody hillside is deforested by a mud slide, slowly the bare terrain is reinvaded by pioneering plants, vegetation comes back, first with nettles and brambles, later with brushes and small trees, which over decennia grow into a mature forest, the so-called climax vegetation. Cycles of (rapid) growth followed periods of consolidation and accumulation. This is a classic boom-and-bust sequence, probably more or less echoed by Kuhn's account of revolutionary science occasionally breaking through vested paradigms. Ecology has fine-tuned this model and added two more stages. The conservation stage leads to accumulating biomass and nutrients bound in increasingly complex networks, which makes it more stable and resilient to external influences. Its "interconnectedness" grows, limiting its flexibility. When external influences become too big, the system turns out to have become fragile and vulnerable. Sudden interference by agents such as forest fires, drought, insect pests or intense pulses of grazing can push the system over a threshold. This triggers a sudden release of all this biomass and nutrients. This is the phase of chaos and disorder. The released materials will in a following phase get reorganised so that they become available for the next phase of exploitation. Pioneering species come in and can rely on seed banks and can tap in on the nutrients left by the devastating events in the release phase. During the reorganisation innovation occurs, new species can blow or fly in from outside, new interspecies relationships come into existence. Micro scale variables, such as the specific amount of micro organisms in the soil may decide on what trees finally will get a foothold. This is - in a nutshell - how ecologists see ecosystems change, in phases that follow each other, sometimes slow, sometimes fast, following a kind of Möbius strip that returns to were it started from, to find the place completely changed (figure 1).

The picture becomes even more intricate and intriguing when one sees how this kind of cycle takes place at different scales of space and time. Take an ecosystem like the High Moors in Eastern Belgium (figure 2). At the smallest visible scale there is the sphagnum mosses, only centimetres long. They grow and die in yearly cycles, their relics forming the underground for the next generation. All this dead plant material has accumulated over thousands of years and is compressed into massive layers of peat. On top of these, patches of trees grow here and there. The scale of trees is up to tens of meters high and many years in life span. A marsh is hundreds of meters wide. The islands of forest in the sea of moor land expand over many kilometres. The whole peat area stretches for tens of kilometres and exists already for millennia. The same scale distribution is found in the climatic events above this landscape: thunderstorms cover vast areas and last an hour or so. Snow cover can last for months (influenced by the particular geographical situation of this landscape), climate warming may take a few decennia, while the last ice age that left its traces lasted for ages. A fire raging through this system may destroy a number of square kilometres but its effects can be felt for tens of years, taking out some species and favouring others. Insects can react very fast to such an external factor, trees much slower. The cycles at higher levels (at larger scales) are slower and bigger.

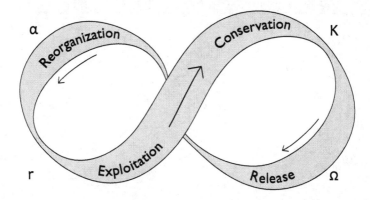

figure 1

Graphic representation of the four phases in an evolving ecosystem.
(Adapted from Gunderson 2002).

r   exploitation phase   birth
K   conservation phase   growth
Ω   release phase   death
α   reorganisation phase   renewal

During the slow sequence from exploitation (r) to conservation (K) the
connectedness and stability increase and a capital of nutients and biomass is
slowly accumulating. This is the 'classical' growth of an ecosystem towards a
dynamic equilibrium. (The r and K are drawn from the traditional ecological
parameters where r stands for the growth rate and the K for the maximum
population that is reached.) Resources become more and more bound within
existing vegetation, preventing other competitors from using them. But this
tightly bound accumulation of biomass and nutrients becomes increasingly
fragile, "over-connected", resulting in a loss of flexibility. Sudden internal or
external events such as a forest fire, drought, insect pests or intense grazing
can initiate a collapse of the system, releasing these acumulated nutrients
into the Ω phase. In the subsequent phase of reorganisation (α) soil processes
minimise nutrient loss and reorganise things for a next phase of exploitation.
Some organisms, the pioneering species, jump through the window of
opportunity to expand, imported from endemic or exotic seed banks.

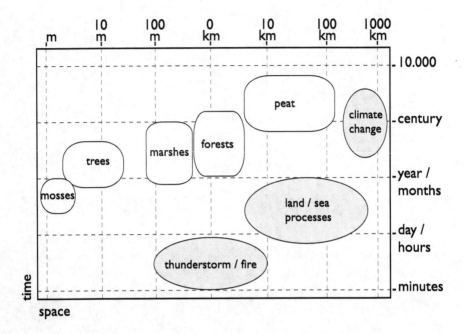

figure 2

Hierarchy of vegetation, landscape structures and atmospheric processes for High Moors system. This diagram shows scales of space and time in shifting 'powers of ten'. (After Gunderson 2002)

At the lowest level mosses dominate the picture. They are host to insects and microorganisms. They may be small but are the basis building block of this system. They lead short lives and when dying they form peat, which slowly raises the soil over thousands of years. The yearly cycle on the level of the moss drives the slow cycle in deep time.

To get an inkling of an idea about the complexity of the real system: one should also make similar diagrams charting herbivores and carnivores, migrating and sedentary animals, micro-organisms, geology, human activities &c. On top of that, one has to picture the interactions between parasites or predators and their hosts or prey. To finish off with a dynamical evolution in population densities and in the interactions between all these components. So this one hierarchy unfolds into a panarchy.

The picture thus painted results in a 'panarchical' view on complex systems. 'Panarchy' as an extension to the classical 'hierarchy', as we know it, a pyramid with lots of grass at the bottom and a few cow on top. Gunderson and Holling give these descriptions:

> The cross-scale, interdisciplinary, and dynamic nature of the theory has led us to coin the term panarchy for it. Its essential focus is to rationalize the interplay between change and persistence, between the predictable and unpredictable
>
> ...
>
> The term was coined as an antithesis to the word hierarchy (literally, sacred rules). Our view is that panarchy is a framework of nature's rules, hinted at by the name of the Greek God of nature, Pan.[351]

Panarchy is a system in which four-phase cycles as described above take place on different scales of time and space, while they exert influence on cycles at levels above and beneath. A critical change on a lower level can cascade up into a vulnerable stage in a slower and larger cycle on a higher level. Or input can come down into the renovation phase from a larger, slower cycle at a higher level. (figure 3) Take as an example an ecosystem in which periodical bush fires clean the forest floor from debris. These fires are rapid and the local trees are evolutionary adapted to be resistent to this heat, which passes now and then. Even more, some seeds are adapted to this periodical burning and need it to sprout. When humans came to live in these forests, they prevented bush fires. Leaves and branches accumulated, small shrubs could grow (not being burned away by the periodical sweep of fire) so that when the next fire got out of hand, it burned hotter and longer than normal, destroying trees and seeds beyond rejuvenation. These fires change the ecosystem fundamentally. So their nature was changed triggered by an external variable, human habitation.

This is not the place to teach an ecology course, I give these examples just to make clear how ecology can help in coming to grips with complex systems. And it may be clear by now that I like to perceive and describe science as a complex adaptive system, a system in which many different individuals, joined in networks (sometimes closely knit, sometimes loosely spread out) generate responses to unsolved problems (knowing that every answer just generates new questions), within a wider web of social, economical and political variables, knowing all this happens at different scales in space and time.

---

351 Gunderson 2002b p 5, 21

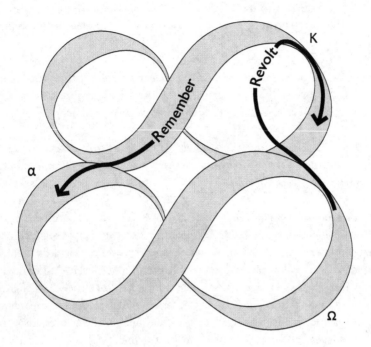

figure 3

Connecting loops between levels (adapted from Gunderson 2002)
Various levels interact and loop back into each other, taking events towards higher or lower levels in the panarchy. Gunderson labels two possible interactions as 'revolt' and 'remember'. When a level enters its Ω phase of creative destruction and experiences a collapse, this 'revolt' can cascade up to the next larger and slower level by triggering a crisis, particularly if that level is in its K phase where resilience is low. Just a handful of escaped muskrats can grow into a permanent burden for the water ecology. The 'remember' cascade feeds potential that has been accumulated in a larger, slower cycle downwards into a cycle in its reorganisation phase α. For its options of renewal a patch of burned forest can draw upon the seed bank, the physical structures and surviving species. They are the biotic legacies that have been accumulated during the growth of the forest.
Translated into the system of eighteenth century science, a 'revolt' is exemplified by how the confusion about the 'green matter' not only obstructed deeper insight in plant physiology but also cascaded up into theories about vitalism and spontaneous generation. A 'remember' loop shows how the new chemistry cascades down into plant physiology, which strugglied with the phlogiston concept. So it delivered the right input to develop a more powerful theory.

An important difference between human cultural systems such as science and a natural ecosystem such as a forest is that the human system is even more complex. It does not just rely on nutrients and energy to built its network relationships, but also on information and social hierarchies embedded in political, religious and cultural frameworks. (Still, it is humbling to keep in mind that the natural and the cultural are sometimes only divided by a very thin line. A disrupted society and a globally expanding transportation system did transform an African virus that infected chimpanzees into a global AIDS epidemic, with a little help from human sexual and drug taking behaviours.)

The flow of events in an ecosystem can serve as a metaphor for what happens in social or cultural systems. After decades of exploitation the car industry had grown into an overconnected and rigid structure, suddenly threatened by internal (complacency, lack of innovation) and external (growing concern about energy and environment) factors that brought it on the brink of collapse. New impulses may be able to stimulate a reorganisation of available resources, material as well as intellectual, cultural and economical. A comparable breakdown occured during the global bankingcrisis. Banks still exist two years later, but it will have to be seen if and how a bank of the future will resemble its twentieth century predecessor. When the communist state systems of Eastern Europe imploded around 1989, there was no resilience left and society became fundamentally reorganised.

In what follows I would like to explore these four levels or scales on which change manifests itself in science. And how ecology could help to understand them. And how that meshes with what some philosophers and historians of science have been trying to do over the last decennia.

In between the ears

At the lowest level, science happens in the brain of an individual. That's why Ronald Giere took the cognitive approach in an effort to get around the shortcomings of the classical approaches to understand science philosophically. His way into the problem was via the then newly developing 'cognitive sciences', an amalgam of research bridging artificial intelligence research and the neurosciences. How do we know the world, how do we represent it in a body of knowledge, how do we judge statements about reality to be effective (not to speak about true or false)? Giere subscribes to naturalism, the view that human activities, science included, are to be seen as entirely natural phenomena, as are the activities of molecules or animals.

Giere spends most of his time examining how theories and experiments and knowledge in general fit reality. He explores the interface between the scientific knowledge and the world out there. What is the relationship between a model or a theory and the world it describes? How do scientists decide which theory or model fits the world best? Therefore he stays very close to the old debates e.g. on subtle differences between empiricism and realism. Here and there he hints at the place where a way out could be found:

> *Once the inadequacies of those traditions are recognized, the way is clear to begin the more fruitful task of explaining how, beginning with the biological and cultural resources of our ancestors, we humans have managed to construct modern science."*[352]

It pushes us straight to an intriguing question. How is it possible that every day, all over the world, for centuries, people (some of them later known as scientists) have been very successful at probing the world and finding out how it works? And that they, based on these principles, made wonderful things and machines such as microwave ovens, corkscrews or antibiotics? And that all these clever philosophers don't seem to be able to explain this success in all their sophisticated writings?

Giere describes the comings and goings in a laboratory of nuclear physics, only to conclude that it does not really fit with the Kuhnian account of scientific development. Lakatosian or Laudanian approaches seem to be a little closer to the mark, he says, to continue with probably one of his most intriguing paragraphs:

> *Closer still is an evolutionary, or ecological, account in which variant theoretical approaches coexist in stable equilibrium, with no approach clearly dominant. The variant approaches become active rivals only when changes in the scientific environment alter their relative fitness. What any such evolutionary account requires is a mechanism by which the ensuing competition takes place. The mechanism is individual decisions.*

Realise that he uses the ecological metaphor here only with respect to the way theories and concepts 'compete'. In doing so, he describes only part of the 'ecological system' of science, predominantly in an evolutionary way. And his example is illuminating. He finds inspiration in an analysis of the scientific 'revolution' in geophysics when the plate tectonics theory of Wegener became accepted in the 1960's following a 40-year standstill after the initial proposals of Wegener. The non-acceptance of the theory in the 1920's was not due to the fact that science boils down to a social construct.

---

352 Giere 1988

It simply means that the decision of a scientific community is a function of the decisions of its members, and that decisions of individuals are, in part, a function of their individual cognitive resources, some of which are derived from their experiences with the world.

The scientific community is not a homogeneous community. Giere suggested that it is the *variation* in cognitive resources among individual members of the scientific community that plays a major role in the *evolution* of the field. (It is Giere who puts the two key words in italics.) Not the Kuhnian concept of 'normal science', the Lakatosian term of 'research program' nor the Laudanian 'research tradition' can explain what happened between 1920 en 1960. These four decades saw an accumulation of new data, generated by new techniques, that provided new insights in magnetism, the ocean floor and more, but were not generated with the aim to prove plate tectonics or continental drift. And they did not have a big impact on the geophysical community because the majority did not learn about the findings, did not fully understand them or did not trust them. Few geologists, or other scientists, possessed the 'cognitive resources' to deal with these issues. In panarchy-terms: biomass was accumulating and getting overconnected in some part of the ecosystem, and spilled over into a higher level, when it had become big enough to do so and triggered a fast evolving new entity there. The ecological view helps to clarify how events on two distinct levels, the one of the individual scientist and the one of the peer group of one scientific discipline, interact and influence each other. Sometimes this interaction is a slowing down of developments resulting in a delay lasting decennia, in other cases it may fasten the development of new concepts or theories.

It also sounds like an attractive perspective to take when looking at the events in plant physiology in the second half of the 18th century. An ecological approach to knowledge implies that rationality is the sign of an intellectual adaptability in face of changing circumstances. One has not to think and behave consistently with one's earlier intellectual presuppositions. It might well be very 'logical' or fruitful to change one's mind, if one has some good reason for it. Adapting to a new situation may ask for a move in a direction where one doesn't know the results yet. Doing something unpredictable can be a rational way of behaving in certain circumstances. Around 1750 quite a few people started looking differently at the world and IngenHousz is a prime example.

More cognition and psychology

Paul Thagard tries another approach to come to grips with the essence of science: the what and how of conceptual revolutions.[353] He sees the changes in science as major transformations in conceptual and propositional systems. His analysis draws parallels between the way scientists develop and change concepts, and the way children do. There he finds a framework by which he can fit science into the real world experience of every day where people of all sorts try to understand the world around them in a way they can live and survive; and learn about it so they can do that tomorrow in a better or more effective way. Everybody looks for facts, everybody tries to explain them and do this in a coherent way. The difference between children and scientists in doing this is only gradual.[354] The scientists do it within a rigorous scheme and within a group of peers, under constant scrutiny. Children (and most adults) are much more free in making up their own story, as long as it is convincing and workable enough for them, without loosing social acceptance or status. This meshes nicely with the view I elaborated earlier in which trustworthy knowledge comes in a spectrum, ranging from what everybody does all the time to survive in a complex world beginning at the moment you're born, to the dedicated and gigantic undertakings such as the Human Genome Project.

Thagard is a philosopher of science who tries a new approach inspired by a scientific perspective from a different discipline (artificial intelligence and machine learning), but he only concentrates on one dimension in the whole complex reality of science, i.e. the construction of theories, being built in hierarchies of concepts with relationships that can change whenever new hypotheses demand it. No matter how useful to illuminate the internal cohesion of theories and what changes in these relationships between concepts when science revolves, it concentrates on one fragment only of the whole of science.

He makes a clear analysis of the way concepts in chemistry concerning oxygen (formerly known as a mysterious non-entity, a lack of phlogiston) changed in the course of twenty years between 1770 and 1790, driven by, among others, Lavoisier. His account reinforces the description of the conceptual changes in the thinking of IngenHousz as seen above. IngenHousz started from a point where he understood that plants were doing something - which we now call photosynthesis. He could explain it perfectly, but some time later he arrived at the point in which the new chemical concept turned out to be a crucial puzzle piece to solve his photosynthesis puzzle, if he only laid the puzzle out in a different way. Here the panarchical perspective shows how individual scientists relate to the higher level of the theory they utilise to understand the world. An individual relates to a cultural or conceptual level that transcends the level of just

---

353 Thaggard 1992
354 Gopnik 2009

one group. It indicates how we are all embedded in a specific view of the world that hands us the framework in which to understand things. This framework is delivered through the interaction with peers, and through communication via various channels and through, usually slow, learning processes. In IngenHousz' time public demonstrations, letters, books and readings at learned societies allowed people to participate in the scientific knowledge.

Science is (no) lonely business

On the next level up from the individual, science manifests itself as group work. In its most obvious form it is a social activity just because *Homo sapiens* is a social species. Still, a human being can do a lot of things on her or his own. Science though is a fundamentally social endeavour. Scientists never work alone, even when they are at the lab bench on their own. They build on the work, expertise and knowledge of others, and their results are worthless if they do not communicate them to the rest of the world. They are, like IngenHousz and many of his contemporaries, often alone but simultaneously linked up in networks. IngenHousz & Co met at coffee shops, read their letters at academic societies, and corresponded in their Republic of Letters, the Facebook of those times. Randall Collins worked out that idea in *The sociology of philosophies*. He tries to describe a global theory of intellectual change, of which science makes up a big part.

> *Science is always two networks: a genealogy of tools and instruments and a human network of people exchanging words and imagery. It is both empirical and conceptual.*
> *Universal because it comprises the operations of treating things as universals. Empirical because arising within experience and applicable to experience, taking place in space and time and the social network.*[355]

The network idea goes very well with the ecosystem view of science. Links between researchers and research groups are made of the empirical, instrumental and conceptual information they exchange. Their links are the information they exchange. This exchange is more than just handing on information, like in acts of teaching or training, often it is a polemical discussion or a downright argumentative power play of public debate, fought out in letters, publications and public demonstrations.

The group is also what David Hull investigated as the fundamental unit in science. However, he looked specifically how ideas are propagated in 'scientific communities', how ideas survive in certain groups and how they die out,

---

355 Collins 1998

in a Darwinian evolutionary way.[356] He specifically looked closely at biologists performing systematics and at the intradisciplinary clash between the 'cladists' and the 'pheneticists'. But he did not so much look at these groups as a sociologist but more as a biologist and wanted to understand the process of adapting new ideas and dropping old ones. He looked at the "evolution of concepts". And he looked at all this in a naturalistic way which he describes referring to Giere as "the view that theories come to be accepted (or not) through natural processes involving both individual judgement and social interaction." He is realistic and gives a description of science that scientists would recognise. These are uneasy about terms like 'true' or 'objective', as they know that what they do is fallible, that all knowledge is temporary and in for improvement. And that sometimes, sometimes, they succeed in producing something better than what they already had. Hull sees science as a major adaptation of our species. He appreciates science in the same way as I described above, as survival tool in the struggle for existence. He admits he does not give much attention to the psychology of scientists and concentrates on their social organisation, because he did not find "these psychological characteristics especially useful in understanding the general characteristics of the scientific enterprise". His book is already huge and detailed, and it is understandable that adding the psychological make-up of all the scientists involved would make the book explode beyond readability. But it shows the way science, and philosophy, work: you need specialisation. Somebody has to study the squirrels, somebody has to track the foxes, somebody else has to measure the production of biomass in the pine trees. But to understand the forest, we will have to put everything together in a framework that does justice to all components and to the complexity of the whole.

Whatever its selfapplied limitations, Hull's account is extremely valuable as he stresses the processes and interactions. He shows how interactions within and between communities can be decisive for the evolution of science. The idea of science as a process is central in his book and its title. But within his evolutionary perspective he may have missed the ecological, environmental substratum in which this evolution takes place. Still, he illustrates how theoretical concepts such as objectivity, truthfulness, trustworthiness and rationality are the glue that makes scientist stick together.

Rationality, between individual and group

Rationality might be seen as one of the 'bridging' variables that spill over from the individual to the group. Some philosophers of science have tried to bridge the gap between individual and group, the gap between logical and social accounts of science. Thomas Nickles attempted to answer the question and

---

356 Hull 1988

ended up by making the problem more complex: explaining science is more than a purely logical-epistemological problem, it is also not a purely psychological or sociological problem: it is a "three-dimensional problem"[357]. It is individuals that do the rational reasoning, and follow some rules in doing so, but it is the group that decides if that's good enough to stay with them. Rationality is one of the ways in which individuals can inspire others, in the way IngenHousz, on basis of his argumentation, could convince his peers that plants produce oxygen in the sun. The way he came to his innovative conclusions was on the basis of simple logic. Logic and rationality were at the core of the discussion.

That was the aim at the 1978 Leonard Memorial Conference in Philosophy: looking for a future for the philosophy of science in light of scientific discovery. Nickles, in his introductory essay,[358] setting the tone for the rest of the symposium, proposes some interesting things. He likes to talk about the logics of discovery in plural as he sees various sets of heuristics, routines and rules. And they are used together or in parallel. This sounds a little bit like a rationalised Feyerabend coming along. Anything works, when and if it works. The creative and unpredictable ways by which people look at the world in a new manner is hard to predict by a rigid Logic of Discovery. Here he enters the battle ground with other philosophers who claim that the process of genuine creative discovery is a psychological problem, not a philosophical one. There is however no good reason why the philosophers of the mind, backed up by cognitive and neuroscientists, could not have a say about this subject. Where do new ideas come from? Studying the meandering reasoning of IngenHousz illustrates how this could work. The crossword puzzle metaphor of Susan Haack is also very illuminating here. Based on limited information, the puzzler tries to fill in words from different angles. What at first may have looked like a good solution turns out to be incompatible with others. There is logic to the whole procedure, but it is an erratic random walk steered by unconscious knowledge, popping up in associative ways.

Nickel makes an appeal to philosophers of science to join historians in the descriptive task of making the actual cases of creative discoveries intelligible. He feels the need for a multidimensional approach to understand the ways of thinking and the patterns of behaviour of scientists as to understand how they end up thinking what they are thinking. It should also make it possible to understand science as a method and a developmental pattern, a form of inquiry, rather than as a collection of established results. As said before: science is essentially a process and should also be understood as such.

Still, Nickels et al. keep racking their brains about the supposed rationality of discovery science while they are in fact talking about the rationality of a scientist. Rationality is no abstract or independent entity out there, it is the

---

357 Nickles 1989
358 Nickles 1978

result of what an individual human being is doing. Something which he, she and most other people would call science. However, they seem to lose sight of the fact that any scientist is equipped with a human brain and all the cognitive capacities that brings, including being crazy, wild or creative, sometimes in a very rational way. It seems to me to be more of a semantic discussion of 'rationality' than a description of what happens when somebody comes up with a new idea. Nickel himself goes some of the way in looking for "rational ways to break through scientific constraints". He concludes that no logic of discovery is necessary. Burian says as much, in his concluding remarks:

> *Such theories (about scientific knowledge) cannot afford to lose sight of our biological constitution, our peculiar sensory apparatus, our fallibility, the formative role played by our conceptual cultural and linguistic heritages, and allied considerations. ... The best available theories of physical and biological nature, of learning, of social enterprises, and of values should be brought to bear on our attempts to understand the scientific enterprise and on our attempts to gain knowledge of the world.*[359]

I always remember what the late Dr Paul Janssen, the greatest drug discoverer of the 20th century[360], said about his method for discovery: always ask what's new, always look for what you don't expect, and start from there. He was not interested in what was to be expected, but instead got really curious when something happened that deviated from the predictable. His motto was Pasteur's dictum: in the field of observation, chance only favours the prepared mind. He used nature to break through the constraints of established scientific thought. IngenHousz found something which at first seemed to be completely inexplicable. So inexplicable that Senebier refused to see it as a fact. But the respiration of plants at night was just one of the innovative and surprising findings that made IngenHousz' theory - in the end - stronger and more convincing.

Struggling with complexity

Few philosophers are as permeated by biology as William Wimsatt. His work centres on the philosophy of the less exact sciences - biology, psychology, and the social sciences - the history of biology, and the study of complex systems. "We are limited beings in a complex world", says Wimsatt[361], and he adds: "False models are the means to truer theories." Key in his view on gathering trustworthy knowledge is the legitimacy, even the essentiality, of making errors.

359 Nickles 1978
360 Magiels 2008
361 Wimsatt 2007

No knowledge without errors that can and will be corrected, as part of a process of learning. Wimsatt turns the fallibility of the procedures into an epistemic advantage.

Another of his key concepts is robustness: "Things are robust if they are accessible (detectable, measurable, derivable, definable, producible, or the like) in a variety of independent ways". Objects in this world have many dimensions, can be approached, observed and measured by different senses and means. Together all these yield knowledge of that object and the more perspectives can be reconciled, the more robust the final result. Trustworthy knowledge is the accumulative result of many efforts. The concept of robustness features not coincidentally also prominently in ecological discussions. In ecology, robustness is used to describe the features of a system that can resist disturbance. Conceptually it could be used as a synonym for stability.[362] An ecosystem (or any living organism) needs robustness because otherwise if would fail to survive in a challenging and changing world. Knowledge is a complex system with inbuilt robustness, because instable concepts or objects are useless ingredients for making knowledge. A bit of knowledge is thrustworthy if you can rely on it, if you can trust it is still valid tomorrow and the day after, and not just for you but also for your wife, kids and neighbours. If knowledge would be volatile, we would not call it trustworthy. A poem by Mandelstam (*Ode to Stalin*) or a painting by Picasso (the *Guernica*) contain knowledge on terror and war, but do not count in the category of trustworthy knowledge. At best they are convincing knowledge, or moving, and probably not convincing or moving for anybody anytime. Knowledge can take many forms (a shopping list, a horoscope, a film by Herzog or a nursery rhyme, the genome database, a text in Linear A, a railway timetable, a philosophy lecture or software manual) and not all are trustworthy. A railway timetable, in whatever country one may choose, is not really to be trusted. Some would say I stretch the concept of knowledge too far, as they see knowledge as trustworthy by definition. Speaking about untrustworthy knowledge would be opening the door to relativism. I don't think that has to be the case. As a biological realist I see knowledge as all things organisms know about the world. So, even animals or plants have knowledge. The trustworthiness of some knowledge can depend on time and place, as it comes in degrees. What was trustworthy in the Middle Ages may not be trustworthy anymore now. Yesterday's version of Wikipedia may be more or less trustworthy as today's. But we can understand why something is trustworthy in specific circumstances. This is not opening the door to relativism, on the contrary. It offers criteria to judge knowledge on its trustworthiness.

This speaks against the relativity of knowledge or the insurmountable multiperpectivalism leading to inescapable subjectivism as advocated by the constructionists or the sociologists of knowledge, in which truth or objectivity

362 Jen 2005

have ceased to exist. Wimsatt precisely argues that out of the interaction among different (multidisciplinary) perspectives robust knowledge grows. Sociologists such as Latour & Woolgar[363] or Barnes, Bloor & Henry[364] go the other way and reduce all knowledge to nothing but a construction by the observer(s), determined by social forces, as expressions of economic, political or cultural self-interest. Their claims have been critically treated[365] elsewhere, but one thing can and should be said here: they cannot explain the complexity of the systems they claim to explain, provided they have ever attempted to do so in the first place.

The more an object of concept is 'triangulated' (a term Wimsatt borrowed from Donald T. Campbell), determined in multiple ways, the robuster it becomes. A table can be detected in different sensory modalities such as vision, touch, smell or hearing, which roughly coincide and are consistent through time, making the table a robust element in our world. But, speaking in panarchical terms (something Wimsatt is not doing), this can lead to rigidity, the incapacity to react towards some unexpected change. Life on earth is so interconnected that a change in some variables will be disastrous for almost all organisms. Turn off the light of the sun, and the whole biological machinery will grind to a halt, apart from some specialised bacteria that can do without light or oxygen. Sometimes a concept in a theory may come under so much pressure from anomalous observations that it will collapse and disappear, its robustness preventing it from adaptation. On the other hand, an object observed only in one modality or by one person at one moment, can't gain a lot of credibility or robustness. Just like that one UFO-report, in bad weather during the night with no other witnesses.

We humans are an adaptive species that developed heuristics to be able to survive in this complex world. Our knowledge of this world, and the tool of science that we developed to built that knowledge, are intrinsically complex systems. Change in such systems is patchy, clumsy and hardly ever fundamental. One does not break down a theory to build a new one from scratch, just as organisms have never been rebuilt from scratch in the course of the evolution. "This is as true for theories or for any complex functional structures - biological, mechanical, conceptual, or normative - as it is for houses" says Wimsatt.[366] His take on the world as a multi-level complex whole leads him to distinguish two very different enterprises that are often conflated in philosophical discussions of reduction. On the one hand, there are intra-level cases, in which one theory replaces its predecessor. On the other hand, there are inter-level cases, in which theories at different but adjacent levels are stitched together. These two types engender very different dynamics. In the case of succession, theories may be

363 Latour 1979
364 Barnes 1996
365 Boghossian 2006
366 Wimsatt 2007 p 137

more or less similar, and replacement occurs when there is less rather than more similarity. Elimination is an option, and given that such similarity mappings are intransitive, it is probably inevitable. The inter-level case is more central to articulating the implications of a hierarchy of organization. Crucially, instead of replacement, what we see is co-evolution of theories and the development of models that span more than one level of organization. The aim is in the end to enhance explanation. Sometimes we can explain what we see by appealing to higher levels of organization; sometimes we need to look lower.

> No such beast as eliminative reduction is to be found anywhere in the history of science, and there is no reason, in terms of scientific functions served, to expect it in the future. It, and its aims, are largely misconceived philosophical inventions. Robust higher-level entities, relations do not disappear wholesale in lower-level scientific revolutions - our conceptions of them transmutate and add new dimensions in interesting ways, yes, but disappear, no.[367]

Curiously, Wimsatt uses here the same concept of transmutation as IngenHousz was using to explain the mechanism behind the vegetable production of dephlogisticated air. On top of that he gives the phlogiston/oxygen dispute as a case in point where a concept disappeared in a succession of theories on the same level. Both oxygen and phlogiston were an extract or species of air. There was no intrinsically different take on things in both approaches. That would only become possible when Dalton's atomic theory came along and things could be resolved on another level. Which brings us back to the end of the eighteenth century.

The case of IngenHousz

This abbreviated but not-so-random walk through the philosophy of science of the last couple of decades converges in the ecological perspective that I would like to apply here to the case of IngenHousz. Many philosophers seeked to develop new perspectives because they felt the classic one did not deliver what they had wished. Feyerabend was right when he said "Anything goes!" to stimulate as many alternative approaches as possible. Indeed anything goes, but not everywhere and not all of the time. Here again, we see how so many approaches and explanations in the philosophy of science have been partial truths. Once proposed, they stimulate fruitful inquiry. Consequently, they have been exposed as partial, and expansion as well as extension proceeded. Partial affection for theory by those who form them and the psychology of the adherents make those approaches contentious. Critics become extreme, caricatures are erected. Looking for integration of all these perspectives is the purpose of this ecological approach.

---

367 Wimsatt 2007 p 168

In the ecological approach they all get a niche in the philosophical eco-system. They were all worked out, seperately: individual creativity, rationality and emotionality, conceptual change, group mechanisms and network effects, theory and experiment as protocols to a world view, society at large as both limiting factor and culture medium. All of them highlighted separate entries to the same labyrinth. In the 'ecosystem of knowledge' (figure 4) one can see the different levels at which various approaches in the philosophy of science have been active. The individual scientist has always been the subject of the biographies of scientific heroes or villains, from the forgetful genius to the dangerous madman.

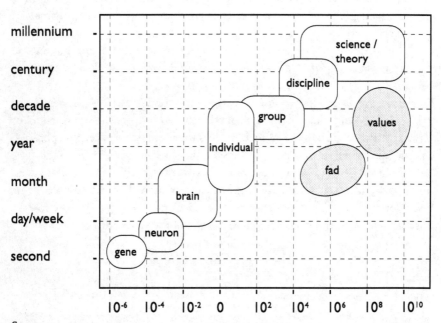

figure 4

The ecosystem of science seen in a diagram similar to the one used earlier to chart a biological ecosystem. This 'hierarchy of knowledge' steps up with a factor of 10^2, beginning with the chromosome, over neuron and brain to the individual, which is a member of groups, that function within scientific disciplines, interacting within greater units (biochemists are affiliated with both chemistry and biology), to fit under the global umbrella of scientific knowledge, where all theories fit together. The lower the level, the higher the turnover rate. It takes less time for an individual to change his or her mind than to get a new theory accepted (as e.g. plate tectonics).

A major difference between change in organic systems and cultural systems such as science, is that cultural change is potentially much faster. Any good idea can spread within one generation, as it does not have to be transmitted to a next generation and filtered through natural selection.

On the lower levels one can find the cognitive and psychological approaches, concentrating on what happens in the brains of individual scientists: changing models, concepts or ideas, creativity, curiosity and stubbornness.

Neurologists and geneticists have already been pointing to the even lower levels of neuron and gene where undoubtedly some of the secrets of human knowledge can be found. How much nature and nurture is there in a scientific mind? Higher up in the hierarchy comes the group, where the sociologists have been active. Their approach extends into a higher level, where groups form scientific disciplines. Across the frontiers of the various disciplines all scientists share a similar approach to gathering knowledge. At this level, science was the subject of the philosophers who concentrated on theories and their logic, on general principles such as falsifiability or compatibility, rationality and objectivity, induction or deduction, the epistemic virtues that make scientific knowledge scientific knowledge.

The 'panarchic view', with actors and factors at various levels and scales influencing each other in various ways, may help to knit all threads together in one messy fabric. Messy because complex systems are far from smooth or ordered. Chaos grows from order and this can always fall apart again. On every level in the hierarchy of knowledge changes take place in a flow of events that can be compared to the one described earlier in ecological systems (figure 5). Three differences between ecological and cultural systems should be reckoned with.[368] Humans have foresight, which ecosystems lack. We look to the future. We let our actions be influenced by expectations (which can be accurate or widely off the mark). This hasn't helped prevent the collapse of several human societies in the past.[369] Another difference is communication. Organisms transfer, test and store experience in a changing world through their genes. Ecosystems do this through forming self-organised patterns. Human systems are also self-organising patterns, but with the unique additional power to communicate and consolidate ideas and experience. These can be incorporated in stories, myths, lab protocols, laws, constitutions or theories. Global interconnectness is speeding up these processes, with the effects of a double-edged sword: reinforcing myths to aggravating oppositions, as well as creating platforms that help resolve crises. Another difference is technology. Tools have been shaping human history in fundamental ways, from the wheel to the computer. Over the ages, we have been travelling roughly one hour a day between home and work, but the distance travelled in that time has increased sharply. This has changed the sight of our landscapes, the way we live and the economy at large. Technology influences what we know (the scanners do let us understand our brains as never before) and how we live (think of mobile phones, containers of microwave ovens).

---

368 Gunderson 202
369 Diamond 2005

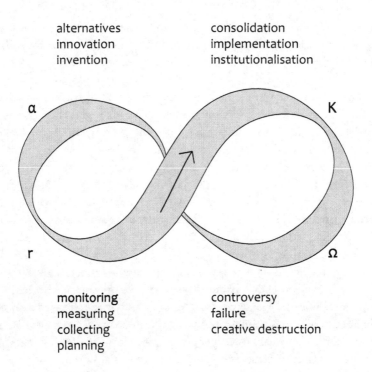

alternatives
innovation
invention

consolidation
implementation
institutionalisation

α

K

r

Ω

monitoring
measuring
collecting
planning

controversy
failure
creative destruction

figure 5

A generalised adaptive cycle applied to changing knowlegde.
In what Kuhn would have called "normal science" knowledge is slowly amassed and built into a generally accepted theory, going from phase r to phase K. Under the influence of new and inexplicable data or persisting anomalies the theory becomes brittle and can collapse. But not everything disintegrates. Instruments and technology remain; basic concepts can survive while others disappear; money, skills and contacts reconfigure; new input from other fields of knowledge may help to integrate into a new whole. This cycle leads from Ω to α, ready to start a new cycle. This process of natural re-cycling may explain the evolutionary aspect of changing knowledge, in contrast to the revolutionary science of Kuhn that fell short of understanding the continuity in the process.
However, human systems seem to have, compared to ecological systems, a greater power for both rigidity and novelty. The conservative abilities of governement bureaucracy are an illuminating example of the former, whilst the capacity to ban all ozon-destroying chemicals from industrial processes is a nice example of the latter. The multi-layered complexity of science has implications for science policy too. Which is the most appropriate scale to interfere with science in order to manage it? Much will depend on what one tries to accomplish. What are the right measures at the right level to manage innovation and avoid rigidity?

As a human activity, science shows all these properties too. It is a human cultural system developed to gather trustworthy knowledge. The story of Jan IngenHousz and his role in the discovery of photosynthesis demonstrates this complex game of science. The fate of IngenHousz and of photosynthesis research can only be understood if one takes into account the individuals involved (Priestley, Franklin, Senebier, Lavoisier, Shelburne, Pringle and many others), as well as their social, cultural and historical context, (the Imperial Court in Vienna, the Royal Society in London, the Enlightenment, before and after the American Independence and French Revolution &c.). And all of these are interwoven in social interactions and interconnected by the theoretical, instrumental and practical requirements (phlogiston theory, eudiometry, the new chemical system, medicine and vaccination) of their search for trustworthy knowledge about some aspect of reality (what plants do to air).

Change is inimical to life, and therefore also to knowledge. This change takes place on all scales on which scientific knowledge develops. The scale of the individual is on average 1.75 meters and 75 years. The scale of the group transcends that of the individual and changes slowly over generations. The scale of theory extends beyond that of the group, as one theory overspans several scientific groups or disciplines and extends in time over the dimension of the group. Society at large is the biggest scale, encapsulating the three others, changing most slowly and gradually spanning the whole globe and even beyond.

The individuals

Every individual is different. We know that about humans, but it is just as true about animals. Not every squirrel is the same. Animals have personalities, just like people. Pet-owners will immediately agree, but in the wild it has been proven to exist too. Garamszegi and colleagues measured the varying intensity of courtship behaviours by male collared flycatchers.[370] Half of the birds seemed to be afraid of entering the nest box (which they knew from a base-line test) when a piece of white paper was attached to it. The other half continued displaying their courtship procedures as before. In a next step, the males were confronted with a male competitor. The males who were undeterred by the paper vigorously attacked the newcomer, whereas those intimidated by the paper were reluctant to attack at all. In a last test the researchers placed traps inside the nest boxes. The aggressive, risk-taking birds were twice as likely to be trapped than the non-risk-takers. Apart from the implications this may have for all studies based on trapped animals and the idea that these form a representative sample of the population, it proves how animals have individuality. This must be the result

---

370 Garamszegi 2009

of evolutionary pressures. Some individuals are more explorative than others. Some are inventive, others are shy, still others lazy or easily scared. This is very natural. The conservative and progressive forces in a group of animals keep each other in check and stabilise the group while leaving open the possibility for change. The same is true for people. And people performing science have their own individual peculiarities, which can make them into a successful scientist. Success in this case should not be measured by the Nobel-standard, but in the output they produce.

Some scientist love to crunch the numbers or do the repetitive work in laboratory or field station - and are good at it, as it is a thing not everybody likes nor is able to do properly. Others love the bold ideas and the grand theories. Both ends of the spectrum and everything in between are necessary for the progress of science. Rosalind Franklin was at one end of the spectrum. She had been doing the hard work with roentgen crystallography to try and determine the structure of the DNA-molecule. Over and over again she did her tests, looking forever more confirmation. All that radiation would cost her her life before she could receive her part of the Nobel prize which went to the two guys whose names everybody knows. Watson and Crick took the big intellectual leap and came up with the breakthrough hypothesis.[371] People come with individual differences and so do scientists. How long does a scientist continue reproducing his or her experiment? When does he or she decide that the hypothesis is tested thoroughly enough? This decision will be partly determined by what one has been taught, in the discipline of one's discipline, but also by one's innate (un)certainty. One will finish off early, another keeps on doubting. The former one will be publishing first, the latter might make less mistakes. Personal style is shaped by education, background, work situation, &c.

What about IngenHousz? He was a man of a good family with a good education. He was first and foremost a physician. As a medical doctor he was trained in being pragmatic. When you try to cure people, you either succeed or you don't, and there is no time for fine theoretical contemplation. He was very much on his own. Although he was married, he and his wife were more frequently separated than together, because of his travels. The couple stayed childless, so no heirs to boast about the deeds of their father. He was sociable, though did not like polemics. He preferred to get on with work rather than discuss with competitors such as Senebier, an attitude he shared with Franklin, who supported him in keeping up that attitude. He was no society man, not interested in pomp and circumstance, although he was a well-known member of the Royal Society. He was independent (materially and psychologically). He was very critical of what we would now call pseudo-science, like mesmerism and other manifestations of quackery. Still, he was an aristocrat, a royalist and religious conformist. Photosynthesis was only one of his passions: his attention was divided across chemistry and electricity (conductivity of metals, electrical machines, lighting rods...), medicine and public health (eudiometry, asthma therapy, inoculation...), physics (Brownian motion and hot air balloons...) and agriculture. Very typical of the Enlightenment, but too much for one person to excel and triumph in.

To take just two of his closest rivals, Priestley and Senebier were different characters. Both were clergymen, the former protestant, the latter catholic. Priestley was married, had a large family and was rather radical in his political and religious standpoints. He was very rhetorical, an outspoken public person, drawing attention with and on everything he did, describing himself as a "meteor". He was a polymath while he had no formal scientific training or education, going some way in explaining why his experimental style was hardly systematic, but haphazard and qualitative. (Remember he described how he hit upon his discoveries by accident.) No wonder he had problems with reproducing his results. Senebier was a man of letters, a librarian. He was verbose and literary. His discussion with IngenHousz on matters of photosynthesis was tinged by theological arguments rather than by hard reproducible data. Three characters, three individuals, interacting in unpredictable ways. But all three were reasoning, thinking logically, and rationally, which was why they were able to communicate and discuss in the first place.

It does not come as a surprise to see how the individual characteristics of a researcher play a role in his or her attitude towards new ideas. Individual styles influence how one reacts towards external or internal input. Priestley would stick to the phlogiston theory until his death, and some have argued that this fitted well with his personality. IngenHousz was a very different kind of personality and as a medical practitioner, forced by the inherent obligations of his profession, to deliver results. Priestley was an old fashioned philosopher of nature, who could indulge in various theoretical reflections, not pressured by the need for reality check. Senebier went with the new chemical theory too, but did not accept the theory that plants respire at night.

The population

These are just three of the main protagonists in those early days of photo-synthesis research. In his publications, IngenHousz was referring to many more peers such as Franklin, Herschel, Lavoisier, Spallanzani, Bonnet or Fontana, openly indicating what he was learning from whom. He was often and honestly giving full credits to his great masters & models Priestley and Fontana. Critiques from Priestley, Senebier or Cavallo were the inspiration to replicate old experiments or devise new ones. He was corresponding intensely with Van Breda, Molitor and Franklin, amongst many others, although not all these letters have been conserved or found. The richest collection left behind is the one with his colleague and friend Van Breda, who was taking care of the Dutch translations of IngenHousz' works. Another curious item left in the archives is a little piece of paper with a list of names destined as mailing list of people to which "copies to give". It is divided in three sections "germany and italy", "holland" and "england". The first lists 32 people and opens with the Emperor, further containing high placed officials at the court, Archduke Ferdinand, the Grand Duke of Tuscany, colleagues and friends such as Landriani, Scherer, Molitor, Jacquin and Van Swieten. He wants the following people in Holland to get a copy: his brother, Van Breda, Van Marum, Dyman and Deckers. The England-list is an abbreviated who's who of eighteenth century science: Beddoes, Rumford, Dimsdale, Gibbs, Hunter, Nichols, Darwin and several people in the Philosophical Society in Bath and in Pensylvania, hand in hand with the political high society: the King, Lord Stanhope, Lord Lansdowne, Marquis Circello (ambassador of Naples), the Archbishop of York and last but not least his amanuensis in Calne Miss Fox.[372]

In such a group of ambitious and curious men and a few women, cooperation and competition is a natural thing. IngenHousz was humble and reserved, and was not inclined to polemicise on who was right or who was first, as he demonstrated in many of his letters. He complains about the attitude of some of his contemporaries who claimed to be the first to discover the beneficial processes of plants (such as Priestley, Senebier and Van Barneveld). In the case of Priestley this has been exquisitely demonstrated by Howard Gest. He shows how Priestley admitted in private that IngenHousz indeed had been the first to describe the beneficial power of plants. He never did so in public. In the meantime, he repeatedly claimed in public to have observed and published before IngenHousz and he kept on repeating this till 1800. Never did he give an accurate reference to IngenHousz' work, never did IngenHousz' name appear in the index of Priestley's works. IngenHousz himself, on the contrary, systematically referred to Priestley, with much respect. His personal opinion on Priestley did evolve over the years. The respect for his scientific accomplishments continued, but his appreciation of Priestley's political and religious viewpoints turned sceptical. A thorough study of the correspondence between IngenHousz, Franklin, Lavoisier, Priestley and others might lead to more material by which to decide on IngenHousz' political and philosophical standpoints. As can be seen in the letters studied here, IngenHousz refrained from making much public fuzz about the attitude of his rivalling colleagues, especially in the priority dispute. But they continued as they did, obfuscating IngenHousz' rightful reputation as the discoverer of photosynthesis in the eyes of the historians and the general public.

---

372 Breda Collectie IngenHousz A11

The population of natural philosophers who were IngenHousz' peers can be split in several groups that were mutually inspiring. Chemistry was a discipline that produced knowledge, which was very useful to beginning plant physiologists who could use it to explain the phenomena they were studying. Electricity did have touch points with chemistry, such as in the application of an electric spark in the advanced version of the eudiometer. This eudiometer was an instrument that played a role in forming different groups of researchers: aerial medicine, pneumatic science, photosynthesis researchers and the new chemistry. In the mean time its measurements were influenced by the practices of the astronomers who inspired other disciplines with their mathematical rigour. Medical doctors, such as IngenHousz, used whatever knowledge they could get in whatever practical application they could think of. Pneumatic medicine, as conceived by IngenHousz and put into practice (and business) by people such as Beddoes, is a prime example. A discovery in one discipline could bootstrap into another, delivering the right input to cause fast progress. Some individuals, such as IngenHousz, linked different groups. Some were geographical go-betweens, such as Magellan who linked English and French chemistry. Looking at the close relationship between IngenHousz and the Herschels, it sounds plausible that the precision in making the astronomical lenses inspired IngenHousz in obtaining a similar accuracy in doing his experiments.

The interactions

The link between all the people mentioned above was their common subject of study, the chemical processes and exchanges between plant and atmosphere. IngenHousz operates at the emerging frontline of modern science, which is born in the cradle of experimental philosophy. IngenHousz was profoundly influenced by epistemological virtues such as clarity, reproducibility and quantifiability. He was calculating error margins on his experimental results, a mathematical method he might have picked up from the astronomers-mathematicians such as Herschel. From the seventeenth century mathematics and astronomy had been intimately linked, while natural philosophy and mathematics were quite seperate areas of endeavour. IngenHousz left the era of haphazard experimentation behind and started consciously formulating hypotheses that were to be confirmed through rigid, controlled and systematic experimentation. Information about all this was openly communicated, with the explicit aim for methods and results to be critically analysed and if necessary refuted or replicated by others.

As mentioned before, IngenHousz was straddled across a classic paradigm shift. In less than two decennia he changed from being a follower of Priestley and the phlogiston theory, to an adept of the new chemical system of Lavoisier.

This makes him a fine exemplar of a "scientific evolutionary", finely rendering redundant any questions of incommensurability. Guided by his own reasoning power, he crossed the grey translation zone between two massively important theoretical systems. Some got lost in the translation, others easily navigated the new conceptual landscape. It is a pity that Freddy Verbruggen when he wrote his doctoral dissertation in 1973 did not know about IngenHousz. With the help of the concept of cognitive dissonance he tried to understand the resistance in individual scientists against a paradigm shift from phlogiston to oxygen.[373] Theories change or evolve, while individuals have to take position on their personal scale of living, in their minds. IngenHousz, it should be remembered, was no professional chemist. That might be one of the reasons why he did not have much to lose professionally when changing his standpoint. Whatever the reasons, all these interactions (partly driven of course by individuals' psychological and social set-ups and hang-ups) make up the theme that was unifying all concerned. IngenHousz himself was not only linked up with the plant physiologists, but also with the medical doctors, with the electricians and other groups of investigators. And because of his travelling, he did get into contact easily with peers all over Europe. More than ever before, science was becoming a transnational endeavour. IngenHousz was not the only one travelling, although he was probably the only one who hardly ever was at home. Priestley travelled to Paris, Fontana to London. Hassenfratz went to Vienna, Van Marum visited Paris and later London, Landriani went to Haarlem and Vienna, to name just a few. And when they were not travelling, they were writing. The world of science was a real Republic of Letters. Photosynthesis research was very much like Abbri[374] described the developments in chemistry: as research that transcended national boundaries. The developments in the embryonic sciences were multi-national and multi-cultural. Trustworthy knowledge is not bound by language, culture or nationality. To the contrary, it is fed and enriched by different perspectives.

The theoretical evolutions in the period described evolved at different speeds in different layers. A theory is not one monolithic bloc that comes as a total package to take or to leave. The science of airs developed so quickly in the 1770s that it became one major focus of chemical investigation and the ferment of reorganisation of the entire chemical theory in the 1780s.[375] Theoretical developments in one field influence those in another, even when they are at different hierarchical levels. Pneumatic science would now be called a part of chemistry. Similarly, the theory about photosynthesis evolved, but at a different speed than the theory on phlogiston. The new insights and dito nomenclature from chemistry cycled down to the level of plant physiology. The use of the eudiometer took place on another level, with a logic of its own, still influencing

---

373 Verbruggen 1973
374 Abbri 1989
375 Bensaude-Vincent 2000

the two other theoretical lines. In another niche of IngenHousz' occuptions, the practice of inoculation evolved into that of vaccination after a slow merger of an old folk wisdom and a empirically tested method. Within the realm of scientific knowledge, various entities exist, interlinked but also partially independent. Sometimes they need one another to make progress, sometimes they obstruct one another. Sometimes one fertilizes the other, sometimes one drives the other up a dead end street.

The biotope

The fact that all this happened in the eighteenth century could be considered very typical for the era. As mentioned before, science was a source of public wonder and debate. Curiosity was the new currency. Many of IngenHousz' activities were linked to immediate practical issues of private and public health, safety and agricultural improvements. The latter was no luxury problem as the population in Western Europe was growing very fast. The number of people living in the most developed parts of Western Europe doubled. In twenty years time many thousands of families would migrate because the Malthusian pressure became too high. Increasing the yields of the fields was an important issue.

Factors external to the development of scientific knowledge played a role too. The discussion between IngenHousz and Senebier crystallised around the fact that Senebier could not accept that God-made creatures, such as plants, would be able to produce something noxious for men. The philosophers of the Enlightenment were active on the borderline between ratio and religion. For some it were two extreme positions that were mutually exclusive,[376] for others the two were both essential and compatible perspectives.[377] IngenHousz was clearly a person who was able to reconcile his belief in God with the rationality necessary for his research. That allowed him to pronounce rather radical critique on current practices and beliefs. He was an advocate of new medical methods such as variolation and proponent of technical innovations such as lighting rods, both of which encountered fierce opposition from religion. It shows how the way IngenHousz lived and worked, communicated and travelled, was very much rooted in the Europe of the late eighteenth century. Speaking and writing in four languages helped him survive in this complex era on a complex continent. He was also lucky to have become Imperial Physician with a livelong stipend. That made him free and independent, geographically as well as spiritually. His financial status allowed him to print and reprint the drawings of his experimental settings and instruments. These very expensive engravings were no limiting factor for a person of his means.

376 Israels 2001
377 Sorkin 2008

However, the French Revolution would force him to flee Paris on 14 July when the people stormed the Bastille. IngenHousz' relationship with the French queen of Austrian descent, travelling under the Austrian flag, forced him to flee. He could only just in time leave the continent that would go under in turmoil for many years, obstructing his postal communication with his family, friends and peers on the continent. The historical descriptions of the French Revolution have described it as that cataclismic event on Bastille Day, others have embedded it in the long slow turn-over in the course of one century that fits in with the other developments in the world. In ecological terms, both are true. Slow and gradually evolving systems are influenced by short bursts of change on a lower and faster level. The accumulation of poverty, unrest and opposition violently disrupted a system that had lost its robustness and collapsed into a new configuration. Science however could keep going, although some of its front men either lost their head or their fatherland, such as Lavoisier and Priestley, respectively. As Lissa Roberts, unknowingly about a possible ecological perspective, described it: "While changes in attitude, policies, and practice in the macro-sphere resonated down to the level of chemistry, changes in chemistry circulated back to inform the larger context in which it was situated. Each realm exerted a constructive impact on the other."[378]

The philosophy of living science

Trying to chart this complex story of just a few decennia in the discovery of photosynthesis, led me, as a biologist, to the scientific discipline specialised in living complexity. Ecology is the study of the complexity of living systems in their interaction with the rest of the world. That sounds like a rather good description of science: a complex but systematic manner in which human beings interact with the world in which they live, in order to try to obtain some reliable, trustworthy knowledge about it.

Philosophy, sociology and history of science each have been trying to understand this particular human behaviour. As some philosophers of science have been indicating, the traditional ways of doing this were flawed, by being too limited or fragmented, repeating the mistake of the six blind elephantologists. Frederic Holmes said something about Lavoisier that is just as applicable to IngenHousz: "The overall result of Lavoisier's activity was a true revolution but not one that can be reduced to one or two events or features lifted from his scientific odyssey."[379] Or like Stephen Jay Gould wrote about Darwin:

---

378 Roberts 1995
379 Holmes 2000

> *If the sum of a person's achievement must be sought in a subtle combination of differing attributes, each affected in marvelously varying ways by complexities of external circumstances and the interplay of psyche and society, then no account of particular accomplishment can be drawn simply from prediction based on inherited mental rank. Achieved brilliance must be (1) a happy combination of fortunate strength in several independent attributes joined with (2) an equally fortuitous combination of external circumstances (personal, familial, societal, and historical) so that (3) such a unique mental convergence can solve a major puzzle about the construction of natural reality. Explanations of this kind can only be achieved through dense narrative. No shortcuts exist; the answer lies in a particular concatenation of details—and these must be elucidated and integrated descriptively.*[380]

A "dense narrative" is the best thing Gould could offer to picture the complexity of scientific progress. But how to write that narrative? As has been illustrated here, an ecological approach may offer a workable toolbox to come to grips with this dense, multifaceted reality of the scientific enterprise. To understand the complex reality of science at work, we will need more than just philosophy or sociology. Maybe ecology may bridge the gap between the two. We will have to integrate methodology, epistemology, history, sociology, logic and psychology to grasp what happens when people try to understand the world they live in, in all its vibrating complexity. It is like a multidimensional puzzle: all elements fit together, but they are all different. Still you need them all, to be able to finish the puzzle.

The advantage of the ecological perspective is that it offers a framework that can capture the dynamics of the multilevel phenomenon of science. And especially in showing how the various variables interact within and in between different levels. It allows to integrate the many approaches in the philosophy and history of science, that have stood apart for so long and that have even sometimes denied each other's right of existence. In the ecological approach they are all needed, because that is the only way to be able to understand the complexity of science. This complexity can't be 'reduced' to a sociological, psychological or epistemological phenomenon. At the same time, this ecological framework is more than a simple addition of some Popper & Kuhn, a little Lakatos, plus Latour, Hull, Nickles and some Haack or Nersessian thrown in. The ecological framework offers a means to integrate these various viewpoints in showing how they interact.

380 Gould 1996

This has been a first exploration of the possibilities of this approach. It deserves more work in order to proof the value of this ecological model. What questions that could be useful to philosophers of science are asked by ecologists? My guess is that on the four levels of the system of science a lot of work has been done and a lot of knowledge is already there. But that progress could be made at the points in time and place where these levels interact, when concepts change in the minds of individuals, circulate in groups through these individuals influencing one another, in such a way that new hypotheses arise, new experiments are designed and theories change, collapse or transmutate. And at the same time, these developments can be coolly allowed, warmly welcomed, harshly resisted or absolutely forbidden by the society in which all this is embedded. Sailing between the cliffs of change and stability, our knowledge of the world slowly but surely evolves towards more reliable and more trustworthy knowledge.

rewind

A case of complexity: ecological questions for the philosophy & history of science

Each individual human being on this earth needs trustworthy knowledge about what's going on around her or him in order to survive. Some of it we can learn ourselves, by trial and error. Easier it is to learn from and with others. The best way of amassing this knowledge is by doing this in cooperation with others, having learnt that the individual cognitive possibilities are limited and often biased. Building and collecting this knowledge has taken on a particular institutionalised form: science, a wonderfully effective way of getting to know why things are as they are and how things work. Hypotheses, corroborated by empirical facts, checked and double-checked against reality (and not against the intuition or the opinion of a prophet or guru - as Raymond Tallis so succinctly explains[381]) within a framework of instruments, experiments and peer-reviewed communication, coagulate into theories. And all this happens in constant interaction with a dynamically evolving environment of society and culture.

Jan IngenHousz was such an individual. His life and his contribution to the discovery of photosynthesis are a prime candidate for an ecological approach to science's philosophy, history and sociology.

---

381 Tallis 1995

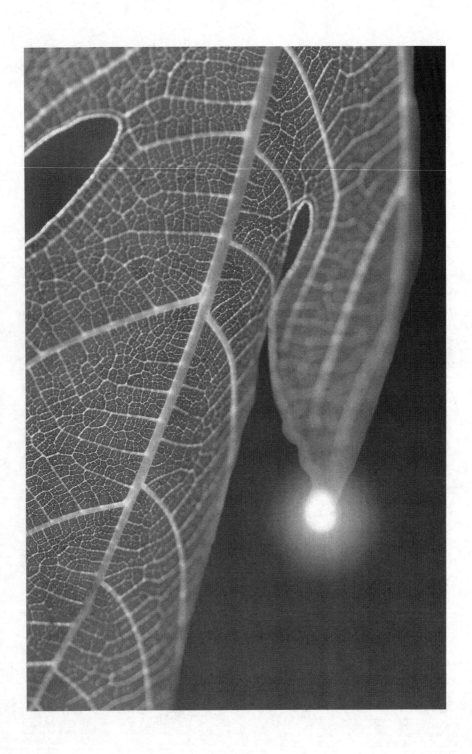

Photosynthesis,
a bright future for the light of the sun

When IngenHousz set up his experiments with photosynthetic plants in the summer of 1779, he took a first glimpse at a natural process that was already millions of years old. This process is intimately linked with the story of life on earth. Life does depend ultimately on oxygen. But there has been a time when the atmosphere did not contain any or hardly any oxygen. Four billion years ago the air on earth probably contained oxygen at one part in a million. Today it contains 208,500 parts per million. The atmospheric concentration of oxygen of slightly less than 21% is the result of one of the most dramatic cases of on-going pollution the earth has ever seen. A first crisis froze the world into one gigantic snowball, when the first photosynthetic bacteria sucked all carbon dioxide from the atmosphere, taking all greenhouse effects away and plunging the planet in a deep freeze.[382] Around 700 million years ago the first multicellular plants announced a second snowball period. Later in the Devonian period (410 to 360 million years ago) the first land plants not only consumed massive amounts of carbon dioxide but also started weathering the rock, which happened to be another carbon dioxide sink. Good for another 50 million year ice age. Every time vast feedback mechanisms shifted the global thermometers back to warm. But it shows how photosynthesis had and still has profound influences on life on earth. When oxygen levels started rising for the first time, this started an environmental crisis because all organisms living in these prehistoric times were bacteria of all sorts thriving in the oxygen-poor atmosphere. Oxygen was a toxic new element and acted as the first weapon of mass destruction. All organisms had to try and cope with this massive pollution by the waste product of the pioneer photosynthetic organisms. Those who survived are our ancestors. One of their descendants was Douglas Adams who wrote, thanks to the oxygen in the air, the following opening sentences in his infamous *The hitchhiker's guide to the galaxy*:

> *Far out in the uncharted backwaters of the unfashionable end of the Western Spiral Arm of the Galaxy lies a small unregarded yellow sun.*

---

382 Ward 2009

Orbiting this at a distance of roughly ninety-two million miles is an utterly insignificant little blue-green planet whose ape-descended life forms are so amazingly primitive that they still think digital watches are a pretty neat idea.[383]

It would be hard to summarize the biological characteristics of our planet in fewer colours. The yellow of the energy of the sun, the blue of the water of the oceans and the green of chlorophyll, the biochemical motor that keeps everything running. The latter is of crucial importance, eminently described by Nick Lane.[384] "The world is so dominated by the machinery of photosynthesis that it is easy to miss the wood from the trees." Without oxygen we would be nowhere and this fact is so all pervading that it is easily overlooked. Just like the invisible but essential presence of oxygen in the air was overlooked for ages.

With the help of light

Photosynthesis literally means "synthesis with the help of light". In oxygenic photosynthesis (other forms exist that do not use oxygen, but I will leave that for now) the synthesis is made of hydrogen and carbon that are built into chains of sugars. It is exactly these sugars that all the other herbivores, carnivores and omnivores depend on. The hydrogen in the sugar comes from water, the carbon from the carbon dioxide. It took until 1941 before Ruben and Kamen, building on the work of Hill, could figure out that the oxygen, which is produced by the plants is a by-product, comes from the water.[385] Therefore, in chemical shorthand:

$$H_2O + CO_2 \quad \rightarrow \quad \text{sunlight \& chlorophyll} \rightarrow (CH_2O) + H_2O + O_2$$

Water plus carbon dioxide
with sunlight and green leaves
gives sugars and water and oxygen.

The essential tool plants use to forge this reaction is chlorophyll, heaped up in the cells of the plants colouring them green. (Plants look green because they absorb the other wavelengths and reflect the green. The green is a side effect, just like the oxygen.) Splitting molecules into atoms and protons, which plants have to do to reassemble water and carbon dioxide, requires a lot of energy. That energy is captured from the light of the sun. The system by which plants do this is very complicated. It takes place in two large assemblies of proteins,

383 Adams 1979
384 Lane 2002
385 Rabinovitch 1979

xlviij *AVANT-PROPOS.*
royale de Londres, le 4 juin 1778, & publié
dans les Tranf. Philof. vol. LXXII, pag. 426.
 Voici à préfent un petit précis (pour autant
qu'il regarde la végétation) du nouveau fyf-
teme de ceux qui rejettent la doctrine du phlo-
giftique.
 Ceux-ci difent que l'eau eft décompofée par
les forces de la nature dans les végétaux, fur-
tout par l'influence de la lumière folaire. La
bafe de l'air inflammable ou l'hydrogène con-
tenu dans l'eau, fe combine avec la fubftance
charboneufe pour former de l'huile, tandis que
la bafe de l'air vital ou déphlogiftiqué ou l'oxygène, autre principe de l'eau, s'unit également
avec la fubftance charboneufe, & forme de l'air
fixe ou acide carbonique, qui entre dans la
compofition des acides végétaux. Une partie de
cet oxygène s'uniffant au *calorique*, eft chaffée
au dehors & fort par les feuilles dans l'état d'air
vital, fur-tout lorfqu'elles font expofées au
foleil. Ainfi, dans ce fyftème, l'eau & la fubftance charboneufe font prefque les feuls principes de la végétation.
 M. de la *Metherie* expliquè d'une manière bien
différente les phénomènes de la végétation.
Suivant lui, l'air qu'on trouve dans les plantes
y eft apporté, foit par les trachées qui l'ab-
forbent, foit par l'eau qui en contient toujours
une quantité plus ou moins confidérable. Cet
air, qui eft celui de l'atmofphère, eft compofé
d'air pur & d'air impur ou phlogiftiqué : il eft
enfuite modifié par le travail de la Nature. L'air
phlogiftiqué eft, en grande partie, décompofé,
& on en retrouve peu dans la plante. Une
portion de ces airs eft changée en air inflammable,

In the introduction to his 1789 book IngenHousz describes in one paragraph the essence of photosynthesis: water is decomposed in plants with the help of sunlight. The hydrogen from water is recombined with carbon to form plant substances. Oxygen, the other principle of water, is expelled, combined with "calorique", the mysterious new parameter in the new chemistry. So, water and carbon are the sole principles of vegetation.

metal ions and green pigments called chlorophylls. These sit huddled together inside the chloroplasts, the photosynthetic factories that lay like rice kernels in the leaves of plants. These were the photosynthetic production units that Senebier collected in his vessels after grinding the leaves. When a photon of light strikes the chlorophyll, it is channelled into a specialised molecule called P680. This releases a high-energy electron, which travels a circuitous route before being used to turn carbon dioxide into sugars. P680 returns to its resting state, waiting for another photon to strike. But before it can do so, it needs to replace its lost electron. This is where a specialised catalytic system kicks in. This system pulls electrons from water to replenish the lost electron in P680. To do so it splits water into hydrogen and oxygen and reduces the hydrogen to a hydrogen proton by pulling off its electron. It is a water-grinding machine that after four light-powered rounds churns two water molecules ($2H_2O$) into one molecule of oxygen (O_2), four electrons ($4e^-$) and four hydrogen ions ($4H^+$). It involves some sophisticated molecular machinery, boosting energy-carrying molecules in cascades of reactions to build a high energy chemical package by which sugar molecules can be forged together. This process has been described elsewhere,[386] but IngenHousz was, unknowingly, very close to a description of this process when he wrote in section XXVI of his 1779 book:

386 Lane 2002

Water itself, one of the simplest and the most unalterable substances known, seems to be changeable into dephlogisticated air, or at least to contain some things which may be transformed into this air by the influence of the day-light; for the green vegetable substance, which serves as a kind of laboratory, in which this salubrious air is produced, is formed from the water itself.[387]

Photosynthesis is the motor that keeps the complicated ecosystem of life on earth running. Plants are the primary producers on which all other life depends. According to one widely quoted estimate the photosynthesis in all the terrestrial and aquatic ecosystems on earth produces about 187 billion tons of carbohydrate per year, roughly 31 tons of carbohydrates per year for every person on the planet.[388] Our twenty-first century society thrives thanks to the primary production in the carboniferous era, some 350 million years ago. We are driving our cars on the leftovers of the photosynthesis of that long gone era. The direct effect of our carbonivorous economy is a tremendous output of carbon dioxide, because that is what is set free after burning.

Plants do not only produce oxygen. They also stabilise the climate, by grabbing carbon dioxide from the atmosphere to build their stems and leaves, but also by eroding the soil. Growing roots can split and break up rock. Moreover, around the roots the soil becomes acidic, which dissolves minerals from the stone. This eroding force of plants turns out to be very important for the climate because the weathering of rock can take away lots of carbon dioxide, as recent models have calculated.[389] Through the reaction between atmospheric CO_2 and a mineral, such as olivine, sand and carbonates are formed. These are washed from slopes and disappear for a long time into the oceans. Over the last tens of millions of years many new mountain ridges were pushed up, like in the Andes or the Himalayas, which offer the right circumstances for plant enforced erosion and CO_2 removal. The CO_2 sink removed carbon dioxide faster than sources such as volcanic eruptions could deliver. That is why the carbon dioxide concentrations over the past 24 million years were buffered at lower concentrations. Still, there is a mechanism that compensates for the cooling effect this has. Less CO_2 means less carbon for plants to grow with. Fewer plants implies less erosion and a diminution of the carbon sink. Plants play a stabilizing role in the complex bio-geo-global system.

IngenHousz would have been pleased to hear how his description of the interactions between roots and leaves, atmosphere and soil are confirmed. His vision on the interrelatedness between plants, animals, air and stones is being demonstrated every day in the growing knowledge of and concern about the fate of the earth, climate and population.

---

387 IngenHousz 1779
388 Ensminger 2001
389 Pagani 2009

The truth about plants

More of IngenHousz' findings or conjectures are still standing firm. The growth of plants is determined by the concentration of CO_2, the temperature and the amount of light. With little light e.g. increased CO_2 concentrations would not help much as there is not enough energy available to boost the production. IngenHousz performed specific experiments to find out whether the time of the day had any influence on the oxygen production of plants; if there was any difference between sunny or cloudy days; if the concentration of fixed air could have any influence. Modern research has confirmed many of his findings.

On cloudy days plants photosynthesise evenly throughout the day. On sunny days though photosynthesis is lower at the end of the afternoon than in the morning and at the beginning of the afternoon. On these sunlit days the photosynthetic apparatus of the plants in the top leaves is turned off to avoid damage by too much light, lowering the total photosynthetic capacity of the plant. This setback is only temporary. During the night the plant resets its normal photosynthetic pathways. Temperature is another variable that influences photosynthesis. While higher temperatures will boost plant growth in cooler regions, in the tropics they may actually impede growth. A two-decade study of rainforest plots in Panama and Malaysia recently concluded that local temperature rises of more than 1°C have reduced tree growth by 50 per cent. There can always be too much of a good thing. Higher concentrations of carbon dioxide boost the photosynthesis but this effect levels off at higher concentrations. This effect is being used in glasshouses to boost productivity in the afternoon. At some point higher CO_2 concentrations do longer increase photosynthesis. Photosynthesis is not influenced by the number of fruits hanging from the plant.

So, IngenHousz' guess that plants could be harmed by too much light has been proven right. Just as his assumption that different levels of fixed air influence the photosynthetic activity, with a clear upper treshold. The climate rumour that plants will start growing harder when the CO_2 levels rise, is therefore an unfounded fairy tale. One thing IngenHousz did not know, was the difference between C_4 and C_3 plants. Some plants have mechanisms for concentrating CO_2 in their tissues, known as C_4 photosynthesis. So, higher CO_2 will not boost the growth of C_4 plants, as they adapted to boost themselves.

IngenHousz supposed that water was not a crucial factor, as he knew many (exotic) plants that could survive with almost no water. He took organisms exceptionally adapted to dry environments as the rule. But water is a limiting factor indeed and with enough water any plant can benefit. Plants lose water through the pores in leaves that let CO_2 enter. Higher CO_2 levels mean they do not need to open these pores as much, reducing the loss of water. However, it is extremely difficult to generalise about the overall impact of the fertilisation effect on plant growth. Experiments in which plots of land are supplied with enhanced CO_2, while comparable nearby plots remain at normal levels, suggest that higher CO_2 levels could boost the yields of non-C_4 crops by around 13 per cent. Climate change in this respect could have other unsuspected effects. The difference between C_3 and C_4 plants that can use higher or lower concentrations of CO_2 may make them fluctuate in numbers, which in turn has an impact on the herbivores that are feeding on them. All this just to prove how there is constant interaction between organisms and their environment, also the non-living environment. Exactly what IngenHousz described.

And what about IngenHousz' concern about the noxious effects of plants at night? And about worries of the pneumatic doctors about the public health effect of bad air in cities, swamps or rooms where crowds were gathered? Plants indeed produce carbon dioxide just as all other organisms, burning sugars to fire their metabolism. We, humans, use on average some 0.60 kilogram of

SECTION XLII.

Une quantité très-modérée d'air fixe mêlée avec un air respirable, accélère-t-elle ou retarde-t-elle la végétation des plantes qu'on y enferme ?

LES plantes déjà formées prospèrent assez bien dans un air respirable, infecté d'un peu d'air fixe, sur-tout au soleil, probablement parce qu'à l'aide de la lumière solaire elles détruisent ou absorbent bientôt cet air. Je trouve dans mes notes, que dans plus d'une expérience il paroissoit que les plantes s'en trouvoient mieux. Je n'oserois cependant pas insister sur la réalité de cette vertu de l'air fixe, parce que cet effet apparent n'a pas été constant

P 2

In his 1789 book IngenHousz explores the role of fixed air (carbon dioxide). Does the concentration of carbon dioxide in the atmosphere influence the photosyntesis and the growth of plants? A question that is stil relevant after all those years.

oxygen per day. In a room of 30 square meters there is some 8 kilograms of oxygen. That's plenty, you would say, to keep several adults alive, knowing air is constantly leaking in through cracks of doors and windows. But the oxygen inhaled becomes carbon dioxide exhaled, produced by burning all those carbohydrates. There is very little carbon dioxide in the air (0,04%), so any extra source of carbon dioxide will have a relatively big impact on the carbon dioxide concentration. The $30m^3$ sleeping room contains about 22 grams of CO_2 at the beginning of a night to which two sleepers will add 550 grams on a night. Without ventilation the CO_2-concentration will have risen to 1% at sunrise.[390] It was Max von Pettenkofer who developed a method to measure the concentration in the air in 1857 and in so doing, revitalised the old movement of the aerial medicine that began a century earlier.[391] Pettenkofer was convinced that a 1 in 1000 concentration would be the healthy limit for CO_2. So, we should beware of ourselves. We produce lots of carbon dioxide. Our own breathing is a massive source of greenhouse gas, which is especially noticeable when in closed spaces. In submarines and space shuttles, this gas is therefore carefully scrubbed out. In sleeping and other rooms, good ventilation is a must, not so much for want of vital air, but for getting rid of the fixed air, as IngenHousz already suspected.

Knowing the primitive state of the knowledge on both electricity and plant physiology, it is no wonder IngenHousz was slightly mistaken about the relationship between the two. The stimulation of plant growth by applying electric current to seeds or plants seems to be more of a myth than reality, as IngenHousz demonstrated after some series of experiments. In the scientific literature nobody seems to be taking it seriously anymore, though some quacks in new-age circles still seem to flirt with the idea. But electricity and plants have other things in common. The air is always charged positively, the earth negatively. Trees standing firmly in the ground are negative too and therefore attract positively charged particles. Simultaneously plants evaporate (negatively charged) water droplets. That is the source of a big stream of rising air, causing storms up to 160 kilometres per hour, strong enough to break trees. It partially explains why thunderstorms often originate over forests, not over the sea. Plants however, profit from these electrically driven currents. They help disperse negatively charged pollen and spores over long distances.

Here comes the sun

Forms of photosynthesis without oxygen and variants that work without chlorophyll exist, such as the halobacterium using bacteriorhodopsin as their

390 Knip 2009
391 Trout 2009

photoreceptive pigment.[392] This shows how nature has been developing various mechanisms to capture the energy of the sun. It makes people think to try and do that too. The first steps towards imitating nature's cleverest invention are being made. Biologists and energy researchers are finding each other probing the most important chemical reaction in the world. To the former it is the key to understanding all complex life forms on this planet; to the latter it could be the solution to the world's energy problems. Together they could learn how to harness sunlight to produce limitless quantities of hydrogen and other energy-rich fuels from water, cleanly and cheaply. But that's still a dream because we do not know enough by far about the trick plants are performing day in and day out, seemingly without effort and without ill effects. Until we know how they do this we will not to be able to build artificial chlorophyll 2.0 and start photosynthesising ourselves. Researchers have not been able yet to mimic the water-splitting reaction, which is essential to provide a never-ending supply of free electrons to replenish the electrons stripped out by the sunlight. Some experts reckon it could take another ten years before they can build such a system. It has been found during the last decennium that the core of the catalytic system consists of four manganese ions, a calcium ion, several oxygen atoms and at least two water molecules, all held in place in a protein scaffold. But how this works inside plants still remains a mystery to explain, let alone reproduce it in the lab. Some 2.5 billion years after photosynthesis started shaping this planet and creating the conditions for multicellular life to develop, we are struggling hard to understand how chlorophyll is doing this seemingly simple task of splitting water. Trying to break a water molecule apart asks for a huge blast of energy. To do it with heat energy alone one would need to raise the temperature of the water with a thousand degrees. Cracking this secret of plants will provide an endless environmentally-friendly source of energy in a way no other renewable energy source could ever provide.

When IngenHousz started his photosynthesis research he could not have dreamed this even in his wildest dreams. Priestley kicked off the century-long research program with experiments on entire plants that improved the air. IngenHousz discovered that the same activity takes place in leaves or other green parts of plants and that it is driven by sunlight. Senebier showed these effects are also produced in shredded leaves. A century or more later, the discovery was made that not only plants could tap the energy of the sun but that bacteria master the same trick. The next step will probably be to let the photosynthetic process run in artificial systems. The work of IngenHousz represents one of the first and fundamental stepping stones on the long road to this point. He was the first to give air to the idea of the earth as a giant ecosystem in which all parts play a role. When we are now concerned about the fate of humanity as a (spare) part in this whole, the roots of that idea can be found in the works of

---

392 Ensminger 2001

this doctor from Breda and Bath and many places in between. Remembering him, in a historical reconstruction of his life, is not a fancy luxury. It shows us how we have come to get a better understanding of the world we live in, while humbly realising that there is still so much we don't know. As Govindjee, Allen and Beatty said in their review of the history of photosynthesis research:

> We hope future generations of students will remember too, that their view of the world will have had an origin and an evolution, and that their contributions may, if they are fortunate, one day become part of someone else's history. That is both the limit of ambition, and yet the noblest aspiration, for any scientist.[393]

This story has shown that we are all heirs to millennia of curiosity. As Oliver Sacks wrote:

> Many scientists, no less than poets or artists, have a living relation to the past, not just an abstract sense of history and tradition, but a feeling of companions and predecessors, ancestors with whom they enjoy a sort of implicit dialogue. Science sometimes sees itself as impersonal, as "pure thought," independent of its historical and human origins. It is often taught as if this were the case. But science is a human enterprise through and through, an organic, evolving, human growth, with sudden spurts and arrests, and strange deviations, too. It grows out of its past, but never outgrows it, any more than we outgrow our own childhood.[394]

And it is as Leszlek Kolakowski said in his Jefferson Lecture: "We learn history not in order to know how to behave or how to succeed, but to know who we are." This is how we are, and photosynthesis is a hidden but central core of our being. Learning this story may also teach us what it is to learn to gather trustworthy knowledge, an equally deeply embedded biological characteristic, which is so evident it is often overlooked. We need that knowledge to survive and we experience how important that is when unreliable knowledge leads us astray or even into disaster. Science, as I have argued, is only the dedicated expression of that trait that drives us to gain knowledge. IngenHousz and the complex cultural, social, psychological and philosophical web in which he was functioning, stacked in multiple interrelated panarchical layers, shows us how easy it is to do yet how difficult to understand. It shows us how we are and how we know, how we succeed and how we fail. How one thing leads to the other. How sunlight led to insight.

393 Govindjee, Allen & Beatty 2004
394 Sacks 1993

Acknowledgments

As I mentioned in the first lines of this book, it all started with a seemingly simple question by a dear friend. Unfortunatley Toon Bartels did not live to see this story unfold, but I remember him in all these words. Equally sad is the fact that Paul Janssen is not here anymore to read this story. He was very interested in this research from the beginning but left us too early. His thoughts on science and history and his search for trustworthy knowledge reverberate through all these pages.

This work would not have been possible without the enthousiasm and support of Gustaaf Cornelis who daringly took the responsability of being my promotor and who guided me safely through the world of academic research.

My gratitude also goes out to the FWO. It was a grant from the Fonds voor Wetenschappelijk Onderzoek Vlaanderen that allowed me to work a full year and full-time on this project. They offered me the material freedom to enjoy twelve months of intellectual freedom.

I wish to thank the Centre of Logic and Philosophy of Science CLFW at the VUB, sponsoring the production of this book, and all its members who were my sounding board for many years, while I was developing my ecological approach. They were always there when I needed another bunch of willing volunteers on which to try out my story.

I cannot thank enough a handful of people who gave me invaluable input and cherished comments, each from their point of view. Geert de Blust and Douwe de Goede with their ecological know-how, Eric Schliesser as a historian and philosopher of science and Filip Buekens as philosopher of language, science and logic.

This is also the place to thank Norman and Elaine Beale, who are themselves in the last stages of a detailed biography of Jan IngenHousz, the beloved doctor in their hometown Calne. It was a pleasure to be able to exchange information and help each other out when necessary. Equally inspiring was the meeting with A. J. Ingen Housz, giving some cherished information on 'Ome Jan'.

I will remember warmly all the people at the various archives, libraries and museums where I spent so much time, browsing through their papers: Teylers Museum in Haarlem, the Gemeentemuseum and Archief in Breda, the Royal Society in London, the NoordHollands Archief in Haarlem, the Library of Reggio Emilia.

Special and very warm thanks for Bruno Cavalchi and Stefano Meloni of the Lazzero Spallanzani Centro Studi in Scandiano, who offered me a rich and inspiring introduction to the world of science in eighteenth century Italy.

I cannot express enough my appreciation for the editing work done by Carla Aerts, making my English more English. Equally thankful I am to Koen de Waal who took the picture on the cover.

Even while the contact with some of the top specialists in this field was short and factual, it was cordial and inspiring. Thanks therefore to Fernando Abbri, Marco Beretta, Bernadette Bensuade-Vincent and Floriane Blanc.

Some friends have been inspiring during all those years, each in their own way. Jo Lissens for thinking along and asking questions, Johan Braeckman for his inspiration, Josse De Pauw for his always present daruma-san and Jerome Sarment for delivering the perfectly blended inspiration in words and fluids.

If I was able to do this, it was ultimately due to my parents. Just biologically, of course, but more important intellectually, because they tought me how to think and act as a critical and curious person. My gratitude to them goes beyond words.

And first and foremost I wish to embrace my wife Anne, for her relentless support and never ending belief in what I was doing. She lost me quite some time to the eighteenth century, but I always knew she was waiting for me in the here and now. It has been a great pleasure and comfort to have been able to do this with her at my side.

Chronology

The Republic of Letters		Jan Ingenhousz
Jan Baptist Van Helmont	1577-1644	
publication of Van Helmont's *Ortus medicinae, vel opera et opuscula omnia*	1648	
Robert Boyle *Sceptical Chymist*	1661	
Robert Boyle *The origin of forms and qualities*	1666	
Stephen Hales *Vegetable Staticks,* description of 'pneumatic trough'	1727	
	1730	Jan IngenHousz, born in Breda
Stephen Hales *Vegetable Staticks* 2nd edition	1731	
Voltaire *Lettres Philosophiques* Joseph Priestley, born in Birstall	1733	
David Hume *A treatise of human nature*	1739	
War of the Austrian Succession	1740-1748	
Jean Senebier, born in Geneva	1742	
Leyden jar by Van Musschenbroeck	1746	

Hume *Enquiry concerning human understanding*	1748	Ingenhousz studies medicine in Louvain till 1753
Charles Bonnet *Recherches sur l'usage des Feuilles dans les Plantes*	1754	Studies in Leyden
Earthquake distroys Lisbon.	1755	Studies in Edinburgh
Joseph Black *Experiments upon Magnesia Alba, etc*	1756	Studies in Paris
Benjamin Franklin arrives in London.	1757	Ingenhousz returns to Breda
	1757-1764	Medical practice in Breda
Lawrence Stern *Tristam Shandy*	1759	
Erasmus Darwin elected FRS	1761	
Franklin returns to Philadelphia.	1762	
	1764-1768	Medical practice in London, performs inoculation with Dimsdale & others, visits Edinburgh, meets Priestley
Priestely elected FRS Cavendish describes 'inflammable air' (hydrogen).	1766	
Nicolas Théodore de Saussure, born in Geneva Priestley minister at Mill Hill Chapel, Leeds	1767	
Thomas Cook's first voyage Priestley invents soda water.	1768	Inoculation Maria Theresia & family, Vienna
James Watt builds his first steam engine.	1769	Inoculation Archduke & family, Firenze. Meets Fontana

Lavoisier "Sur la nature de l'eau … et la possibilité de son changement en terre."	1770	
Priestley puts sprig of mint in jar with water 17 August, and publishes "Observations of different kinds of air"	1771	Ingenhousz travels to Switzerland, Paris, Holland, England, Vienna
John Pringle president RS Joseph Priestley publishes "Observations on Different Kinds of Air" in *Philosophical Transactions*	1772	More inoculations in Firenze
Priestley (in Bowood, 73-80) as chaplain and librarian of Earl of Shelburne Priestley wins Copley Medal.	1773	Experiments on *torpedo* in Livorno
Priestley *Experiments and Observations on different kinds of air* (Vol I): vegetation restores vitiated air. Lavoisier isolates oxygen. Lavoisier meets Priestley in Paris in October. Senebier *Essai sur l'art d'observer*	1774	Vienna, experiments on diminution of air upon mixing of common air and nitrous air
Lavoisier *Easter Memoir* Priestley *Experiments and Observations on different kinds of air* (Vol II): dephlogisticated air	1775	Marries Agatha Jacquin in Vienna
14 July, American Declaration of Independence 3 Dec, Franklin arrives in Paris Adam Smith *The wealth of nations* Edward Gibbon *Decline and fall of the Roman Empire*	1776	Read to RS, 15 Feb "Early methods of measuring the diminution of Bulk, taking place upon the Mixture of common Air and nitrous Air; together with experiments on Platina"

Lavoisier's experiment on formation of fixed air through respiration (first critique on phlogiston theory of combustion)	1777	
Lavoisier presents his new theory to the French Academy Priestley experiments on plants in June	1778	4 June Baker lecture "Electrical experiments to explain how far the phenomenon of the electrophorus may be accounted for by Dr Franklin's theory"
19 Feb., Priestley experiments with plants, publishes Vol. I of *Experiments and Observations relating to Various Branches of Natural Philosophy; with a continuation of the Observations on Air.*	1779	Elected Fellow of Royal Society Starts in June a series of 500 experiments during summer, England Publication *Experiments upon Plants.* Travels first to Bath, later to the continent
Van Marum teaches about IngenHousz' theory in Holland. Priestley leaves Bowood and moves to Birmingham. Maria Theresia dies in Vienna, her son Joseph II comes to the throne.	1780	Voyage to Vienna via Paris, where he meets Franklin and Fontana. *Experiences sur les plantes*, Paris *Versuche mit Pflanzen*, Leipzig First letters to Van Breda, Delft *Proeven op Plantgewassen*, Delft
Priestley *Experiments and Observations relating to various branches of natural philosophy*, London Willem van Barneveld *Proeve van onderzoek omtrent de hoeveelheid van bedarf, 't welk in onzen dampkring ontstaat, nevens deszelfs verbetering door den groei der plantgewassen*	1781	Dutch ed in *Vaderlandsche Letteroefeningen*, vol 2, Delft "Uitslag der proefnemingen op de planten, strekkende ter ontdekking van derzelver zonderlinge eigenschap om de gemeene lugt te zuiveren op plaatsen waar de zon schynt, en derzelve te bederven in de schaduw en gedurende de nacht."
Senebier *Mémoires physico-chimiques* (3 volumes) Death of John Pringle, buried in Westminster Abbey	1782	Read to Royal Society, 13 June "Some farther considerations on the influence of the Vegetable Kingdom on the Animal Creation"

Montgolfiers launch the first hot air balloon. Lavoisier *Réflexion sur phlogiston* Senebier *Recherche sur l'influence de la lumière* Erasmus Darwin *A system of vegetables*	1783	Paper Royal Society 13 June 1782 published in London Hassenfratz visits IngenHousz in Vienna and assists in eudiometric experimentation.
Cavendish decomposes water by an electric spark. Report on mesmerism by Franklin, Lavoisier et al.	1784	*Journal de Physique* May "Observations Physiques" June "Sur l'économie des plantes" July "Sur l'origine de la matière verte" 7 Aug demonstration by Volta of explosion eudiometer *Vermischte Geschrifte*, Vienna
Franklin returns to Philadelphia. Lavoisier demonstrates publicly the making of water from hydrogen and oxygen.	1785	*Journal de Physique* May, "Sur la construction et l'usage de l'eudiometre" Decembre, "Au sujet de l'influence de l'électricité sur la végétation" *Nouvelles Experiences* Vol I, Paris
Lavoisier *Réflexions sur le phlogistique*	1786	Ingenhousz *Versuche mit Pflanzen*, Vienna
Lavoisier, Guyton de Morveau, Fourcroy and Berthollet, *Méthode de nomenclature chimique*	1787	*Experiences sur les plantes* ed 2 vol 1 (was written in Paris in 1780)
Senebier *Experiences sur l'action de la lumière* Laws of cricket are established. Prussian intervention in Dutch Republic	1788	Travels from Vienna to Paris *Journal de Physique*, May "Au sujet de l'influence de l'Electricité atmosphérique sur les Végétaux"

14 July, storming of the Bastille Lavoisier *Traité elementaire de chimie* George Washington elected President of the USA	1789	*Experiences sur les plantes* ed2vol2 (was written in Vienna in 1781) 15 July departure to Brussels, via Breda to London Last letters to Van Breda, Delft *Nouvelles Experiences* Vol II
Benjamin Franklin dies 17 April. Van Marum visits London. Edmund Burke *Reflections on the revolution in France*	1790	*Versuche mit Pflanzen,* *hauptsächlich über die Eigenschaft,* *die Luft im Sonbnenlicht zu reinigen* *u.s.w.* 3 volumes (1786-90) translated by Scherer, Vienna
14 July, 'Church and King' Riots, Priestley leaves Birmingham.	1791	
Jean-Henri Hassenfratz 3 mémoires in *Annales de Chimie* "Sur la nutrition des Plantes" Erasmus Darwin *The economy of vegetation*	1792	
Louis XVI is executed in Paris in Jan, Marie Antoinette in Oct. Great Britain declares war on France.	1793	
Priestley migrates to America. Lavoisier beheaded on 8 May.	1794	
Annexation of Austrian Netherlands to French Republic	1795	*Miscellanea physico-medica*, Vienna
Edward Jenner develops smallpox vaccine.	1796	Ingenhousz elected Honorary Member of Board of Agriculture. *Proeve over het voedzel der planten* *en de vrugtbaarmaking van* *landeryen*, Delft *An essay on the Food of Plants and* *the Renovation of Soils*, London

De Saussure in *Annales de Chimie* 1797
"Essai sur cette question:
La formation de l'acide carbonique
est-elle essentielle a la végétation?"

Malthus 1798 Ingenhousz
Essay on the principle of population *Ueber die Ernäherung der Pflanzen,*
Leipzig

Napoleon becomes First Consul. 1799 7 Sept Ingenhousz dies
in Bowood House, Calne

Senebier 1800 Agatha, his widow, dies in Vienna.
Physiologie végétale 5 vol.

Priestley dies 1804
in Philadelphia on 6 February.
De Saussure
Recherches chimiques
sur la végétation, Geneva

Fontana dies in Firenze. 1805

Abolition of the slave trade 1807

Senebier dies in Geneva. 1808

Jean-Baptiste Lamarck 1809
Philosophie zoologique

Bibliography

archive and manuscript sources

Stadsarchief Breda, Collectie Ingen Housz

A.

1. Terminologia plantarum J. Ingen Housz
2. Notes in the field of medicine J. Ingen Housz
3. Alphabetical list of diseases and their therapies J. Ingen Housz
4. Recipes of Dr. Pringle. "Formulae quibua utebatur praecipue celeber Prinlius".J. Ingen Housz
5. Alphabetical list of Dutch expressions and their Latin equivalent.
   "Liber Phrasium Johannis Ingen Housz 1746"
   "Phrases Iosephi Ingen Housz 1780"
   "Phrases Ludovica Ingen Housz 1795"
6. Register with summary of letters by IngenHousz, 1774 Febr. 24 - 1793 March 15.
7. Diary, financial notes by Jan I.H. 1767-1789, July 4.
8. Handwritten copies of dissertations by Dr. Jan Ingen Housz
9. Handwritten (in different handwriting) copies of dissertations by Jan Ingen Housz
10. Extracts of various works, by Jan Ingen-Housz.
11. Scientific notes by Jan Ingen Housz.
12. Various notes by Dr Jan Ingen Housz.
13. Summaries of and notes on letters of Dr. Jan Ingen Housz
14. Poems, offered to Jan Ingen Housz at graduation in Leuven 24 July 1753
15. Documents of death and burial of Dr. J. Ingen Housz from A.J. Ingen Housz Breda, 1799
16. "Papers of Dr. Gibbes", belonging to papers of Jan Ingen Housz
16a. List of addresses in London, by Dr. Jan Ingen Housz.
38. Document of admission of Dr. Johan Ingen Housz, body physician to His Imperial Majesty, as Foreign Honorary Member of the Board Df Agriculture, Londen. 1796 febr. 23

B.

1. Notes by Jan Ingen Housz on scientific and other topics in Paris (15 July 1786 - 8 July 1789), Breda and 's-Hertogenbosch (Sept. 1789) and London (27 Sept 1789 - Febr. 1791); and about expenses in London in 1778 and 1779 and Paris in 1788 and 1789

2. Scientific notes by Ingen Housz about plant physiology concerning dispute with Dr. Priestley, (c.1788)

3. Fragment of French essay by Jan Ingen Housz about experiments with platinum (ca. 1790)

4. Two letters by Jan Ingen Housz from London to his wife in Vienna, 17 Juli and 24 July 1798

5. Three letters to Jan Ingen Housz from his wife Agatha from Vienna, 24 Aug. 1798, 21 Nov. 1798 en 13 Aug. 1799.

6. Three letters from Jan Ingen Housz to his nephew Joseph Jacquin in Vienna, 17 and 24 July 1798, 1 June 1799.

7a. Letter from Jan Ingen Housz from London to banker Stametz in Vienna, 24 juli 1798.

7b. Payment order to Archbishop of Toulouse, 1785.

8. Letter from Joseph Jacquin to his uncle Jan Ingen Housz, 20 mei 1796.

9. Letter from Henri Crumpipen, vice-president of the Governement Board in Brussels, 16 juli 1789.

10. Letter from Williams at Nantes to Jan IngenHousz in Paris, 14 oktober 1777.

11. Document of participation in East-India-Stock, by Jan Ingen Hausz, 14 October 1773.

13. Official statement by Royal Oberstkammerer-amt in Vienna for body physycian Johan Ingen Housz, 6 april 1792.

14-16. Documents on financial and other transactions between Jan and Louis Ingen Housz

17. Instructions by Jan Ingen Housz to make "fijne dag-hemden en "groffe nagthemden"

18a. Testament of Jan Ingen Housz made in Vienna 4 March 1785. Authentic copy, 1800.

18b. Testament of Jan Ingen Housz d.d. London 10 June 1796

19- 21. Financial statements and documents on heritage of Jan Ingen Housz in Vienna 1802, 1803

22. Newspaper clippings:
a) Nieuwe Rotterdamsche Courant 3 and 4 June 1905 about Jan Ingen Housz.
b) Neue Freie Presse (Vienna) Abendblatt 13 June 1905 International Botanical Congress Vienna.

c) Neue Freie Presse (Vienna) Morgenblatt 15 June 1905
d) Frankfurter Zeitung 20 June 1905 about International Botanical Congress in Vienna.
e) short note on congress, (c. 1905).

MB414 (contains eightteen pages of personal notes of InghenHousz)
Senebier, J 1782, *Mémoires physico-chimiques, sur l'influence de la lumière solaire pour modifier les êtres des trois règnes de la Nature, & sur-tout ceux du règne végétal.* Tome Premier, Chirol, Genève

Breda's Museum
Conserves some books that have been owned by IngenHousz himself and in which he made notes in the margin.

IngenHousz, J 1779, *Experiments upon plants, discovering the great power of purifying the common air in the sun-shine, and of injuring it in the shade and at night too which is joined, a new method of examining the accurate degree of salubrity of the atmosphere.* Elmsly & Payne, London.

IngenHousz, J 1780, *Experiences sur les vegetaux, specialement Sur la Propriété qu'ils possèdent à un haut degré, soit d'améliorer l'Air quand ils sont au soleil, soit de le corrompre la nuit, ou lorsqu'ils sont à l'ombre; auxquelles on a joint Une methode nouvelle de juger du degré de salubrité de l'Atmosphère.* Didot le Jeune, Paris

IngenHousz, J 1785, *Nouvelles experiences et observations sur divers objects de physique.* Barrois le jeune, Paris

IngenHousz, J 1787, *Expériences sur les végétaux: spécialement sur la propriété qu'ils possèdent à un haut degré, soit d'améliorer l'air quand ils sont au soleil, soit de le corrompre la nuit, ou lorsqu'ils sont à l'ombre; auxquelles on a joint une méthode nouvelle de juger du degré de salubrité de l'atmosphére.* Nouvelle édition, revue et augmenté, T. Barrois Le Jeune, Paris

IngenHousz, J 1789, *Expériences sur les végétaux: spécialement sur la propriété qu'ils possèdent à un haut degré, soit d'améliorer l'air quand ils sont au soleil, soit de le corrompre la nuit, ou lorsquils sont à l'ombre; auxqueles on a joint une méthode nouvelle de juger du degré de salubrité de l'atmosphére.* Tome Seconde, T. Barrois Le Jeune, Paris

Teyler's Museum Haarlem

Correspondence IngenHousz - Van Breda
Envelope 2243: Juny 1780 - Dec 1782
Envelope 2244: 5 Feb 1783 - 28 Dec 1783
Envelope 2245: 5 Jan 1785 - March 1789
Envelope 2246: 23 March 1790 - 2 Jan 1798
Envelope 2247: some copies of letters of Van Breda to IngenHousz

Noord Hollands Archief Haarlem

Correspondence IngenHousz - Van Marum
5 Sept 1789 - 14 Oct 1799

Royal Society London

Experiments on the torpedo (read 10 Nov 1774) (L&P.VI.71); published in *Phil. Trans.* 65, 1.

Lighting a candle by a very moderate spark (read 9 July 1778) (L&P. VII. 60.) published in *Phil. Trans.* 68, 1022

The Bakerian Lecture for 1778 - Experiments to explain how phenomena of electrophorus can be explained on Dr. Franklin's theory (L&P.VII.48.) published in *Phil. Trans.* 68, 1027

Of a new kind of inflammable air (read 25 Mar 1779) (L&P.VII.100 and 109B-appendix) published in *Phil. Trans.* 69, 376

Improvement in electricity - 'The Bakerian Lecture' (read 3 June 1779)(L&P. VII.105); published in *Phil. Trans.* 69, 661

On some new methods of suspending magnetical needles. (read 17 June 1779) (L&P.VII.115); published in *Phil. Trans.* 69, 537

On the degree of salubrity of the air at sea (read 13 April 1780)(L&P.VII.151.); published in *Phil. Trans.* 70, 354

The influence of the vegetable on the animal kingdom (read 13 June 1782) (L&P.VII.259); published in *Phil. Trans.* 72, 426

printed sources

Abbri, F 1989, 'The chemical revolution: a critical assessment', *Nuncius*, 4 (2), p 303-319

Abbri, F 1991, *Science de l'air - Studi su felice Fontana*, Brenner Editore, Cosenza

Adams, D 1979, *The hitchhiker's guide to the galaxy*, Pan Macmillan, London

Anderson, E (ed.) 1985, *The letters of Mozart and his family*, Macmillan, London

Anonymus, 1784, 'Recherches sur l'influence de la lumière', in *Journal des Savants*, Tome III, Mars 1784, Valade, Paris

Appenzeller, T 2004, 'The case of the missing carbon', in *National Geographic*, February 2004

Arneth, A Ritter von, 1876, *Geschichte Maria Theresia*, Bd.VII, Wien

Bailey, B, Melis, A, Mackey, KRM, Cardol, P, Finazzi, G, van Dijken, G, Mine Berg G, Arrigo, K, Shrager, J & Grossman, A 2008, 'Alternative photosynthetic electron flow to oxygen in marine Synechococcus', *Biochimica et Biophysica Acta (BBA) - Bioenergetics*, Vol 1777, 3, p 269-276

Bandinelli, A 2007, 'Scientific communication during a major change in the approach to empirical research: Annales de chimie vs Observations sur la physique/Journal de physique (1789-1803)', presented at *6th International Conference on the History of Chemistry*, Leuven, 28 Aug-1 Sept 2007

Barcott, B 2009, 'Ice paradise', in *National Geographic*, April 2009

Barnes, B, Bloor, D & Henry, J 1996, *Scientific knowledge. A sociological analysis*, Athlone, London

Barnes, C 1893, 'On the food of green plants', *Botanical Gazette* 18, p 403-411

Beale, N & Beale, E 1999, *Who was Ingen Housz, anyway?* Calne Town Council, Calne, UK

Beale, N & Beale, E 2000, 'Looking for Dr Ingen Housz. The evidence for the site and nature of the burial, in Calne, of the famous Dutch physician and scientist of the eighteenth century', in *Wiltshire Archeological & Natural History Magazine*, 93, p 120-130

Beale, N & Beale, E 2005, 'Evidence-based medicine in the eighteenth century: the Ingen Housz-Jenner correspondence revisited', *Medical History*, 49, p 79-98

Beaudreau, SA & Finger, S 2006, 'Medical Electricity and Madness in the 18th Century: the legacies of Benjamin Franklin and Jan Ingenhousz', *Perspectives in Biology and Medicine*, 49, 3, p 330-345

Benedict, F 1912, *The composition of the atmosphere, with special reference to its oxygen content*, Carnegie Institution, Washington

Bensaude-Vincent, B 1993, *Lavoisier, Mémoires d'une revolution*, Flammarion, Paris

Bensaude-Vincent, B 2000, 'Pneumatic medicine viewed from Pavia', in Bevilacqua, F & Fregonese, L (eds) 2000, *Nuova Voltiana*, 2, p 15-33

Beretta, M 1998, *Lavoisier - la rivoluzione chimica*, La Scienze, Milano

Beretta, M 2000, 'Pneumatics vs. 'Aerial Medicine': salubrity and respirability of air at the end of the eighteenth century', in Bevilacqua F and Fregonese L (eds.), *Nuova Voltiana, Studies on Volta and his times*, vol. 2, Pavia

Best, M, Neuhauser, D & Slavin, L 2003, 'Evaluating mesmerism, Paris 1784: the controversy over the blinded placebo controlled trials has not stopped', in *Quality and Safety in Health Care*, 12, p 232-233

Black, J 1993, *Eigtheenth Century Europe 1700-1789*, Macmillan, New York

Black, J 1996, *An illustrated history of eighteenth century Britain 1688-1793*, Manchester University Press, Manchester

Boghossian, P 2006, *Fear of knowledge*, Claredon, Oxford

Boylston, Z 1726, *An historical account of the smallpox inoculated in New England, upon all sorts of persons, whites, blacks, and of all ages and constitutions: with some account of the nature of the infection in the natural and inoculated way, and their different effects on human bodies: with some short directions to the unexperienced in this method of practice.* Chandler, at the Cross&Keys in the Poultry, London

Broadie, A 2003, *The Cambridge companion to the Scottish Enlightenment*, Cambridge University Press, Cambridge

Brooke, JH 2003, 'Science and religion', in Porter, R (ed.) 2003, *Eighteenth Century Science*, The Cambridge History of Science Vol. 4, Cambridge University Press, Cambridge

Brougham, HP 1837 (chairman), *Penny cyclopaedia of the Society for the Diffusion of Useful Knowledge*, Volume X, Knight, London

Brown, R 1827, *A brief account of microscopical observations, made in the months of June, July and August (1827) on the particles, contained in the polen of plants and n the general existence of active molecules in orgaic and inorganic matter*, The Ray Society, London

Browning, R 1993, *The war of the Austrian succession*, Dt Martin's Press, New York

Bud, R & Warner, DJ (eds) 1998, *Instruments of science. A historical encyclopedia*, Garland Publishing, London

Buffon, 2007, *Oeuvres*, Gallimard, Paris

Burian, RM 1978, 'Why philosophers should not despair of understanding scientific discovery', in Nickles T (ed.) *Scientific discovery, logic, and rationality*, Reidel, Dordrecht

Buys, E 1776, *Nieuw en volkomen woordenboek van konsten en weetenschappen*, sesde deel, Amsteldam MDCCLXXIV, Baalde boekverkooper

Capuano, F & Cavalchi, B (eds) 1998, *Spallanzani e la respirabilita dell'aria nel tardo 700: Strumenti e misure della chimica pneumatica*, Pioppi, Scandiano

Carpanetto, D & Ricuperati, G 1987, *Italy in the age of reason 1685-1789*, Longman, London

Cavallo, T 1781, *A treatise on the nature and properties of air, and other permanently elastic fluids, to which is prefixed an introduction to chymistry*, Printed for the author, London

Cavendish, H 1783, 'An account of a new eudiometer', *Phil Trans Roy Soc*, 73, p 106-135

Chevreul, E 1856, 'Recherches expérimental sur la végétation', in: *Journal des Savants année 1856 septembre*, septième article, Imprimerie Impériale, Paris

Chevreul, E 1857, 'Recherches expérimental sur la végétation', in: *Journal des Savants année 1857 juillet*, septième article, Imprimerie Impériale, Paris

Chevreul, E 1857b, 'Recherches expérimental sur la végétation', in: *Journal des Savants année 1857 aout*, huitième article, Imprimerie Impériale, Paris

Cohen, IB (ed) 1941, *Benjamin Franklin's experiments*, Harvard University Press, Cambridge, Mass., p 139-148

Cohen, IB & Schofield, R 1952, 'Did Divis erect the first European protective lightning rod, and was his invention independent?', in *Isis*, Vol 43, 4, p 358-364

Collins, R 1998, *The sociology of philosophies, a global theory of intellectual change*, Harvard University Press, Cambridge

Conant, JB 1950, *The overthrow of the phlogiston theory, the chemical revolution of 1777-1789*, Cambridge

Crosland, M 2000, 'Slippery substances: some practical and conceptual problems in the understanding of gases in the pre-Lavoisier era', in Holmes, FL & Levere, TH (eds), *Instruments and experimentation in the history of chemistry*, MIT Press, Cambridge Massechussets

Darnton, R & Roche, D (eds) 1989, *Revolution in print - The press in France 1775-1800*, University of California Press, Berkeley

Daston, L & Galison, P 2007, *Objectivity*, Zone Books, New York

Darwin, E 1789, *The Botanic Garden, a poem with philosophical notes*, Parts I & II, Johnson, London

Davy, H & Davy, J (ed.) 2001, *The collected works of Sir Humphry Davy*, Thoemmes, London

De Saussure, T 1797, 'Essai sur cette question: La formation de l'acide carbonique est-elle essentielle à la végétation?', *Annales de Chimie*, Tome XXIV, Guillaume, Paris

De Saussure, T 1804 (AN XII), *Recherches chimiques sur la végétation*, Nyon, Paris

Desmond, A & Moore J, 1991, *Darwin: the life of a tormented evolutionist*, Warner, New York

Diamond, J 2005, *Collapse. How societies choose to fail or succeed*, Penguin, New York

Donovan, A.L. 1983, *Philosophical chemistry in the Scottish Enlightenment. The doctrines and discoveries of William Cullen and Joseph Black*, Edinburgh University Press, Edinburgh

Ducheyne, S 2006, 'J.B. Van Helmont and the Question of Experimental Modernism', *Physis: Rivista Internazionale di Storia della Scienza*, vol. XLII, p 305-332

Ellenbroek, F 2006, *The Biological Evolution of the Arts*, Natuurmuseum Brabant, Tilburg

Ensminger, PA 2001, *Life under the sun*, Yale University Press, New Haven

Evans, F 1956, 'Ecosystem as the basic unit in ecology', *Science*, 123, p 1227-1228

Finger, S & Zaromb, F 2006, 'Benjamin Franklin and shock-induced amnesia', *American Psychologist*, 61, 3, p 240-248

Fissell, M & Cooter, C 2003, 'Exploring natural knowledge', in Porter, R (ed.) 2003, *Eighteenth Century Science*, The Cambridge History of Science Vol. 4, Cambridge University Press, Cambridge

Fontana, F 1767, *Osservazioni sopra la ruggine del grano*, Giusti, Lucca

Fontana, F 1775, *Recherches physiques sur la nature de l'air nitreux et de l'air dephlogistiqué*, Nyon l'Aîné, Paris

Fontana, F 1775b, *Descrizione, e usi di alcuni stromenti per misurare la salubrita' dell' aria*, Cambiagi, Firenze

Fontana, F 1779, 'Experiments and observations on the inflammable air breathed by various animals. By the Abbe Fontana, Director of the Cabinet of Natural History Belonging to His Royal Highness the Grand Duke of Tuscany; Communicated by John Paradise, Esq. F. R. S.', *Philosophical Transactions* vol 69, p 337-361

Fontana, F 1779b 'Account of the Airs Extracted from Different Kinds of Waters; With Thoughts on the Salubrity of Air at Different Places. In a Letter from the Abbe Fontana, Director of the Cabinet of Natural History Belonging to His Royal Highness the Grand Duke of Tuscany, to Joseph Priestley, LL.D. F. R. S.' *Philosophical Transactions*, vol 69, p 432-453

Forbes, RJ 1969, *Martinus Van Marum. Life and work*, Tjeenk Willink, Haarlem

Forti, G (ed.) 1972, *Photosynthesis, two centuries after its discovery by Joseph Priestley*, Junk, The Hague

Franklin, B 1757, Letter to John Pringle, December 21. In *Writings of Benjamin Franklin*, vol 7, MacMillan, New York, p 298-300

Franklin, B 1759, *Some account on the success of inoculation for the small-pox in England and America together with plain instructions by which any person may be enabled to perform the operation, and conduct the patient through the distemper*, W Strahan, London

Franklin, B 1775, 'Letter of Franklin to Mazzei', in *The William and Mary Quarterly*, second series, Vol 9, 4 October 1929, p 323

Fitzmaurica, EGP 1912, *Life of William, Earl of Shelburne*, Vol 2, Macmillan, London

Galison, P 1987, *How experiments end*. University of Chicago Press, Chicago

Galison, P 1997, *Image and logic. A material culture of microphysics*, University of Chicago Press, Chicago

Gallo, DA & Finger S 2000, 'The power of a musical instrument: Franklin, the Mozarts, Mesmer and the Glass Armonica', in *History of psychology* 3, vol 4, p 326-343

Garamszegi, L , Eens, M & Török J, 2009, 'Behavioural syndromes and trappability in free-living collared flycatchers, *Ficedula albicollis*', in Animal behaviour, 77:4, p 803-812

Gascoigne, J 2003, 'Ideas of nature', in Porter, R (ed.) 2003, *Eighteenth Century Science*, The Cambridge History of Science Vol. 4, Cambridge University Press, Cambridge

Gensel, L 2005, 'The medical world of Benjamin Franklin', *J Roy Soc Med* 98, p 534-538

Gest, H 1997, 'A 'misplaced chapter' in the history of photosynthesis research; the second publication (1796) on plant processes by Dr Jan Ingen-Housz, MD, discoverer of photosynthesis - A bicentenniel 'ressurection', *Photosynthesis Research* 53, p 65-72

Gest, H 2000, 'Bicentenary homage to Dr Jan Ingen-Housz, MD (1730-1799), pioneer of photosynthesis research', *Photosynthesis research* 63, p 183-190.

Gest, H 2002, 'History of the word photosynthesis and the evolution of its definition', *Photosynthesis Research* 73, p 7-10

Giere, RN 1988, *Explaining science, a cognitive approach*, Chicago University Press, Chicago

Giobert, JA 1793, *Des eaux sulpureuses et thermales de Vaudier*, Zea, Turin

Glynn, I & Glynn, J 2004, *The life and death of smallpox*, Cambridge University Press, Cambridge

Godfroi, MJ 1875, *Het leven van Dr Jan Ingen-Housz, geheimraad en lijfarts van Z.M. Keizer Jozef II van Oostenrijk*, Gebroeders Muller, Printers to the Association of Arts and Sciences of North Brabant, 'sHertogenbosch

Golinski, J 1992, *Science as public culture Chemistry and enlightenment in Britain, 1760-1820*, Cambridge University Press, Cambridge

Golinski, J 1998, *Making natural knowledge - constructivism and the history of science*, Cambridge University Press, Cambridge

Gopnik, A 2009, *The philosophical baby: what children's minds tell us about love, truth and the meaning of life*. Farrar, Straus and Giroux, New York

Gould, SJ 1996, 'Why Darwin?', in *New York Review of Books* 43, April 4 1996

Gould, SJ 1999, *Rocks of ages. Science and religion in the fullness of life*, Ballantine, New York

Govindjee, Allen, JF & Beatty, T 2004, 'Celebrating the milennium: historical highlights of photosynthesis research, Part 3', Photosynthesis Research, 80, p 1-13

Govindjee & Krogmann, D 2004, 'Discoveries in oxygenic photosynthesis (1727-2003): a perspective', Photosynthesis Research 80, p 15-57

Grene, M 1987, 'Historical realism and contextual objectivity: a developing perspective in the philosophy of science', in Nersessian, N (ed.) 1987, The process of science. Contemporary philosophical approaches understanding scientitifc practice, Martinus Nijhoff, Dordrecht

Grison, E 1996, L'étonnant parcours du républicain J.H. Hassenfratz (1755-1827) Du Faubourg Montmartre au Corps des Mines, Les Presses Mines, Paris

Gunderson, LH & Pritchard, L Jr 2002, Resilience and the behaviour of large-scale systems, Island Press, Washington

Gunderson, LH & Holling CS (eds) 2002b, Panarchy - Understanding transformations in human and natural systems, Island Press Washington

Haack, S 2003, Defending science - within reason, New York, Prometheus Books.

Hackmann, DS 1971, 'The Design of the Triboelectric Generators of Martinus van Marum, F.R.S. A Case History of the Interaction between England and Holland in the Field of Instrument Design in the Eighteenth Century', in Notes and Records of the Royal Society of London, Vol 26, 2, p 163- 181

Hänngi, P, Marchesoni, F & Nori, F, 2005, Brownian Motors, Annnalen der Physik, Vol 14, p 1-3

Harcken, AH 1976, 'Oxygen, politics and the American revolution (with a note on the bicentennial of phlogiston)', Annals of Surgery, November 1976

Hasquin, H (ed.) 1987, Oostenrijks België 1713-1794 De Zuidelijke Nederlanden onder de Oostenrijkse Habsburgers, Gemeentekrediet van België, Brussel

Hill, R 1972, 'Joseph Priestley (1733-1804) and his discovery of photosynthesis in 1771', in: Proceedings of the 2nd International Congress on Photosynthesis Research, Junk Publishers, Den Haag

Hochadel, O 2002, 'The foggy summer of 1783: the introduction of lightning rods in German-speaking territories' in: Taming the electrical fire, a conference on the history and cultural meaning of the lightning rod, Minneapolis

Holmes, FL 1985, *Lavoisier and the chemistry of life, an exploration of scientific creativity*, University of Wisconsin Press, Madison

Holmes, FL 1998, *Antoine Lavoisier, the next crucial year; or, the sources of his quantitative method in chemistry*, Princeton University Press, Princeton NY

Holmes, FL 2000, 'The evolution in Lavoisier's chemical apparatus', in Holmes, FL & Levere, TH (eds), *Instruments and experimentation in the history of chemistry*, MIT Press, Cambridge Massechussets

Holmes, R 2009, *The age of wonder: how the romantic generation discovered the beauty and terror of science*, Pantheon, London

Hull, DL 1988, *Science as a process - an evolutionary account of the social and conceptual development of science*, University of Chicago Press, Chicago

Hunter, M & Webster, J (eds.) 1997, *Opera buffa in Mozart's Vienna*, Cambridge Studies in Opera, Cambridge, Cambridge University Press

Huta, C 1998, 'Jean Senebier (1742-1809): un dialogue entre l'ombre et la lumière - L'art d'observer à la fin du XVIIIe siècle', in *Review of History of Science*, 51, 1, p 93-105

IngenHousz J 1768, *Lettre de Monsieur Ingen-Housz, Docteur en Médecine, à Monsieur Chais, Pasteur de l'eglise Wallone de la Haye sur la nouvelle méthode d'inoculer la petite variole*, Amsterdam

IngenHousz, J 1775, 'Extract of a letter from Dr. John Ingenhousz, F.R.S. to Sir John Pringle, Bart. P.R.S. containing some experiments on the torpedo', *Philosophical Transactions*, Vol 65, p 1-4

IngenHousz, J 1776, 'Early Methods of measuring the Diminution of Bulk, taking place upon the Mixture of common Air and nitrous Air ; together with experiments on Platina', *Philosophical Transactions*, Vol 66, p 257-267

IngenHousz, J 1778, 'Electrical experiments, to explain how far the phenomenon of the electrophorus may be accounted for by Dr. Franklin's theory of positive and negative electricity; being the annual lecture instituted by the will of Henry Baker', *Philosophical Transactions*, Vol 68, p 1027-1048
also seperately published as *The Baker lecture for the year 1778*, Nichols, London

IngenHousz, J 1779, 'Account of a New Kind of Inflammable Air or Gass, Which Can Be Made in a Moment without Apparatus, and is as Fit for Explosion as Other Inflammable Gasses in Use for That Purpose; Together with a New Theory of Gun-Powder. By John Ingen-Housz, Body Physician to Their Imperial Majesties, and F. R. S.', *Philosophical Transactions*, Vol 69, p 376-418

IngenHousz, J 1779b, *Experiments upon plants, discovering the great power of purifying the common air in the sun-shine, and of injuring it in the shade and at night too which is joined, a new method of examining the accurate degree of salubrity of the atmosphere.* Elmsly & Payne, London.

IngenHousz, J 1780, *Experiences sur les vegetaux, specialement Sur la Propriété qu'ils possèdent à un haut degré, soit d'améliorer l'Air quand ils sont au soleil, soit de le corrompre la nuit, ou lorsqu'ils sont à l'ombre; auxquelles on a joint Une methode nouvelle de juger du degré de salubrité de l'Atmosphère.* Didot le Jeune, Paris

IngenHousz, J 1780b, *Proeven op plantgewassen: ontdekkende derzelver zeer aanmerkelyk vermogen om de lucht des Dampkrings te zuiveren, geduurende den Dag, en in de Zonne-schyn, en om gemeene lucht des nachts, en wanneer zy in de schaduw zyn, te bederven : beneevens eene nieuwe manier om den graad van gezontheid des dampkrings nauwkeurig te toetsen.* Van der Smout & De Groot, Delft

IngenHousz, J 1780c, *Versuche mit Pflanzen: wodurch entdeckt worden, dass sie die Kraft besitzen, die atmosphärische Luft beim Sonnenschein zu reinigen, und im Schatten und des Nachts uber zu verderben, und nebst einer neuen Methode, den Grad der Reinheit und Heilsamkeit der atmosphärischen Luft zu prüfen*, translated by Scherer AJ, Wengandschen, Leipzig

IngenHousz, J 1782, *Some farther considerations on the influence of the vegetable kingdom on the animal creation*, Nichols, London

IngenHousz, J 1782b, *Vermischte Geschriften, physisch-medizinische Inhalts*, translated by N. Molitor, Kraus, Vienna

IngenHousz, J 1784, *Vermischte Geschriften, physisch-medizinische Inhalts*, erster Band, translated by N. Molitor, Kraus, Vienna

IngenHousz, J 1784b, *Vermischte Geschriften, physisch-medizinische Inhalts*, zweyte verbesserte und vermehrte Band, translated by N. Molitor, Kraus, Vienna

IngenHousz, J 1785, "Observations sur la construction et l'usage de l'eudiomètre de M. Fontana et sur quelques propriétés particulière de l'air nitreux", in *Journal de Physique*, Tome XXVI, p 339

IngenHousz, J 1785b, *Nouvelles experiences et observations sur divers objects de physique*, Barrois le jeune, Paris

IngenHousz, J 1786, *Versuche mit Pflanzen: hauptsächlich über die Eigenschaft, welche sie in einem hohen Grade besitzen, die Luft im Sonnenlichte zu reinigen, und in der Nacht und im Schatten zu verderben : nebst einer neuen Methode, den Grad der Reinheit und Heilsamkeit der atmosphärischen Luft zu prüfen*, Vol I, translated by Scherer AJ, Wappler, Wien

IngenHousz, J 1787, *Expériences sur les végétaux: spécialement sur la propriété qu'ils possèdent à un haut degré, soit d'améliorer l'air quand ils sont au soleil, soit de le corrompre la nuit, ou lorsqu'ils sont à l'ombre; auxquelles on a joint une méthode nouvelle de juger du degré de salubrité de l'atmosphére*, Nouvelle édition, revue et augmenté, T. Barrois Le Jeune, Paris

IngenHousz, J 1789, *Expériences sur les végétaux: spécialement sur la propriété qu'ils possèdent à un haut degré, soit d'améliorer l'air quand ils sont au soleil, soit de le corrompre la nuit, ou lorsquils sont à l'ombre; auxqueles on a joint une méthode nouvelle de juger du degré de salubrité de l'atmosphére*, Tome Seconde, T. Barrois Le Jeune, Paris

IngenHousz, J 1789b, *Nouvelles experiences et observations sur divers objects de physique*, Tome Seconde, Barrois le jeune, Paris

IngenHousz, J 1795, *Miscellanea physico-medica*, J Scherer (ed.), Patzowsky, Vienna

IngenHousz, J 1796, *An essay on the food of plants and the renovation of soils*, Appendix to the Outlines of the fifteenth chapter of the proposed general report from the Board of Agriculture / On the subject of manures, Bulmer & Co, London

IngenHousz, J 1796b, *Proeve over het voedzel der planten, en de vruchtbaarmaking van landeryen*, uit het Engelsch vertaald door J. van Breda, Roelofswaert, Delft

IngenHousz, J 1798, *Ueber die Ernährung der Pflanzen und die Furchtbarkeit des Bodens*, u.s.w. A.d. Engl. von G Fischer, Schäfer, Leipzig

IngenHousz, J 1798b, letter to Edward Jenner, 12 Oct 1798, Gemeentearchief Breda IV, 5-38, 55

Ingen Housz, JM, Beale N & Beale E 2005, 'The life of Dr Jan Ingen Housz (1739-99), private counsellor and personal physician to Emperor Joseph II of Austria', *Journal of Medical Biography*, 13, p 15-21

Israel, J 2001, *Radical Enlightenment, philosophy and the making of modernity 1650-1750*, Oxford

Jackson, J 2005, *A world on fire. A heretic, an aristocrat and the race to discover oxygen*, Viking, New York

Jay, M 2009, *The atmosphere of heaven. The unnatural experiments of Dr Beddoes and his sons of genius*, Yale University Press, New Haven

Jeffries, J 2005, 'The UK population: past, present and future', in *Focus on people and migration*, UK National Statistics, www.statistics.gov.uk

Jen, E 2005, 'Stable or robust? What's the difference?', in Jen, E (ed.) *Robust Design - A repertoire of biological, ecological and engineering case studies*, Oxford University Press, Oxford

Jenkins, JS 1999, The English inoculator: Jan Ingen-Housz *J R Soc Med* 92, p 534-537

Johns, A 2003, 'Print and public science', in Porter, R (ed.) 2003, *Eighteenth Century Science*, The Cambridge History of Science Vol. 4, Cambridge University Press, Cambridge

Johnson, S 2008, *The invention of air. A story of science, faith, revolution, and the birth of America*, Riverside Books, New York

Jones, W & Jones, S 1822, *Catalogue of optical, mathematical and philosophical instruments*, Clondinning, London

Kahneman, D, Slovic, P & Tversky, A 1982, *Judgment under uncertainty: heuristics and biases*, Cambridge University Press, Cambridge

Keeman, JN 2006, 'Blaasstenen en lithothomie, een verdwenen kwaal als fundament van de urologie', in *Nederlands Tijdschrift voor Geneeskunde*, 150, p 2805-2812

Kitcher, P 1993, *The advancement of science - Science without legend, objectivity without illusions*, Oxford University Press, Oxford

Knip, K 2009, 'Slaapademen', in *NRC*, *Wetenschap&Onderwijs*, zaterdag 28.02.2009

Knoefel, PK 1979, 'Famine and fever in Tuscany. Eighteenth century Italian concern with the environment', *Physis*, Anno XXI, 7-35

Knoefel, PK 1984, *Felice Fontana - Life and works*, Societa di studi Trentini di scienza storiche, Trento

Kohler, RE 2005, 'A generalist's vison', *Isis*, 96: 244-229

Kuhn, T 1962, *The structure of scientific revolutions*, Chicago

Labaree, W B (ed.) 2001, *The papers of Benjamin Franklin*, Yale University Press, New Haven

Lakatos I. 1981. 'History of science and its rational reconstructions', in Hacking (red) *Scientific Revolutions*, Oxford University Press, Oxford, p 107

Lane, N 2002, *Oxygen. The molecule that changed the world*, Oxford University Press, Oxford

Lanska, DJ & Lanska, JT 2007, 'Franz Anton Mesmer and the rise and fall of animal magnetism: dramatic cures, controversy, and ultimately the triumph for the scientific method', in Whitaker, H, Smith, CUM & Finger, S 2007, *Brain, mind and medicine: essays in eighteenth century neuroscience*, Springer, New York

Latour, B & Woolgar, S 1979, *Laboratory Life: the Social Construction of Scientific Facts*, Sage, Los Angeles

Lavoisier, A 1770, 'Sur la nature de l'eau et sur les expériences par lesquelles on a prétendu prouver la possibilité de son changement en terre', in *Mémoires de l'Académie des Sciences*, année 1770, p 73

Lavoisier, A 1786, 'Réflexions sur la décomposition de l'eau par des substances végétales et animales', *Mémoires de l'Académie des sciences*, année 1786, p 590

Lavoisier, A 1789, *Traité élémentaire de chimie*, Cuchet, Paris

Lavoisier, A 1790, 'De la décomposition de l'air par le souffre, de la formation des acides sulfureux et sulfurique. Et de l'emploi des sulfures dans les experiences eudiométriques', in *Lavoisier* (1764-93), II

Lemonick, MD 2009, *The Georgian star. How William and Caroline Hershell revolutionized our understanding of the cosmos*, Norton, New York

Levere, TH 1970, 'Friendship and influence Martinus Van Marum, FRS', in *Notes and Sources of the Royal Society of London*, Vol 25, 1 (Jun 1790) p 113

Levere, TH 2000, 'Measuring gases and measuring goodness', in: Holmes, FL & Levere, TH (eds), *Instruments and experimentation in the history of chemistry*, MIT Press, Cambridge Massechussets

Levere, TH 2005, 'The role of instruments in the dissemination of the chemical revolution', in *Endoxa, Series Filosoficas*, 19, 227-242

Levere, TH & Holmes FL 2000, 'Introduction: a practical science', in: Holmes, FL & Levere, TH (eds), *Instruments and experimentation in the history of chemistry*, MIT Press, Cambridge Massechussets

Liebig, J 1840, *Die chemie in ihrer Anwendung auf Agrikultur und Physiologie*, Braunschweig

Lightman, A 2002, 'The art of science', in *New Scientist*, 21 Dec 2002, 68-71

Lopez CA 1994, 'Franklin and Mesmer: an encounter', in *Yale Journal of Biology and Medicine* 66, 325-331

Maddox, B 2002, *Rosalind Franklin - The dark lady of DNA*, HarperCollins, London

Magiels, G 2002, 'De wereld, een proeftuin', in *Standaard der Letteren*, 27 juni 2002, p 14

Magiels, G 2007, 'Dr Jan IngenHousz, or why don't we know who discovered photosynthesis?' *1st Conference of the European Philosophy of Science Association* Madrid, 15-17 November 2007

Magiels, G 2008, *Paul Janssen, Pioneer in Pharma and in China*, Dundee University Press, Dundee

Mason, SF 1991, 'Jean Hyacinthe de Magellan, FRS, and the chemical revolution of the eighteenth century', in *Notes and Records of the Royal Society of London*, 45, 2, p 155-164

Matthews, MR 2009, 'Science and worldviews in the classroom: Joseph Priestley and photosynthesis', *Science & Education*, 18, p 929-960

Mayer, JR von 1845, *Die organische Bewegung in ihren Zusammenhange mit dem Stofwechsel*. Drechsler, Heilbronn

McConnell, A 2007, *Jesse Ramsden (1735-1800) London's leading instrument maker*. Ashgate, Aldershot

McEvoy, J 2001, 'Whither the history of science: Philosophical reflection on the historiography of the chemical revolution', *History Unveiled Science Unfettered: A Conference in Honor of Professor James E. McGuire*. Pittsburgh, PA; 19 Jan, 2002, p 5

Michaud, LG 1818, *BIOGRAPHIE UNIVERSELLE, ancienne et moderne, ou histoire par ordre alphabétique, de la vie publique et privée de tous les hommes qui sont fait remarquer par leurs écrits, leurs actions, leurs talents, leurs vertus et leurs crimes*, Tome 21. Michaud Ed, Paris

Michaud, LG 1823, *BIOGRAPHIE UNIVERSELLE, ancienne et moderne, ou histoire par ordre alphabétique, de la vie publique et privée de tous les hommes qui sont fait remarquer par leurs écrits, leurs actions, leurs talents, leurs vertus et leurs crimes*, Tome 36. Michaud Ed, Paris

Michaud, LG 1825, *BIOGRAPHIE UNIVERSELLE, ancienne et moderne, ou histoire par ordre alphabétique, de la vie publique et privée de tous les hommes qui sont fait remarquer par leurs écrits, leurs actions, leurs talents, leurs vertus et leurs crimes*, Tome 42. Michaud Ed, Paris

Michaud, LG 1858, *BIOGRAPHIE UNIVERSELLE, ancienne et moderne, ou histoire par ordre alphabétique, de la vie publique et privée de tous les hommes qui sont fait remarquer par leurs écrits, leurs actions, leurs talents, leurs vertus et leurs crimes*, Tome 20. Nouvelle Edition, Mme Deplaces, Paris

Michaud, LG 1860, *BIOGRAPHIE UNIVERSELLE, ancienne et moderne, ou histoire par ordre alphabétique, de la vie publique et privée de tous les hommes qui sont fait remarquer par leurs écrits, leurs actions, leurs talents, leurs vertus et leurs crimes,* Tome 34. Nouvelle Edition, Mme Deplaces, Paris

Michaud, LG 1860b, *BIOGRAPHIE UNIVERSELLE, ancienne et moderne, ou histoire par ordre alphabétique, de la vie publique et privée de tous les hommes qui sont fait remarquer par leurs écrits, leurs actions, leurs talents, leurs vertus et leurs crimes,* Tome 39. Nouvelle Edition, Mme Deplaces, Paris

Michaud, LG 1860c, *BIOGRAPHIE UNIVERSELLE, ancienne et moderne, ou histoire par ordre alphabétique, de la vie publique et privée de tous les hommes qui sont fait remarquer par leurs écrits, leurs actions, leurs talents, leurs vertus et leurs crimes,* Tome 40. Nouvelle Edition, Mme Deplaces, Paris

Miller, G 1957, *The adoption of inoculation for smallpox in England and France,* University of Pennsylvania Press, Pittsburgh

Murray, J 1809, *A system of chemistry,* 2nd ed., Creech, Bell & Bradfute, Edinburgh

Nash, LK 1952, 'Plants and the atmosphere', in: Conant J.B. (ed), *Harvard Case Histories in experimental science,* Harvard University Press, Cambridge Massachusetts

Nersessian, N 1984, *Faraday to Einstein: constructing meaning in scientific theories,* Martinus Nijhoff, Leiden

Nersessian, N 2008, *Creating scientific concepts,* MIT Press, Cambridge, Massachusetts

Nickles T, 1978, Scientific discovery and the future of philosophy of science, in Nickles T (ed.) *Scientific discovery, logic, and rationality,* Reidel, Dordrecht

Nickles, T 1989, 'Justification and experiment', in Gooding, D, Pinch, T & Schaffer, S (eds) 1989, *The uses of experiment,* Cambridge University Press, Cambridge

Obrist, HA & Vanderlinden B 1999, *Laboratorium,* Dumont, New York

Odum, EP 1971, *Fundamentals of ecology,* Saunders, Philadelphia

Pagani, M et al. 2009, 'The role of terrestrial plants in limiting atmospheric CO^2 decline over the past 24 million years, *Nature* 460, 85-88 (2 July 2009)

Paine, S 2005, 'Apocalypse then', in *New Scientist*, 21 May 2005, 56-57

Piattelli-Palmarini, M 1994, *Inevitable illusions. How mistakes of reason rule our minds*, Wiley, New York

Peacock, G 1855, *Miscellaneous works of the late Thomas Young*, Vol II, John Murray, London

Pearson, R 2005, *Voltaire Almighty. A life in pursuit of freedom.* Bloomsbury, London

Piccolino, M 2007, 'The taming of the electric ray: from a wonderful and dreadful 'art' to 'animal electricity' and 'electric battery', in Whitaker, H, Smith, CUM & Finger, S 2007, *Brain, mind and medicine: essays in eighteenth century neuroscience*, Springer, New York

Porter, R (ed.) 2003, *Eighteenth Century Science*, The Cambridge History of Science Vol. 4, Cambridge University Press, Cambridge

Porter, R 2003b, *Flesh in the age of reason*, Allen Lane, London

Priestley, J 1767, *The history and present state of electricity with original experiments*, J Dodsley, London

Priestley, J 1769, *History of electricity*, 2nd ed., p 500

Priestley, J 1772, 'Observations on Different Kinds of Air' *Philosophical Transactions* vol 62, p 147-264

Priestley, J 1772b, *Impregnating water with fixed air, in order to communicate to it the peculiar spirit and the virtues of pyrmont water and other mineral waters of similar nature.* J Johnson, London

Priestley, J 1774, *Experiments and observations on different kinds of air*, Johnson, London

Priestley, J 1775, 'An account of further discoveries in air', *Philosophical Transactions* vol 65, p 384-390

Priestley, J 1775b, *Experiments and observations on different kinds of air*, Vol II, Johnson, London

Priestley, J 1775c, *Experiments and observations on different kinds of air*, Second edition corrected, Johnson, London

Priestley, J 1775b, *Philosophical empiricism: containing remarks on a charge of plagiarism respecting Dr. H-s, interspersed with various observations relating to different kinds of air.* Johnson, London

Priestley, J 1779, *Experiments and observations relating to various branches of natural philosophy*, Vol I, Johnson, London

Priestley, J 1781, *Experiments and observations relating to various branches of Natural Philosophy with a continuation of observations on air.* Vol 2, Pearson and Rollason, Birmingham

Priestley, J 1800, *The doctrine of phlogiston established: and that of the composition of water refuted*, printed by Kennedy, Northumberland, Pennsylvania

Priestley, J 1787, Birmingham, 24 November 1787.

Rabinovitch, EI 1945, *Photosynthesis and related processes*, Vol I Interscience Publishers, New York

Rabinovitch, E & Govindjee, R. 1969, *Photosynthesis.* Wiley Interscience, New York

Rabinovitch, E 1971, *An unfolding discovery*, Proc Nat Acad Sci USA, vol 68, no 11, p 2875-2876.

Reed, HS 1949, 'JAN INGENHOUSZ plant physiologist; with a history of the discovery of photosynthesis', *The Chronica Botanica* New York

Roberts, L 1995, 'The death of the sensuous chemist: the 'new' chemistry and the transformation of sensuous technology', in *Studies in History and Philosophy of Science*, 26, p 503-529

Rocca, J 2007, 'William Cullen (1710-1790) and Robert Whytt (51714-1766) on the nervous system', in Whitaker, H, Smith, CUM & Finger, S 2007, *Brain, mind and medicine: essays in eighteenth century neuroscience*, Springer, New York

Roe, SA 2003, 'The life sciences', in Porter, R (ed.) 2003, *Eighteenth Century Science*, The Cambridge History of Science Vol. 4, Cambridge University Press, Cambridge

Sachs, J von 1875, *History of botany (1530-1860)*, English translation 1860 by Garnsey HEF, Oxford University Press, Oxford

Sacks, O 1993, 'The poet of chemistry', in *New York Review of Books*, Nov 4 1993, 40, p 18

Schaffer, S 1984, 'Priestley's questions: an historiographical survey', *History of Science* XXII, p 151-183.

Schaffer, S 1986, 'Scientific discoveries and the end of natural philosophy', in *Social Studies of Science*, 16, p 387-420

Schaffer, S 1990, 'Measuring virtue: eudiometry, enlightenment and pneumatic medicine' in: Cunningham, A & French, R (eds) 1990, *The medical enlightenment in the eighteenth century*, Cambridge, Cambridge University Press

Schierbeek, A 1954, 'Een tot op heden niet gepubliceerde brief van de eerste markies van Lansdowne over de laatste levensdagen van Jan Ingenhousz en een oproep om te komen tot het aanbrengen van een gedenksteen in de kerk te Calne', in *Nederlands Tijdschrift voor Geneeskunde*, 98, p 36

Schliesser, E 2005, 'Wonder in the face of scientitif revolutions: Adam Smith on Newton's 'proof' of Copernicanism', *British Journal for the History of Philosophy*, 13(4), p 697-732

Schofield, RE 1966, *A scientific biography of Joseph Priestley*. MIT Press, Cambridge Massachusetts.

Schofield, RE 1997, *The Enlightenment of Joseph Priestley - a study of his life and works 1733 to 1773*, Penn State Press, Pennsylvania

Seguin, A 1791, 'Mémoire sur l'eudiométrie', *Annales de chimie*, 9

Senebier, J 1782, *Mémoires physico-chimiques, sur l'influence de la lumière solaire pour modifier les êtres des trois règnes de la Nature, & sur-tout ceux du règne végétal*, Tome Premier, Chirol, Genève

Senebier, J 1783, *Recherches sur l'influence de la lumière solaire pour métamorphoser l'air fixe en air pure par la végétation*, Chirol, Genève

Senebier, J 1788, *Experiènces sur l'action de la mumière solaire dans la végétation*, Barde, Manget & Co, Genève

Senebier, J 1800, *Physiologie végétale, contenant une description des organes des plantes*, Paschoud, Genève

Senebier, J 1802 (An X), *Essay sur l'art d'observer et de faire des experiènces*, Tome I, II & III, Paschoud, Genève

Sibum, HO 1995, 'Reworking the mechanical value of heat: instruments of precision and gestures of accuracy in early Victorian England', in *Studies in History and Philosophy of Science*, part A 26, p 73-106

Sibum, HO 2000, 'Experimental history of science', in Svante Lindqvist (ed.) *Museums of Modern Science*, Canton, Massachusetts: Science History Publications 77-86

Sibum, HO 2003, 'Experimentalists in the Republic of Letters', *Science In Context*, 16 (1-2), p 89-120

Silverman, K 1984, *The life and times of Cotton Mather*, Harper&Row, New York

Smit, P 1980, 'Jan Ingen-Housz: some new evidence about his life and work'. *Janus*, LXVII, p 125-139.

Smyth, AH 1907, *The writings of Benjamin Franklin*, MacMillan, New York

Sorkin, D 2008, *The religious Enlightenment*, Princeton, New York

Standage, T 2005, *A history of the world in six glasses*, Walker & Co, New York

Strathern, P 2000, *Medeleyev's dream: the quest for the elements*, Penguin, London

Taleb, NN 2004, *Fooled by randomness: the hidden role of chance in life and in the markets*. Penguin, London

Tallis, R 1995, *Newton's Sleep, two culture and two kingdoms*, MacMillan, London

Thagard, P 1992, *Conceptual revolutions*, Princeton University Press, Princeton

Theunissen, B 2004, *Diesels droom en Donders' bril*, Nieuwezijds, Amsterdam

Thimann, KV 1971, Photosynthesis Bicentennial Symposium, introduction by the chairman, *Proc. Nat. Acad. Sci USA*, Vol 68, No 11, p 2875

Thordarson, T & Self, S 1995, "Atmospheric and environmental effects of the 1783-1784 Laki eruption: a review and assessment", in *Journal of Geophysical Research*, American Geophysical Union, 108, D1, 4011, p 717-28

Trout, DL 2009, 'Max Josef von Pettenkofer (1818-1903). A biographical sketch', *Journal of Nutrition*, August 28 2009

Uglow, J 2002, *The lunar men. The friends who made the future 1730-1810*. Faber and Faber, London

Van Barneveld, W 1781, *Proeve van onderzoek omtrent de hoeveelheyd van bedarf, 't welk in onzen dampkring ontstaat, nevens deszelfs verbetering door den groei des plantengewassen*. Verhandelingen Utrechtsch Genootschap

Van der Korst, JK 2003, *Een dokter van formaat. Gerard van Swieten, lijfarts van Maria Theresia*. Bert Bakker, Amsterdam

Van der Pas, J 1964, 'The Ingenhousz - Jenner correspondence', *Janus*, 51, p 202-220

Van der Pas, P 1973, 'Jan IngenHousz', in : Gillispie CC (ed.) *Dictionary of Scientific Biography*, New York, Scribner's Sons, Vol VII, p 11-16

Van der Pas, P 1978, 'The Latin translation of Benjamin Franklin's papers on electricity', *Isis*, vol 69, No 1, p 82-85

Van der Pas, PW 1968, *The discovery of the Brownian motion*, XII International Congress of the History of Science, 26 August 1968

Van der Pas, PW 1971, 'The discovery of the Brownian motion', *Scien. Historiae*, 13, p 17

Van Marum, M 1791, 'Lettre à Jean Ingenhousz, contenant la description d'une machine électrique, construite d'une manière nouvelle et simple, et qui réunit plusieurs avantages sur la construction ordinaire', *Journal de Physique*, 38, p 447-459

Van Helmont, JB 1662, *Oriatrike, or physick refined: the common errors therein refuted and the whole are reformed and rectified*. Lodowick-Loyd, London
Verbruggen, F 1973, *Cognitieve dissonantie en de phlogistoncontroverse*, doctoral dissertation, University Gent

Ward, P 2009, *The Medea hypothesis: is life on earth ultimately self-destructive?*, Princeton University Press, Princeton

Wiechmann, A & Palm, LC 1987, *Een elektriserend geleerde. Martinus van Marum 1750-1837*, Enschede & Zonen, Haarlem

Wiesner, J 1905, *Jan Ingen-Housz. Sein Leben und sein Wirken als Naturforscher und Artz*, Verlagsbuchhandlung Carl Konegen, Vienna

Whitaker, H, Smith, CUM & Finger, S 2007, *Brain, mind and medicine: essays in eighteenth century neuroscience*, Springer, New York

Wilson, B 2007, *The making of Victorian values. Decency & dissent in Britain: 1789-1837*, The Penguin Press, New York

Wimsatt, WC 2007, *Re-engineering philosophy fior limited beings - piecewise approximations to reality*, Harvard University Press, Cambridge, Massachusetts

Zuidervaart, HJ 2006, 'An eighteenth-century medical-meteorological society in the Netherlands: an investigation of early organisation, instrumentation and quantification', *British Society for the History of Science*, 39(1), p 49-66

NAME INDEX

SUBJECT INDEX

DATE DUE